Safety and Health Essentials
for Small Businesses

Safety and Health Essentials
for Small Businesses

William F. Martin

James B. Walters

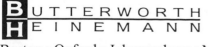
BUTTERWORTH
HEINEMANN

Boston • Oxford • Johannesburg • Melbourne • New Delhi • Singapore

Library of Congress Cataloging-in-Publication Data
Safety and Health Essentials For Small Businesses
William F. Martin; James B. Walters
ISBN 0-7506-7127-0

British Library of Congress Cataloging-in-Publication Data
A catalogue record for this book is available form the British Library.

The publisher offers special discounts on bulk orders of this book.
For information, please contact:
Manager of Special Sales
Butterworth-Heinemann
225 Wildwood Avenue
Woburn, MA 01801-2041
Tel: (781) 904-2500
Fax: (781) 904-2620

For information on all Butterworth-Heinemann publications available, contact our World Wide Web home page at: http://www.bh.com

10 9 8 7 6 5 4 3 2 1

Contents

Authors

William F. Martin, PE holds a civil engineering degree from the University of Kentucky and a master's degree in environmental health engineering from the University of Texas. He served 22 years as a commissioned officer in the U.S. Public Health Service. He held positions with the Indian Health Service, U.S. Coast Guard, Federal Water Pollution Control Administration, and the National Institute for Occupational Safety and Health. A registered professional engineer in Texas and Kentucky, he has presented and published numerous technical papers both foreign and domestic. He served on the Superfund steering committee made up of EPA, OSHA, NIOSH, and the U. S. Coast Guard. He served as the NIOSH Hazardous Waster Program Director with primary responsibility for coordinating all Institute Superfund activities including research projects and the production of comprehensive health and safety guidelines, worker bulletins, and training materials. Mr. Martin has consulted on environmental engineering and hazardous waste health and safety with Valentec International Corporation, Environmental Systems & Services, Inc., and Greenglobe Engineering, Inc.

James B. Walters, CSP, PE, holds a BS degree from Union College in Kentucky and a Masters degree in Public Health Administration from the University of North Carolina at Chapel Hill. Prior to retirement, he occupied the position of Occupational Safety and Health Manager, Office of the Director, Division of Training and Manpower Development, National Institute for Occupational Safety and Health in Cincinnati, Ohio. His 38 years of federal employment with the U.S. Public Health Service, Food and Drug Administration, and the U.S. Consumer Product Safety Commission included 32 years in the areas of program development, on-site training and curriculum development aspects of home, recreational institutional, product, and occupational safety. During this time, he was employed at the local, state and national levels. He co-authored and assisted with the development of various types of safety and health training materials directed at the total work force. This included college and university faculty, private industry, trade associations, vocational teachers and administrators, elementary, high school and college students, prison officials, small business managers, and officials representing the World Health Organization. He conducted courtesy on-site safety and health inspections, presented workshops, and provided specialized safety and health training for small business managers and employees. When requested, utilizing OSHA standards and acceptable safety and health practices, he planned and conducted safety and health inspections of occupational facilities in state prisons, prepared reports and provided expert testimony for the U.S. Department of Justice. Mr. Walters served on the Safety Advisory Committee of the Hand Tool and Power Tool Institute and the NIOSH special work group responsible for developing a national strategy for the prevention of occupational traumatic injuries.

Preface

This health and safety handbook was developed for the small business with less than 100 employees. Many of the small businesses do not employ a fulltime health and safety professional. They often assign safety and health duties as extra work to an employee that has other primary responsibilities. This handbook is written in straight forward non-technical language assuming the reader does not have extensive training in industrial hygiene, safety, or occupational health. The authors encourage companies to seek professional health and safety input and suggest that in house staff members be trained in health and safety. However, years of experience has lead the authors to develop this handbook for those employees given health and safety responsibilities with little or no training for their new role.

This handbook is not a comprehensive compliance and/or regulation document but an aid to small businesses to protect their workers, reduce liability, and provide easily understood guidelines for meeting the Occupational Safety and Health Administration (OSHA) regulations. Extensive use has been made of the many publications that OSHA, National Institute for Occupational Safety and Health (NIOSH) and Environmental Protection Agency (EPA) make publicly available. The authors have listed some publications at the end of each chapter that would aid personal development of an employee working in health and safety.

More than half (56%) of the U.S. workforce in private industry is employed in business establishments with fewer than 100 employees. Of all U.S. private industry establishments, 98% employ fewer than 100 employees, and more than 87% employ fewer than 20 employees. Prevention of occupational injury and illness is often difficult in these establishments because small businesses generally have few safety and health resources, cannot usually hire staff devoted to safety and health activities, and often lack the ability to identify occupational hazards and conduct surveillance. The NIOSH and OSHA recognize these special challenges to safety and health in small business establishments.

In 1994 in the United States, 6.5 million private industry establishments were operating, and they employed more than 96.7 million employees. These private industry establishments represent all major divisions of the economy.

Table 1 shows the number and percentage of employees in establishments with fewer than 100, 20, and 10 employees in 1994.

TABLE 1

Number and percentage of employees in establishments with fewer than 100, 20, and 10 employees, by industry division in 1994*.

Industry/division	No. of employees	<100 employees		<20 employees		<10 employees	
		Number	%	Number	%	Number	%
Agriculture[†]	551,507	478,147	87	336,800	61	215,612	39
Mining	607,721	>272,655	>45[‡]	>100,747	>17[‡]	>47,306	>8[‡]
Construction	4,709,379	3,716,851	79	2,150,840	46	1,336,954	28
Transportation, communication, and public utilities	5,713,515	2,637,978	46	1,024,351	18	551,853	10
Wholesale trade	6,365,973	4,681,038	74	2,254,753	35	1,193,682	19
Retail trade	20,320,266	15,457,780	76	6,929,207	34	3,770,063	19
Finance, insurance, and real estate	7,002,431	4,149,983	59	2,262,835	32	1,388,866	20
Services	33,253,032	16,616,313	50	8,726,606	26	5,370,812	16
Total	96,733,300	>53,404,287	>55[‡]	>25,287,715	>26[‡]	>14,559,801	>15[‡]

Source: Bureau of the Census County Business Patterns 1994.
*Includes salary and wage-earning employees in all private industry
[†] Agricultural services only (SIC 07).
[‡] Indicates that it was not possible to determine the exact percentages in all size categories, as employment information id routinely withheld from County Business Patterns (CBP) data by the Bureau of the Census to assure confidentiality of operations for some establishments. It was still possible to calculate at least the minimum percentage of workers in the desired categories for every industry.

These statistics are given in terms of business establishments-not companies. An establishment is an individual business location such as a fast-food restaurant. In contrast, a company consists of all establishments under common ownership or control, such as a chain of fast-food restaurants owned by a single employer. The Bureau of the Census [1998] reports that the 6.5 million private establishments operating in the United States in 1994 were owned by 5.3 million companies. This figure indicates that most companies own only a single establishment, but multi-establishment companies may have a disproportionate impact. Although 56% of the U.S. private workforce is employed by establishments with fewer than 100 employees, only 38% is employed by companies with fewer than 100 employees.

Statistics about establishment size are generally the most appropriate for this book, since regulations and guidelines usually apply to individual establishments at a given business location. When we refer to resources available to an employer, company size might be more relevant.

Employers, employees, and insurers involved with small businesses can all benefit from a heightened awareness of the risks and hazards within various small business industries. Other accounts have noted that providing this information (possibly in the form of industry report cards) would spur market-adjusting forces to reduce injuries and illnesses as employees are made aware of risks and as employers strive to provide safer workplaces or pay a higher compensating wage.

Incidence of Injuries, Illnesses, and Fatalities

According to the U.S. Bureau of Labor Statistics (BLS), the incidence of nonfatal work-related injuries and illnesses in private industry fell in 1995 to the lowest rate this decade, continuing a 3-year decline in these rates. A total of 6.6 million work-related injuries and illnesses were reported during 1995. This figure translates into an average rate for all private industry of 8.1 cases per 100 full-time equivalent workers (i.e., per 200,000 hr, based on 40 hr per week for 50 weeks per year per worker), compared with a high of 8.9 cases per 100 full-time equivalent workers in 1992. Although this trend is encouraging, nearly 6.1 million injuries alone in 1995 resulted in lost work time, medical treatment other than first aid, loss of consciousness, restriction of work or motion, or transfer to another job. An estimated half million occupational illnesses occurred in 1995. However, this estimate is uncertain, as it is generally more difficult to determine whether an illness is work-related. Therefore, occupational illnesses tend to be under-reported. Work-related fatalities in private industry totaled 5,495 in 1995 and 5,521 in 1996. These figures also indicate that serious hazards are still associated with many workplaces.

Consequences of Occupational Injuries and Illnesses

The consequences of occupational injuries and illnesses are significant. In addition to causing pain, suffering, and loss of productivity for affected employees and their families, occupational injuries and illnesses exact a hefty toll on U.S. industry. The National Safety Council (NSC) estimates that occupational injuries and fatalities in the United States in 1996 cost $121 billion, or $790,000 per death and $26,000 per disabling injury. These data do not include costs associated with occupational illness. Another study places employers' costs of occupational injuries alone at $155 billion annually. Others estimate the total annual costs of occupational injuries, illness, and fatalities in the United States to be greater than $170 billion in 1992, noting that these estimates are conservative because (1) they ignore costs associated with pain and suffering and home care provided by family members and (2) the numbers of work-related injuries and illnesses are likely to be undercounted. Underreporting of occupational injuries and illnesses, especially among small business establishments, has been recognized as a major source of bias in the Federal statistics

Occupational Health Services Provided by Small Businesses

Over the last half century, devising strategies to provide a safe workplace and to protect the health of workers have been common goals of government agencies, labor organizations, and industry groups. Providing for the safety and health of workers was identified as a priority in a 1965 report to the U.S. Surgeon General entitled *Protecting the Health of Eighty Million Americans: A National Goal for Occupational Health*. The report recognized that with a few exceptions, minimal or no occupational health surveillance or services were provided by small business establishments-a fact that was later verified by the National Occupational Exposure Survey. Significant differences found between small business establishments (8 to 99 employees) and large ones (500 or more employees) are presented in table 2.

TABLE 2
Occupational Health Services Provided by Small Businesses

	Small Establishments (%)	Large Establishments (%)
On-site occupational health units:		
With a physician in charge	0.1	30
With a staff person in charge	1.9	78
One or more medical screening tests for all employees	6	54
Pre-placement physicals	21	82
Industrial Hygiene or safety personnel	2	59
Industrial hygiene consultants	14	56

OSHA conducted a similar but broader survey of more than 7,000 establishments in 1990. They determined that of 5.3 million establishments with 85 million employees, only 6.3% had a medical surveillance program. The category of smallest business establishments (1 to 19 employees) had the fewest establishments with a medical surveillance program (3.8 %), followed by those with 20 to 99 employees (14.4%), those with 100 to 249 employees (33.4%), and those with 250 or more employees (55.8%).

Strategies for Increasing Occupational Health Services

Recognizing occupational safety and health challenges for small businesses is an important first step toward determining effective strategies for addressing these concerns. In drafting the national public health agenda in 1990, the U.S. Department of Health and Human Services expressed the need to "establish in 50 States either public health or labor department programs that provide consultation and assistance to small businesses to implement safety and health programs for their employees."

OSHA established a program to consult with employers about establishing effective occupational safety and health programs. The program is active in all 50 States and emphasizes businesses with the most hazardous operations. Primary attention is given to smaller businesses (i.e., those with fewer than 250 workers). The OSHA Targeted Training Program provides grants for safety and health training programs in small businesses and for specialized training in ergonomics, scaffolding, and workplace violence. To reduce the burden of regulatory compliance for small businesses, the Small Business Regulatory Enforcement Fairness Act of 1996 requires Federal agencies to help small businesses comply with government regulations and to develop policies that reduce or eliminate fines for small businesses. In implementing the statutes of the Act, the U.S. Department of Labor reported that 88% of small businesses fined by OSHA between March and December 1997 had fines reduced by a total of $110 million.

On an international level, the importance of setting safety and health priorities for small businesses has also been recognized. Within the European Union, increasing emphasis is being placed on providing education, training, and additional support to small and medium-sized enterprises for improving the recognition, communication, and prevention of hazards in these establishments.

Bibliography

NIOSH, *Identifying High-Risk Small Business Industries*. Center for Disease Control. Publication No. 99-107.

Acknowledgements

This practical health and safety handbook for small businesses would not be possible without the previous work of many individuals, companies, and government agencies. The authors made extensive use of the publications developed by the Occupational Safety and Health Administration (OSHA), the Department of Labor, and the National Institute for Occupational Safety and Health (NIOSH), and the Center for Disease Control and Prevention. A special recognition is given to Robert J. Firenze for his work with NIOSH and the development of health and safety guidelines for vocational education shops and industrial machine operations. Without the initial efforts and encouragement from Michael Forster of Butterworth-Heinemann, this publication would not have been undertaken by the authors.

During the past 20 years, the authors have worked with a host of highly qualified health and safety professionals in the nation's efforts to protect and enhance the quality of life for all workers. A special thanks to Nancy Orr, CIH, Sage Environmental, for her review and suggestions on the draft of this book.

The authors wish to thank Laurie Goodale of Priester and Associates, for her extensive efforts to translate NIOSH and OSHA publications to a desktop publishing format. Thanks to Ann T. Kiefert, M.S., for her technical editing of the final draft. Ms. Kiefert's experience with Florida's environmental regulations, her work on hazardous waste health and safety handbooks, and her graduate studies at Florida State University contributed to her expert input.

1

The Safety and Health Program

THE PURPOSE of the safety and health program is to prevent accidents and illnesses; protect workers, property, and community; control or reduce losses; provide means for management/employee involvement in the safety and health program; and to comply with legal requirements. Managers and supervisors must understand the importance of a systematic approach to identifying, evaluating and controlling those factors in the work environment which are responsible for accidents and their effects. Once the causes and nature of accidents are understood, management can begin developing a safety and health program where supervisors, workers, and others can interact with each other and provide a safe and healthy place to work.

It is essential that a safety and health program receive the full support and commitment of the owner(s). Managers and supervisors must accept responsibility for the safety and health program and furnish the drive to get the program started, oversee its operations, and continuously improve it. In order for the program to succeed, those in control must delegate the necessary authority down to the appropriate levels of the organization and provide the resources to accomplish the goals.

1.1 Impact of Accidents and Injuries on Small Businesses

Each year accidents continue to take their toll on workers in all types of industries. In 1998, annual statistics compiled by the National Safety Council indicate that over 3,800,000 workers were disabled and over 5,100 workers were killed that year as a result of work place accidents. Actual figures are much higher because many accidents are not reported. Table 1.1 lists the number of deaths and disabling injuries in a variety of classes.

Accidents are harmful not only when measured in human terms, that is, injuries and illnesses to workers, but also when measured in terms of their adverse affects on the overall business. Accidents can result in fiscal losses

1

to the business due to the costs of replacing damaged equipment and materials, medical treatment expenses, administrative costs, and any liabilities or penalties incurred.

TABLE 1.1
Deaths and Disabling Injuries by Class, 1998

Deaths and disabling injuries by class occurred in the nation at the following rates in 1998:

Class	Severity	One Every	Number per			1998 Total
			Hour	Day	Week	
All	Deaths	6 min.	11	253	1,770	92,200
	Injuries	2 sec.	2,210	53,200	373,100	19,400,000
Motor-	Deaths	13 min.	5	113	790	41,200
Vehicle	Injuries	14 sec.	250	6,000	42,300	2,200,000
Work	Deaths	103 min.	1	14	100	5,100
	Injuries	8 sec.	430	10,400	73,100	3,800,000
Workers	Deaths	14 min.	4	103	720	37,600
Off-the- Job	Injuries	6 sec.	650	15,600	109,600	5,700,000
Home	Deaths	19 min.	3	77	540	28,200
	Injuries	5 sec.	780	18,600	130,800	6,800,000
Public	Deaths	26 min.	2	55	380	20,000
Nonmotor -Vehicle	Injuries	5 sec.	760	18,400	128,800	6,700,000

While you make a 10-minute safety speech, 2 persons will be killed and about 370 will suffer a disabling injury. Costs will amount to $9,140,000. On the average, there are 11 unintentional-injury deaths and about 2,210 disabling injuries every hour during the year.

1.2 Definition of an Accident

Generally, the term accident refers to an unusual event caused by human, situational, or environmental factors which result in or has the potential to result in injuries, fatalities, or property damage. Three important points contained in this definition are:

1. *Accidents do not have to result in actual injuries.* If, for example, a worker slips and falls on an oil spot but the fall does not injure the worker or cause any damage, the incident is still classified as an accident because it interrupted the production process and had the potential to cause injury and damage.

2. *Accidents are unusual and unexpected events.* It is important to recognize that the potential for accidents is always present. Unless supervisors and workers are aware of and fully alert to the ever present possibility of accidents, and unless they act to discover and eliminate potentially dangerous situations, the potential for acci-

dents will remain.

3. *Accidents do not occur without reason, they are caused.* Some common causes of accidents include the improper use of tools and machines, the failure to use protective equipment, the failure to follow correct procedures, the use of faulty equipment and tools, the poor condition of walking and working surfaces, improper maintenance of equipment, and inadequately guarded machinery.

1.3 Definition of Hazard

Generally, safety professionals refer to accident causes as hazards. A hazard may be defined as any existing or potential condition which, by itself or by interacting with other variables, can result in the unwanted effects of property damage, illnesses, injuries, deaths and other losses. Hazards are generally grouped into two broad categories, hazards dealing with safety and hazards dealing with health.

A safety hazard usually results in trauma. It evolves from a situation in which a worker may be injured or killed because of electrical, thermal, or mechanical conditions. Examples of safety hazards include faulty electrical wiring on tools and equipment, unguarded gears or blades on equipment, slips and falls, grinders without tool rests, and tool rests improperly adjusted or lacking guards. A health hazard is a condition, which has the potential to cause illness. Examples of health hazards include high noise levels, dust, fumes, mists, vapors, smoke, solvents, viruses, and bacteria.

In some instances, a substance or material can be classified as both a safety and health hazard. For example, benzene vapor can cause lung irritation and perhaps more advanced forms of illness if inhaled over long periods of tune. If this substance comes in contact with an open flame, it may ignite and explode. In this case, the benzene vapor, normally considered a health hazard, is also a safety hazard.

1.4 Causes of Accidents

Accidents are caused by either human, situational, or environmental factors. In each accident situation, the cause can be directly or indirectly attributed to the supervisor or worker (human factors); to the operations, tools, equipment and materials (situational factors); or to work place conditions such as noise, vibration, and poor illumination (environmental factors).

1.4.1 Human Factors

An accident can be attributed to human factors if a person caused the accident either by something they did or by something they failed to do. For example, a worker may cause an accident when he sharpens a chisel on a grinder without resting the material or tool on the grinder's tool rest or he may cause an accident when he fails to clean up an oil spot from the floor. Table 1.2 provides five factors that are key to reducing instances of human error.

TABLE 1.2
Five Key Factors For Reducing Human Error

Factor	Example
Knowledge	Supervisors and workers must know the correct methods and procedures to accomplish given tasks.
Ability	Workers should demonstrate skill proficiency before using a particular piece of equipment. Many accidents are the result of a person's inability to use equipment, tools, and safety devices correctly.
Physical Fitness	The physical characteristics and fitness must be taken into consideration as it affects a worker's job task. For example, poor eyesight or a problem with depth perception is a factor which can cause a worker to make faulty judgements. The temporary loss of the use of a hand or fingers may interfere with the manual dexterity required to do the job safely.
Regard for Safe Work Practices	Supervisors and workers should maintain a high and continuous regard for the health and safety of themselves and that of their co-workers. If they are constantly aware of potentially dangerous situations, take corrective action and encourage others to the same, then the workers will be more productive and the hazards will be minimized.
Supervision	The supervisor must constantly be aware of the skill level of each worker and adjust the amount and type of supervision given to each worker accordingly. Furthermore, when the supervisor accepts nothing less than safe work practices, the potential for accidents is minimized.

1.4.2 Situational Factors

Along with human factors, situational factors are a major cause of accidents. Situational factors include the operations, tools, equipment, and materials which contribute to accident situations. Examples of situational factors include:

- unguarded, poorly maintained, and defective equipment
- electric shock from ungrounded equipment
- equipment with inadequate warning signals

- congestion hazards caused by poorly arranged equipment
- poorly located equipment which exposes workers to rotating and/ or moving parts

1.4.3 Environmental Factors

Environmental factors also play a role in work place accidents. Three general categories of environmental factors are physical, chemical and biological.

1. Physical Category. Noise, vibration, fatigue, illumination, radiation, heat, and cold are factors which have the capacity to directly or indirectly influence or cause accidents. For example, if operations on a machine lathe produce high noise levels, the noise may interfere with communications in the work place and may preclude a worker from warning others of a hazard.
2. Chemical Category. This category includes toxic fumes, vapors, mists, smokes and dust. In addition to causing illnesses, these elements can impair a worker's skill, reactions, judgment, or concentration. For example, a worker who has been exposed to the narcotic effect of some solvent vapors may experience an alteration of judgment and move his hand too close to the cutting blade of a milling machine.
3. Biological Category. The biological category includes instances in which a person becomes ill from contact with bacteria and microorganisms. For example, maintenance personnel working around waste systems and sewage facilities or workers handling contaminated cutting oils without adequate protection are at risk of exposure.

1.5 Sources of Situational and Environmental Hazards

Situational and environmental hazards enter the work environment from many sources. The primary contributors are the workers; those responsible for purchasing items for use in the work environment; those responsible for tool, equipment and machinery placement and for providing adequate machine guards; and those responsible for maintaining shop equipment, machinery, and tools.

Workers may contribute to situational and environmental hazards by disregarding safety rules and regulations, by making safety devices inoperative, by using equipment and tools incorrectly, by using defective tools rather than taking the time to secure serviceable ones, and by failing to use control equipment such as exhaust fans when required.

Those responsible for purchasing items for the business are often instrumental in causing situational and environmental hazards. With little consideration given to hazards, machinery may be acquired without adequate guards and other safety devices, especially when such items can be obtained at a bargain. Toxic and hazardous materials are sometimes purchased when less hazardous materials could have been used instead. Furthermore, purchasing agents sometimes fail to acquire and distribute the warning and control information provided by the vendor.

Those responsible for tool, equipment, and machinery placement may contribute to environmental and situational hazards by placing equipment and machinery with reciprocating parts where workers can be injured, by locating control switches so that the operator is exposed to the cutting tools and blades, and by locating work stations with high hazard potential where they unnecessarily expose other workers.

Shop maintenance workers may contribute to accident potential by failing to:

- adequately insulate exposed electrical wire splices, thereby increasing the possibility of an electrical shock
- detect or replace worn or damaged machine and equipment parts
- adjust and lubricate equipment and machinery on a scheduled basis
- inspect and replace worn hoisting and lifting equipment
- replace worn and frayed belts on equipment
- replace guards removed during the repair or leaving repair debris at the work site

1.6 Major Components of a Safety and Health Program

The major components of a safety and health program include:

- Inspection Programs
- Maintenance Programs
- Hazard Analysis
- Housekeeping
- Standards Compliance
- Accident Investigation

- Safety Committees
- Confine Space Entry
- Hazard Communication
- Waste Management
- Standard Operating Procedures
- First Aid
- Program Policy
- Program Objectives
- Measuring Program Effectiveness
- Assigning Responsibility and Authority
- Worker Orientation and Training
- Supervision
- Chemical Management
- Lock/Out Tag Out
- Emergency Procedures
- Record Keeping
- Management of Change
- Medical Surveillance

1.7 Developing a Policy Statement

The first step in developing a safety and health program is the development of a formal policy. See Appendix A for Model Policy Statements. The policy should be available to all personnel, state the purpose behind the safety and health program, and require the active participation of all those involved in the program's operation. The policy statement should reflect the:

- importance the business places on the health and well-being of its workers and the protection of the environment and community in which it operates
- emphasis the organization places on efficient operations with a minimum of accidents and losses
- intention of integrating hazard control into all phases of the facility
- intent of the management to bring its facilities, operations, machinery, equipment, and tools into compliance with health and safety standards and regulations

1.8 Establishing Program Objectives

Critical to the design and organization of a safety and health program is the establishment of objectives that are consistent with the organization's goals and which guide the program's development. Safety and health program objectives include:

- gaining and maintaining support for the program
- motivating, educating, and training those involved in the program to recognize and correct or report hazards
- engineering hazard control into the design of machines, tools, and production facilities
- providing a program of inspection and maintenance for machinery, equipment, tools, and other items
- incorporating hazard control into worker training complying with established safety and health standards

1.9 Providing for an Adequate Budget

In addition to those funds traditionally associated with training and professional development, adequate funds must be allocated for the safety and health program. Small businesses should define their safety and health program needs and prepare short and long-range (three to five years) budget projections.

1.10 Establishing Responsibilities

Responsibility for the safety and health program should be established at the following levels:

1. Managers. Managers are required to safeguard employees' health by ensuring that the work environment is adequately controlled. They must be aware of those operations that produce airborne fumes, mists, smokes, vapors, dusts, noise, and vibration and those operations that have the capacity impair health or cause discomfort among the workers. Managers must be aware that occupational illnesses can begin in the work place but may not be fully realized until years later.

2. Supervisors. Supervisors are in strategic positions to control hazards. Without their full support, the best designed health and safety program will not be effective. Their leadership and influence ensure that safety and health standards are enforced and upheld and that standards and enforcement are uniform throughout the facility. The responsibilities of supervisors include but are not limited to the following:

- train and educate workers in methods and procedures which are free from hazards
- demonstrate an active interest in and comply with safety and health policies and regulations
- actively participate in and support company policy and safety and health committees
- making certain that materials, equipment, and machines slated for distribution to the work areas are hazard free or that adequate control measures have been provided
- making certain that equipment, tools, and machinery are being used as designed and are properly maintained
- keeping abreast of accident and injury trends occurring in their area and to take proper corrective action to improve these trends
- investigating all accidents occurring under their supervision
- ensuring the enforcement of all hazard control rules, regulations, and procedures
- requiring that a Job Hazard Analysis be conducted for each operation
- anticipating hazards associated with process changes

3. Workers. Workers are critical for ensuring the success of the safety and health program. Highly trained workers who actively participate in the program can make the greatest contribution to the program. Worker responsibilities include:

- following safety and health rules and regulations
- recognizing and reporting hazardous conditions or work practices
- using protective equipment, safety devices, tools, and machinery properly
- reporting all injuries or exposure to toxic substances to the supervisor as soon as possible
- participating in training sessions and health and safety committee activities

4. Purchasing Agents. Those responsible for purchasing items for the small business are in a key position to help reduce hazards associated with business operations. Some of the specific responsibilities of those who purchase items include:

- ensuring that tools, equipment, and machinery are purchased with consideration for health and safety and with adequate protective devices
- obtaining adequate information on the health hazards associated with substances and materials
- striving to purchase effective materials with the lowest possible health and safety impact

5. Maintenance Personnel. Maintenance personnel play an important role in reducing accidents. Responsibilities of those in maintenance include:

- performing construction and installation work in conformance with good engineering practices
- complying with safety and health standards
- conducting planned preventive maintenance on electrical systems, machinery, equipment, and ventilation to prevent abnormal deterioration, loss of service, or safety and health hazards
- providing for the timely collection and disposal of scrap materials and waste
- actively participating in and supporting the safety and health committee

6. Safety Committees. Generally, safety and health committees are comprised of representatives from management, maintenance, supervisors, and workers. Their responsibilities include:

- inspecting for the presence of safety and health hazards, advising management of the presence of any such hazards, and providing recommendations for their correction
- examining the business' safety and health practices and the safety information contained in materials and instructions
- evaluating the acceptability of safety devices and personal protective equipment to be purchased
- conducting accident investigations

1.11 Monitoring Program

In order to maintain control over hazards in the facility, small business owners must implement a safety and health monitoring program. The purpose of the monitoring program is to detect potential hazards, provide early and effective countermeasures, and measure the effectiveness of the program. Management must decide what the program should yield in terms of reduced accidents, injuries, or illnesses. Metrics specific to the program should be developed in order to measure its performance.

Summary

In order to ensure a safe work place, an effective safety and health program must be developed. Safety and health programs are successful at reducing the number of accidents and removing hazards only when the causes and nature of accidents are fully understood and the managers and supervisors actively support and promote the program. A reduction in work place accidents will save small businesses from undue financial losses and prevent countless injuries, illness, and deaths.

Bibliography

Dade County, Fla., Safety for Industrial Education and Other Vocational Programs, School Board Policies, extracted from Policies and Regulations of Dade County Public Schools, no date.

DiBerardinis, Louis J., "Handbook of Occupational Safety and Health". Part I. John Wiley & Sons, N.Y., N.Y., 1999.

DiNardi, Salvatore R., "The Occupational Environment – Its Evaluation and Control", Chapters 37 and 41, AIHA Press, Fairfax, VA., 1997.

Firenze, Robert J., "Guide to Occupational Safety and Health Management", Kendall/Hunt Publishing Company, Dubuque, Iowa, 52001. 1973.

Firenze, Robert J., "The Process of Hazard Control", Kendall/Hunt Publishing Company, Dubuque, Iowa, 52001. 1978.

Geller, E. Scott. "The Psychology of Safety", Chilton Book Co., Radnor, PA. 1996.

National Safety Council. (1998). Accident Facts. Chicago: IL, National Safety Council.

National Safety Council. (1998). Accident Prevention Manual, 11th ed. (Administration & Programs), Chicago: IL, 60611.

U.S. Department of Health, Education, and Welfare, National Institute for Occupational Safety and Health, "Evaluation and Control of Workplace Accident Potential," January 1978.

U.S. Department of Health, Education, and Welfare, National Institute for Occupational Safety and Health, Maintaining Facilities and Operations, September 1979.

2

Essential Processes in Hazard Control

IN ANY PROGRAM designed to reduce or eliminate hazards, five essential processes must be addressed. The essential processes include:

1. hazard identification and evaluation
2. management decision making
3. establishing corrective and preventive measures
4. safety and health inspections
5. evaluation of program effectiveness

This chapter identifies and highlights the important aspects of assessing, eliminating, and controlling hazards in work place. The chapter also introduces managers to the method for locating and evaluating hazards, decision making in hazard control, the installation of preventative and corrective measures, the role of safety and health inspections, and safety and health program evaluations.

2.1 Step One: Hazard Identification and Evaluation

The first process in a comprehensive hazard control program is to identify and evaluate hazards located in the workplace. See Appendix B for a Self-Inspection Check List. These hazards are associated with machinery, equipment, tools, chemicals, work methods, and the physical plant. The goal of this first process is to:

- acquire information about what specific hazards exist or could develop, that is, conduct a hazard analysis
- rank hazards according to their potential negative impacts
- estimate the probability of the hazard resulting in an accident or release

Under such a system, efforts can be directed toward correcting problems with the most serious consequences and the highest probability of occurrence first, while leaving other hazards to be corrected at a later date.

Anyone involved in coordinating the early phases of a safety and health program for a business has many ways to acquire information about hazards associated with their operations. Examples are:

- conduct a Shop Operations Hazard Analysis
- conduct a thorough walk-through of the workplace, looking for hazards
- interview those with experience in specific operations, especially including line operators
- review surveys, inspections, and facility inspection reports conducted by local, state or federal enforcement organizations, where conditions in the shop can be evaluated against established safety and health standards and company policy
- review all injury, illness, and accident data associated with the area

2.1.1. Other sources of Information

1. The Occupational Safety and Health Administration (OSHA) for hazard information from OSHA inspections of industrial workplaces similar to those in the industrial work place.
2. The National Institute for Occupational Safety and Health (NIOSH) for safety and health hazard information.
3. Another source for acquiring important hazard information is OSHA compliance information, which describes violations in industrial settings that may appear in the small business.
4. The OSHA website: http://www.osha.gov/.

2.1.2 Conduct a Hazard Analysis

A hazard analysis is a procedure used to review processes, equipment, and procedures in order to uncover and correct hazards. These hazards may have been overlooked in the layout of the facility and in the design and location of machinery or may have developed as the equipment or process had been modified overtime. Specific methods for conducting hazard analysis are more fully described in Chapter 3.

The benefits of operations hazard analysis include:

- identifying specific hazardous conditions and potential hazards
- providing information to help establish effective control measures
- determining the skill level and physical requirements that workers need to perform specific tasks
- discovering and eliminating unsafe procedures, actions, or worker positions

2.1.3 Ranking Hazards According to Potential Negative Impacts

Once hazards have been identified, the next step is to rank these hazards according to their potential negative impacts. By judging hazards according to established criteria, it is possible to determine which hazardous conditions warrant immediate action, those of a secondary priority, and which can be addressed in the future. Without such a system, there can be no consistent guide for corrective action. Even worse, if time is not taken to rank hazards on a "worst first" basis, efforts and resources could be directed toward problems of lower consequence while those that pose a greater risk are overlooked. Table 2.1 provides suggested categories for ranking hazards.

TABLE 2.1
Hazard Evaluation Scheme

Hazard Consequence Category	Explanation
I. Catastrophic Hazard	The hazard is capable of causing death, possible multiple deaths, widespread occupational illnesses, and loss of facilities.
II. Critical Hazard	The hazard can result in death, injury, serious illness, and property and equipment damage if not corrected as soon as possible.
III. Marginal Hazard	The hazard can cause injury, illness, and equipment damage, but the injury, illness, and equipment damage may not be serious or permanent.
IV. Negligible Hazard	The hazard will not result in a serious injury or illness. The potential for the hazard causing damage beyond a minor first aid case or minor property damage is extremely remote.

2.1.4 Estimating Hazard Probability

Once hazards are uncovered and ranked according to their potential negative impact, the next step is to estimate the probability of the hazard resulting in an accident or loss. When the hazards have been ranked according to both sets of criteria, it is easy to determine where controls can be implemented. A hazard rated IA, for example takes priority over a rating of ID. Table 2.2 provides one approach to estimating probability.

TABLE 2.2
Hazard Probability Category

Hazard Probability Category	Qualitative Estimate of Probability
A	Probable
B	Reasonably Probable
C	Remote
D	Extremely Remote

2.2 Step Two: Management Decision Making

Once hazards are identified and ranked according to their negative impact, a report is prepared and submitted to management. The purpose of this approach is to enable management to make informed decisions in order to upgrade their operations and training methods while reducing accidents, injuries and other undesirable situations. When managers receive the information, they can choose to take no action; they can modify the shop workplace, methods, or procedures; or they can redesign the shop workplace or its components.

When managers choose to take no positive steps to correct hazards uncovered, it usually is because of one of three reasons:

1. they do not appreciate the need for and the purposes of hazard control as an integral part of doing business
2. there are insufficient resources such as staff, capitol, and time
3. they are presented with only the best and most costly solutions leaving no intermediate plans to choose from

A decision to not correct a known hazard is not only a poor management decision, it is also unethical.

When management chooses to modify the workplace, provide training, they do so with the idea that their programs are generally acceptable and that, with the reported deficiencies corrected, their programs' performance will be improved. Some examples of modification alternatives include:

- acquiring and installing machine guards
- acquiring and using safety equipment, such as ground-fault circuit interrupters, and combustible gas indicators
- modifying training techniques
- isolating hazardous materials and processes
- replacing hazardous materials and processes with non- or less hazardous ones
- acquiring and using personal protective equipment

Although redesign is not a popular alternative, it is sometimes necessary. When redesign is selected, managers must keep in mind that they are going to have to deal with certain problems including the fact that the redesign usually involves substantial cash outlay and inconvenience. Another problem that should be anticipated is that the new design may contain hazards of its own. For this reason, whenever redesign is offered as an alternative, those making such a recommendation must establish and execute a plan to detect problems in the early stages and eliminate or reduce them before they present new hazardous conditions.

2.2.1 Formats for Decision Making

One method to expedite decision making regarding actions for hazard control is to display hazard information for decision making as illustrated in Table 2.3.

2.3 Step Three: Implementing Controls

After hazards have been identified and evaluated and the information needed for informed decision making has been developed, the next step involves the implementation of control measures. Hazardous conditions can be controlled or eliminated at its source, through its path, or at the receiver(s).

The first and perhaps best solution is to control a hazard at its source. One example of controlling a hazard at its source is to eliminate a harmful material from a process or substitute it with a less harmful material. The second alternative for controlling hazards is to affect the path of the hazard by erecting a barricade or shield between the hazard and the worker. Examples of such controls are machine guards to prevent a worker's hands from making contact with an unguarded hazard such as a saw blade,

TABLE 2.3
Form for Hazard Control Decision Making
Sample Record of Safety and Health Deficiencies and Specifications for Corrective Action

Deficiency No.	Date Recorded	Description of Hazardous Conditions	Specific Location	Identification of Acceptable Standards	Hazard Rating		Recommended Corrective Action and Responsible Staff	Estimated Costs Required for Correction	Date Deficiency Corrected	Actual Costs of Correction
					Conseq.	Prob.				
MS-1	02/01/99	Ungrounded tools and equipment	Through-out shop	OSHA: Subpart S National Electrical Code, Article 250; 4S	1	A	Training and education	$1000	02/01/99	$1,000
							Purchase materials and equipment	$2,000	02/22/99	$2,000
							Modify methods	$2,000	04/03/99	$1,900
								$5,000		$4,900

protective curtains to prevent worker exposure to welding lights, and a local exhaust system to remove toxic vapors from the workers' breathing zone.

The third alternative is to focus control efforts at the receivers, which are the workers. Examples of this approach include removing the worker from the hazardous location by employing automated or remote control options (automatic feeding devices on planers, or shapers); and providing personal protective equipment when all options have been exhausted and it is determined that the hazard does not lend itself to correction through substitution or engineering redesign.

Protective equipment should be used as a last resort when there is no immediately feasible way to control the hazard by more effective means. Personal protective equipment is also useful as a temporary measure, while more effective solutions are being installed. There are, however, major short-comings associated with the use of personal protective equipment. These shortcomings include the fact that:

- personal protective equipment is often uncomfortable and difficult to wear properly
- the protection factors provided by the devices may not be adequate in all work situations
- if the protective equipment fails for any reason, the worker is exposed to the full effects of the hazard
- the devices require constant care, maintenance, and proper fit testing
- the protective equipment may be cumbersome and interfere with the workers ability to perform a task

2.4 Step Four: Safety and Health Inspection and Audits

The fourth step in the hazard control process deals with the inspection and monitoring of activities in order assess the effectiveness of existing controls. It is necessary to provide safety and health inspections and audits in order to provide assurance that hazard controls are working properly, to ensure modifications have not altered the workplace so that hazard controls can no longer function adequately, and to discover new hazards.

2.5 Step Five: Program Evaluation

The fifth element in the hazardous control program deals with evaluating the effectiveness of efforts to improve the overall quality of safety and health within the business. When managers are conducting program evaluation, they should consider the:

1. benefits of the hazard control program
2. impact those benefits have on improving the overall safety and health program
3. costs of locating and controlling hazards

When determining the effectiveness of their safety and health program effort, managers should address the following items:

- improved reporting of near-misses
- improved handling of worker complaints
- number of injuries compared to exposure hours
- cost of medical care
- material damage costs
- facility damage costs
- equipment and tool damage or replacement costs
- number of days lost from accidents and illnesses

Summary

The five essential processes for hazard control are:

1. Hazard identification and evaluation – the process of conducting a hazard analysis and then ranking the hazards according to potential negative impacts and probability of occurrence.
2. Management decision making – the process of providing management with the information they need to take the proper actions regarding hazard control.
3. Establishing corrective and preventive measures – the process of implementing controls to address the hazards identified in the hazard analysis.
4. Safety and health inspections - the process of inspecting and monitoring activities in order to assess the effectiveness of existing controls.
5. Evaluation of program effectiveness – the process of evaluating the effectiveness of efforts to improve the safety and health of the business.

Bibliography

ACGIH, (1995) "Industrial Ventilation: A Manual of Recommended Practice" 22 edition. Lansing, MI

Birkner, L. R., R.K. McIntyre (1997) "Anticipating and Evaluating Trends Influencing Occupational Hygiene" Chapter 50 in The Occupational Environment – Its Evaluation and Control, AIHA, Fairfax, VA.

Burgess, W.A. (1995) "Recognition of Health Hazards in Industry" John Wiley, New York.

Firenze, Robert J., (1978). The process of Hazard Control. Dubuque, Iowa: Kendall/Hunt Publishing Company.

Horowitz, M.R., M.F. Hallock (1999) "Recognition of Health Hazards in the Workplace" Chapter 4, in the Handbook of Occupational Safety and Health, John Wiley Publisher, N.Y., N.Y.

3

Hazard Analysis

THIS CHAPTER describes the purpose and uses of hazard analysis, the steps required to set up and conduct a hazard analysis, and solution alternatives from hazard analysis information. With the techniques of hazard analysis, the small business manager and supervisor can create a more effective and thorough safety and health program. As a general rule, there should be hazard analysis for every existing and new process and operation conducted.

Hazard analysis is a technique that will:

- identify the essential requirements for performing specific tasks safely
- uncover hazards that may have been overlooked in the original design or setup of a particular process, operation, or task
- reduce the potential for accidents, injuries and illness
- increase effectiveness of process, operations and tasks
- provide for the best selection and placement of tools, equipment and machinery
- review training and educational approaches to uncover areas which warrant improvement and identify educational and training needs of workers

Hazard analysis can be used to:

- modify shop processes, operations, and tasks
- produce information essential in establishing effective control measures (adoption of procedural changes, special procedures, safety devices such as machine guards, protective equipment, etc.)
- locate and arrange equipment and machinery so that worker and managers will not be exposed to unnecessary hazards

- identify situational hazards (in facilities, equipment, tools, materials and operational events)
- identify human factors responsible for accident situations (worker capabilities, activities, limitations, etc.)
- identify exposure factors that contribute to injury and illness (contact with hazardous substances, materials or physical agents)
- identify physical factors that contribute to accident situations (noise, vibration, insufficient illumination, etc.)
- determine appropriate inspection methods and maintenance standards needed for safety

3.1 Participants in Hazard Analyses

In order to be fully effective and reliable, a hazard analysis should represent as many different viewpoints as possible. Every person familiar with a process or operation has valuable insights concerning potential problems, defects, and situations which could cause accidents. Department heads, maintenance personnel, and members of the business's safety and health committees all can contribute valuable information and their insights should be included in the hazard analysis.

Equally important is the input of the workers. There are four good reasons for encouraging employee participation when conducting hazard analysis. First, employees know the job and the process best and can offer some of the most practical and effective suggestions for improvement. Second, employees may have personally observed problems or have been involved in near-misses, experiences which can shed light on previously unknown hazardous situations. Third, worker involvement in the analysis process reinforces the legitimacy and importance of safety and health in the shop, helping the worker see where hazards exist and learn what can be done about them. Finally, workers are more likely to conform to procedures they have helped to develop.

Many processes, operations, and tasks conducted in the shop are candidates for hazard analysis because of their potential to cause accident situations. Some, however, are better candidates than others. In determining which processes, operations, and tasks receive priority for analysis, the following considerations should be made:

1. Frequency of Accidents. An operation or task that has repeated accidents associated with its performance is a good candidate for

analysis, especially if different workers have the same type of accident while performing the same operation or task.

2. Potential for Injury. Some processes and operations may have a low accident frequency but a high potential for major injury. For example, tasks on the grinder conducted without the use of a tool rest or tongue guard have a high potential for injury.

3. Severity of Injury. A particular process, operation, or task with a history of serious injuries is a worthy candidate for analysis even if the frequency of such injuries is low.

4. New or Altered Processes and Operations. As a general rule, whenever a new process, operation, or task is created or an old one altered because of machinery or equipment changes, a hazard analysis should be conducted. For maximum benefits, the hazard analysis should be done while the process or operation is in the planning stages.

5. Excessive Material Waste or Damage to Equipment. Processes or operations which produce excessive material waste and/or damage to tools and equipment are candidates for hazard analysis. The same problems which are causing excessive waste and damage may be the ones which could cause injuries.

3.2 Steps In Conducting A Hazard Analysis

Once the processes have been identified for analysis, the following steps are taken:

1. Break the process down into its operations and tasks.
2. Analyze the health and safety aspects of individual operations.
3. Identify potential hazards.
4. Develop recommendations for safe operations.

3.2.1 Step One: Break the Process Down into its Operations and Tasks

In performing the first step, two errors are commonly made. One is breaking the process down into an overly detailed and unnecessarily large number of tasks. The result is an analysis to cumbersome to be of any benefit. The second common error is making the operational breakdown so general that important steps are not recorded. Consequently, hazards associated with these steps remain hidden and adequate control measures are never identified.

In breaking a process down, the manager or supervisor needs to determine which operation comes first in the overall process, which major operations follow, and which major tasks complete the operation. The worker's thorough knowledge of the process should be consulted because they can provide the necessary detail for each step of the process.

In defining the major operations and tasks for a particular process, the manager or supervisor should record the process on paper in the form of a flow diagram. Flow diagrams can be made by using various symbols which define what is taking place through each step of the operation or task. The symbols in Figure 3. 1 are commonly used in constructing flow diagrams. Figure 3.2 is an example of a completed flow diagram.

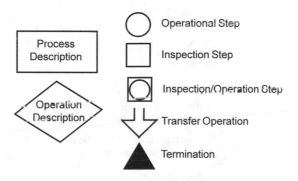

FIG. 3.1 Typical symbols for flow diagrams.

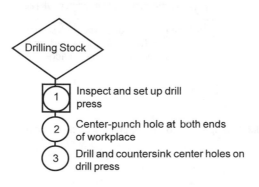

FIG. 3.2 Sample flow diagram.

3.2.2 Step Two: Analyze the Health and Safety Aspects of Individual Operations

    The information for the hazard analysis should be recorded in a format designed to maintain consistency, clarity, and logic. Table 3.1 is an example of a format for conducting analysis of shop operations hazards.

    The following is an example of a completed hazard analysis for a badly rutted grinding wheel. When an abrasive wheel is rutted, worn or out of balance, it can damage the machine, injure the operator and produce poor work. Dressing the grinding wheel means removing a large area of the face to restore the wheel to good condition. If a wheel cannot be balanced by truing and dressing, it should be removed from service. Figure 3.3 is flow diagram for the rutted grinding wheel and Table 3.2 is a completed hazard analysis form for this operation. Figure 3.4 is an example of a stand grinder without adequate controls.

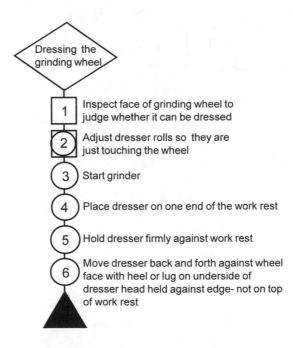

FIG. 3.3 Example of a flow diagram for a badly rutted grinding wheel.

TABLE 3.1
Operations Hazard Analysis Format

Process/ Equipment	Major Operations	Task	Potential Hazard	Conditions Triggering Hazard into Accident	Hazard Consequence Classification	Hazard Probability Category	Procedural Requirements	Safety & PPE Requirements	Special Instructions
(specific hazardous process or equipment)	(operations required to complete the process)	(activity required to complete an operation)	(situation which, when triggered by factors in the workplace, has potential to cause an accident or release)	(conditions brought about by human error, situational or environmental factors which cause a hazardous situation to result in an accident or release)	(assessment of hazardous consequence: I-Catastrophic, II-Critical, III-Marginal, or IV-Negligible)	(assessment of the probability of hazard result in an accident: A-Probable, B-Reasonably Probable, C-Remote, or D-Extremely Remote)	(actions to be taken to eliminate or reduce hazards)	(necessary items required to reduce the possibility of injuries and illness while performing operations and tasks)	(special procedures or actions required to ensure safety and health)

TABLE 3.2
Sample Hazard Analysis

Process/ Equipment	Major Operations	Task	Potential Hazard	Conditions Triggering Hazard into Accident	Hazard Consequence Classification	Hazard Probability Category	Procedural Requirements	Safety & PPE Requirements	Special Instructions
Turning between centers/ Pedestal grinder	Dressing the grinding wheel	1 — Inspect face of grinding wheel to judge whether it can be dressed							Round off wheel edges with hand stone before dressing to prevent edges from chipping
		2 — Adjust dresser rolls so they are just touching the wheel	Misalignment causing wheel to come in contact with dresser rolls when grinder is started	Wheel binding against dresser rolls and possibly degrading when grinder is started	II	A	Adjust work rest so dresser rolls are just touching grinding wheel	Face shield, work apron	
		3 — Start grinder	Flying chips or debris coming from wheel	Starting grinder	II	B	Operator should stand to one side of grinder when starting machine. Upper peripheral guard must be in place	Face shield	Point of operation guards

TABLE 3.2 (Continued)

Process/ Equipment	Major Opera- tions	Task	Potential Hazard	Conditions Triggering Hazard into Accident	Hazard Consequence Classification	Hazard Probability Category	Procedural Requirements	Safety & PPE Requirements	Special Instructions
			Wheel or spindle ends catching clothing	Starting grinder	II	A	Roll up or clip sleeves		Point of operation guards
			Excessive vibration	Not dressing wheel properly	II	A	Shut down equipment	Equipment adequately grounded or GFCI	
			Wheel exploding	Running grinder at excessive RPM with crack or other defect	I	B	Following instructions for RPM		
		④ Place dresser on one end of the work rest					Make sure that lugs of dresser are against edge of work rest		Use dressing wheel approved for job. Never use lathe cutting tool

TABLE 3.2 (Continued)

Process/ Equipment	Major Operations	Task	Potential Hazard	Conditions Triggering Hazard into Accident	Hazard Consequence Classification	Hazard Probability Category	Procedural Requirements	Safety & PPE Requirements	Special Instructions
		(5) Hold firmly against edge or work rest	Excessive pressure on wheel	Operator exerting excessive pressure on dresser rolls	II	A	Exert only sufficient pressure to allow dresser to remove small amount or abrasive wheel at one pass; roll up or clip sleeves; remove rings	Face shield; hair nets	Hands should be free of grease or oil
		(6) Move dresser back and forth against face of wheel, with heel or lug on underside of dresser head held firmly against edge, not on top of work rest	See 5 above	See 5 above	II	A	See 5 above	See 5 above	See 5 above. Round off wheel edges with hand stone after dressing to prevent edges from chipping

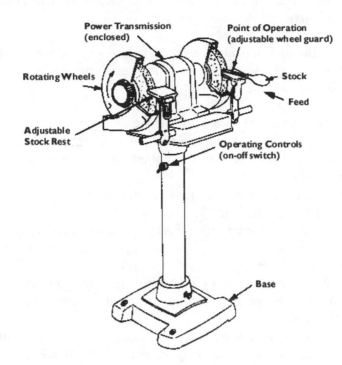

Fig. 3 4 Stand grinder without adequate controls.

3.2.3 Step Three: Identify Potential Hazards

After selecting the process or operation worthy of analysis, breaking the process down into operations and tasks, and obtaining an analysis format for recording information acquired during the analysis, the analyst is now ready to identify the potential hazards and conditions which trigger the hazard into an accident. Potential hazards will not only be associated with human, situational, and environmental factors but also will be associated with the process, operations, and tasks under examination. For each of the tasks identified, the analyst should address the following questions:

1. Is there a danger of striking against, being struck by, or otherwise making injurious contact with an object?
2. Can the worker be caught in, on, or between the object?
3. Can the worker slip, trip, or fall?
4. Can he strain himself by pushing, pulling, or lifting?

5. Is the environment hazardous due to the presence of noise, heat, cold, toxic gas, fumes, vapor, mist, or dust?

To assist the analyst in identifying hazards associated with a particular operation or task, a review of the following basic causes of accidents is of value:

1. Failure to give necessary instruction, giving incorrect instructions, or inspections not made according to an acceptable schedule.
2. Failure of person in charge to properly plan or supervise operations and tasks.
3. Failure to plan or establish safe work practices.
4. Use of unsafe methods.
5. Inexperienced or unskilled worker performing tasks beyond their capabilities.
6. Failure to maintain order and discipline.
7. Failure to enforce safety and health rules and regulations.
8. Improper design, construction, or layout of equipment, tools and machinery.
9. Unsafe housekeeping, poor ventilation, or inadequate lighting.
10. Protective devices, proper equipment and tools required to perform the task safely are not provided.
11. Failure to maintain tools, equipment, and machines.
12. Workers not following instructions.
13. Improperly inspected or maintained equipment or safety devices.
14. Failure to use prescribed protective devices.
15. Workers using defective or insufficient protective devices.

Once the hazards have been identified to the extent that knowledge and experience allows, attention must be given to hazard control strategies in which the analyst considers all possible ways to eliminate potentially hazardous situations. Possibilities include changing tools and equipment or substituting less toxic materials for hazardous ones.

Another approach to controlling the effects of hazards is to reduce the possibility of an injury if an accident does occur. For example, if the potential of a grinding wheel exploding exists, then action must be taken to protect the worker from injury by installing a shield or guard. Another alternative might be to decrease exposure to the hazard by removing from the path of the hazard those workers who are not required to be there to perform a task. Personal protective equipment is selected either as a last alternative or as a secondary protection to be used along with machine guards.

3.2.4 Step Four: Develop Recommendations for Safe Operations

The final step in the hazard analysis process is to develop recommendations for safe operations and work tasks to prevent the occurrence of accidents. The principal solutions are to find a new way to perform the operation or task, remove or alter the physical and environmental conditions that create the hazards, or change procedures.

To find an entirely new way to perform an operation or task, it is necessary to determine the objective of the operation and then analyze the various ways of reaching this goal to determine which is safest. If a new process cannot be found for the operation or task, then a change in physical conditions (tools, materials, equipment, or location) or environmental conditions (noise, or ventilation systems) is needed in order to eliminate the hazard or prevent the accident.

The third solution in eliminating or reducing hazards in the business shop is to investigate changes in the tasks each worker must perform to complete an operation. The analyst should determine what the worker can do to eliminate the hazard or prevent a potential accident and how the worker should do it.

Instructions must be specific and concrete if new procedures and work tasks are to be effective. Vague precautions such as "be careful," "don't make mistakes," and "do what you are told," are not helpful. Instructions should state clearly what to do and how to do it, using such action words as "remove," "adjust," and "insert." Sometimes instructions will call attention to potential malfunctions or hazards, for example, "remove small chips with brush instead of with hands."

Suppose the hazardous situation concerns the possibility that a worker may be injured while sharpening a tool on the grinder because the tool rest is not adjusted properly. The instruction to "Make certain that the cutting tool does not get pulled down into the space between the tool rest and the abrasive wheel" is not helpful. It does not tell the worker what to do to prevent the cutting tool from being caught in the space between the tool rest and the wheel. The instruction should be revised to state that "Before sharpening a cutting tool on the grinder, be sure to adjust the tool rest to within 1/8" or 3 mm. of the abrasive wheel and maintain this distance by adjusting the tool rest as the wheel wears down." In this way, the instruction clearly specifies what to do and how to do it.

Summary

A hazard analysis is conducted for each process and operation in order to identify hazards and reduce the potential for accidents, injuries, and illness. Every person familiar with a process or operation, including department heads, maintenance personnel, and workers should be consulted during the hazard analysis. The four steps for conducting a hazard analysis include:

1. Breaking the process down into its operations and tasks.
2. Analyzing the health and safety aspects of individual operations.
3. Identifying potential hazards.
4. Developing recommendations for safe operations.

Bibliography

Accident Prevention Manual for Industrial Operations, 7th ed., National Safety Council, Chicago, IL 60611.1974.

Hawkins, N.C., S.K. Norwood and J.C. Rock (1991) "A Strategy for Occupational Exposure Assessment", AIHA Publication, Akron, OH.

Horowitz, M.R., M.F. Hallock (1999) "Recognition of Health Hazards in the Workplace" Chapter 4, in the Handbook of Occupational Safety and Health, John Wiley & Sons, N.Y., N.Y.

Lee, Susan B., B. Johnson (1997) "Biohazards in the Work Environment" Chapter 18 in The Occupational Environment - Its Evaluation and Control, AIHA Publication, Fairfax, VA.

U.S. Department of Health and Human Services, National Institute for Occupational Safety and Health, "Evaluation and Control of Workplace Accident Potential", January 1978.

4

Safety Committees

A SAFETY and health committee is a group appointed to aid and advise management on matters of safety and health pertaining to operations. In addition, it performs essential monitoring, educational, investigative, evaluative tasks, and provide an official channel for combining the knowledge and experience of many people in order to accomplish the objectives of reducing hazards and losses.

The committee, depending on its type, is composed of members of representing key components of the organization. Safety and health committees provide a means whereby hazard information and suggestions for hazard control can travel between employees and management. Because the information flow is facilitated, managers can process problems and take action more expediently. To establish and maintain a safety and health program of high quality, the full cooperation of everyone in the business is required. The safety and health committee is a vehicle for obtaining this cooperation while effectively and efficiently distributing the work load.

This chapter highlights the formation and functions of safety and health committees in the small business facility. Like any other kind of committee, the safety and health committee can either provide a valuable service or end up as nothing more than a social circle which is neither productive nor efficient. What makes one committee highly effective while another fails? The answer lies in part with the original purpose of the committee, how well it is structured and staffed, and the support it receives while carrying out its responsibilities. See Appendix C, Codes of Safe Practices. This chapter introduces management to the value of safety committees, the types of committees, and the way to organize and operate committees.

4.1 Types of Safety and Health Committees

There are two basic types of safety and health committees used in small business settings: policy and shop committees. In a smaller business these two committees may need to be combined into one group of six to eight persons, representing employees, supervisors, and maintenance personnel. Usually, the Safety and Health Policy Committee is composed of top level management, at least one supervisor and a workers' representative where applicable. The work of the policy committee includes:

- identifying, defining, and studying problems which have a significant impact on the safety and health of workers
- acting on or evaluating the effectiveness of recommendations from the Shop Safety and Health Committee
- assessing recommendations in the light of appropriations and setting priorities for expending funds to improve safety and health
- reviewing and updating all safety and health regulations
- promoting and evaluating training in hazard recognition and control
- standardizing disciplinary courses of action for noncompliance with safety and health rules and regulations
- reviewing accident and illness reports
- reviewing employee suggestions to improve safety and health

The Safety and Health Policy Committee also gives guidance to the Shop Safety and Health Committee, to which it refers specific tasks. The Shop Safety and Health Committee, unlike the Policy Committee, is the group that promotes safety and health at the production level. It consists of department heads, maintenance personnel, supervisors, and a workers' representative. The work of this committee includes:

- conducting periodic inspections to detect hazardous physical and environmental conditions as well as unsafe procedures and practices
- studying and evaluating accident and injury data
- investigating accidents
- conducting safety and health training and evaluation programs
- conducting hazard analyses of all processes and operations
- reviewing and upgrading instructional materials and training methods
- field-testing safety and health equipment and making recommendations to the Policy Committee

- studying the implications of changes in manufacturing processes, operations, and tasks
- recommending actions to be taken by the Policy Committee to eliminate or reduce hazards
- considering ways of improving the effectiveness of rules, regulations, and procedures to promote safety and health
- promoting first aid training for all staff

4.2 Organizing Committees

Top management must keep in mind two important requirements in organizing the committee. First, management should ensure the committee is small and effective. Second, management should appoint a chairman who is respected and who demonstrates interest in the area of safety and health. The manager must impress upon the chairman that committee members, no matter how good their intentions, should be constantly motivated and encouraged to take on additional responsibilities while still maintaining their planning, training, and evaluative duties. Thus, the person selected to chair the committee must fully understand the importance of and actively support an effective safety and health program.

The ideal size of the Policy Committee is five people, representing management, department heads, and supervisors. The safety and health committee, might range in size from five to ten persons. The committees should be of manageable size so that each member will have the opportunity to participate. Committee membership should be staggered so that the term of office of all committee members does not expire at the same time, destroying the continuity of committee programs.

Committee members will, for the most part, need safety and health training and education before they can be effective. Consequently, managers must see to it that committee members receive basic instruction in the principles and techniques of hazard recognition and control.

To arrive at recommendations, the safety committee needs an established procedure. If a business has only one safety and health committee, recommendations will go directly to the management/owner. In businesses with two committees, the Shop Safety and Health Committee will make its recommendations to the Policy Committee, to which the administration has delegated its authority.

In addition, the following procedures are suggested:

1. When a committee member makes a recommendation, it should be discussed by the entire committee to determine if the recommendation is acceptable, if it needs modification before being submitted to the policy committee, or if it should be rejected.
2. Any recommendation regarding safety and health must be referred directly to the safety and health committee. A recommendation received from outside the committee should be subject to the same procedure described above: committee discussion, leading to acceptance, modification, or rejection.
3. Recommendations which have been accepted by the committee should be submitted in writing to the policy committee.
   a. Recommendations should state what is to be done, by whom and when. Committees should avoid making recommendations to do something "at once," "as soon as possible," "when funds are available," "when convenient" and so forth. Such recommendations are either too demanding or too vague and transfer the responsibility of deciding priorities to the doer.
   b. The written statement should explain why a course of action is recommended, information which also should be included in the committee meeting minutes. Such explanations should make clear how the recommended action will reduce or eliminate hazards and improve safety and health within the company.
   c. How the recommendation is to be carried out should be stated only if the "method" is a condition of the recommendation.

4.3 Conducting Committee Meetings

The chairman of the Shop Safety and Health Committee should establish a date and time for each committee meeting, taking into account the schedules of members. The committee should meet at least once a month and carry out assignments between meetings. The Policy Committee meets as needed, at the discretion of its chairman.

At least five days in advance of a committee meeting, notice of the meeting and the agenda should be sent to each committee member. By receiving the agenda beforehand, committee members have time to think about the topics to be considered and organize their ideas and opinions at a leisurely pace. A good practice is to include a copy of the minutes of the last meeting with the agenda. Whenever possible, meetings of the safety and health committee should last no longer than one hour. In order to accomplish an entire agenda in that time, the committee chairman must take time beforehand to organize the business at hand into a reasonable schedule.

During committee meetings, the chairman must keep the program on course. It is easy for the chairman, as coordinator, to slip out of the leader's role and find himself doing the leg work of committee members. When the chairman steps out of his role, even for a short period of time, he frustrates the members. On the other hand, if the chairman conducts meetings in such a manner that he makes all the recommendations himself, the role of the committee is destroyed. Instead of the members jointly participating in decisions, they are forced into accepting one person's opinions. If such a situation is allowed to continue, the effectiveness of the committee will be drastically reduced and soon committee members will refuse to participate.

The committee should adopt a formal set of rules to govern the conduct of the meeting. Committee meetings are more productive when minutes of proceedings are kept. These minutes need not be elaborate but should be complete and informative and should be reviewed by the chairman after each meeting. Minutes taken by the committee secretary should follow a pattern similar to the following:

- date of meeting
- time meeting opened
- members present
- members absent
- minutes of previous meeting read and approved or disapproved
- unfinished business (including issues involving recommendations not yet resolved)
- recommendations completed since last meeting
- new business
- new recommendations
- listing and discussion of accidents which occurred since last meeting (date of injury, employee, cause, recommendations, etc.)
- other committee remarks

- committee resolutions
- time meeting adjourned and date of next meeting
- secretary's signature
- space for signature of reviewing official and date
- action items
- time meeting adjourned

4.4 Effectiveness of Safety Committees

The effectiveness of safety committees depends on several factors including:

- regular meeting times and regular attendance at meetings indicate interest
- committee members must be sincerely interested in the safety and health program and be willing to cooperate with others to improve that program
- committee recommendations and suggestions must be carefully considered and acted upon
- the committee's work and accomplishments should receive recognition, either by means of public announcement or via letters of commendation.

Summary

The safety and health committee performs an important role in developing and maintaining an effective safety and health program. The committee should be comprised of six to eight persons representing employees, supervisors and maintenance personnel and be led by a chairperson who can keep the members interested and motivated. A formal set of rules for conducting safety and health committee meetings should also be developed.

Bibliography

Colvin, Rosemary, R. Colvin (1999) "Management's Roles and Responsibilities in an Effective Safety and Health Program" Chapter 2 in the Handbook of Occupational Safety and Health, John Wiley & Sons, New York.

Leibowitz, Alan (1997) "Program Management" Chapter 37 in The Occupational Environment - It's Evaluation and Control, AIHA Publication, Fairfax, VA.

State of Maryland, Department of Labor and Industry, Safety Engineering and Education Division, (1967). "Safety Program Safety Committee Manual"

U.S. Department of Labor, OSHA, (1975). Organizing a Safety Committee, OSHA 2231, June.7

5

Accident Investigation Techniques

IN PREVIOUS chapters, the importance of identifying and evaluation hazards in the workplace to reduce or eliminate the potential for accidents was emphasized. While the idea behind this concept is sound and in fact has proven itself to be a critically desirable element of hazard control programs, there will be times when the hazards are not found before an accident occurs. When an accident occurs, management must be prepared to acquire as much information as possible about its cause(s) so that similar accidents can be avoided by establishing new control strategies and additional training.

There are four important reasons for investigating accidents that occur in the work place. The first is to determine the cause of the accident. The second reason to investigate accidents is to identify what action can be taken and what improvements are necessary to prevent similar accidents from occurring in the future. The third reason is to document the facts involved in an accident in the event of compensation and litigation. Many times a business has been placed in an embarrassing situation when an attorney for the business or for the injured party has asked the management to produce facts concerning an accident situation when few or insufficient records are available. The report produced at the conclusion of an investigation becomes the permanent record of facts involved in the accident. This ensures the availability of a complete and detailed record of the incident years after its occurrence.

The fourth reason for investigating accidents is that a thorough investigation is likely to uncover problems which indirectly contributed to the accident. Such information benefits accident reduction efforts. For example, a worker slips on an oil spill and is injured. The oil spill is the direct cause of the accident, but a thorough investigation might reveal other contributing factors such as poor housekeeping, failure to follow maintenance schedule, inadequate supervision, or faulty equipment.

In addition to the four primary reasons for investigating accidents, there are other reasons that are peripherally important. One is that the investigation can provide information on the direct and indirect costs associated with accidents. Direct costs include medical expenses; compensation costs; property damage; and city, state and federal penalties. Indirect costs include the supervisor's and worker's time, liability claims, increases in insurance premiums, and paper work. In addition, the investigation projects a positive image of the business' interest in safety and health and the involvement of staff in the investigative process promotes cooperation which is vital to the overall safety and health program.

As a general rule, all accidents, no matter how minor, are candidates for thorough investigation. Many accidents are considered minor because their consequences were not serious. Such accidents or "incidents" are taken for granted and may not receive the attention they demand. Managers, safety and health committees, and supervisors must be aware that serious accidents arise from the same hazards as minor "incidents." It is usually sheer luck that determines whether a hazardous situation results in a minor incident or a serious accident.

5.1 Fact-Finding Not Fault Finding

Despite what many believe, accident investigation is fact finding, not a faultfinding, process. When attempting to determine the cause(s) of an accident, the novice is tempted to conclude that the person involved in the accident was at fault. But if human error is chosen when it is not the real cause, the hazard, which caused the accident will go unobserved and uncontrolled. Furthermore, the person falsely blamed for causing the accident will respond to the unjustified corrective action with contempt and alienation. Such alienation will discourage future cooperation and undermine respect for the safety and health program. The intent of accident investigation is to pinpoint causes of error and/or defects so that similar incidents can be prevented.

When an accident occurs, it indicates that:

- something has gone wrong in the process, operation, or tasks
- someone failed to perform a task properly
- a hazardous condition existed without adequate safeguards
- a newly developed process or substance has defects and dangers, which only recently have become known

Upon determining the facts, the investigator(s) is in a position to offer suggestions involving:

- improvements in engineering
- changes in processes, operations, and tasks
- improvements in supervision
- training and education

Accident investigations must be conducted immediately following the incident. On-the-scene accident investigation provides the most accurate and useful information. The longer the delay in examining the accident scene and interviewing the victim(s) and witnesses, the greater the possibility of obtaining erroneous or incomplete information. With time, the accident scene changes, memories fail, and people talk to each other. Whether consciously or unconsciously, witnesses may alter their initial impressions to agree with someone else's observation or interpretation. Prompt accident investigation also expresses a positive concern for the safety and well-being of the employees.

5.2 Who Should Investigate?

Accident investigation logically begins with the supervisor. The supervisor is the person who is usually on the scene and can most quickly size up the situation, call for assistance, isolate the hazard, and record names of eyewitnesses. The supervisor knows the workers (their educational level, experience, and personal characteristics); the equipment, tools, and material (how they are operated, their peculiarities, and their potential to cause further damage); and the environment in which the worker and equipment must function. Furthermore, the supervisor has both a personal and professional interest in worker safety and health.

The mere physical presence, knowledge, and interest do not ensure that supervisors will make good accident investigators. But with experience, training, and guidance, supervisors will be able to make valuable contributions to the investigative process. When supervisors participate in accident investigation, their sense of involvement and responsibility increases. Denying supervisors this opportunity tends to undermine their sense of responsibility and involvement in the safety and health program.

The safety and health committee should become directly involved with investigating serious accidents, that is, those that result in injury to workers, and property damage. The safety and health committee can provide additional expertise to support and complement the specific knowledge of supervisors. The committee's involvement indicates widespread interest in hazard control and the safety and health of workers. The committee will submit reports recommending actions and improvements to keep similar accidents from recurring.

It is necessary for the safety and health committee to become involved with corrective action. The committee must routinely follow up its investigations to determine the status and effectiveness of corrective action and to provide the stimulus necessary for the effective functioning of the safety and health program. A lack of concern or involvement on the part of the safety and health committee will be reflected in the workers' and supervisors' attitudes, resulting in an ineffective program.

5.3 How to Investigate Accidents

Conducting an accident investigation is not simple. It is difficult to look beyond the details of the incident to uncover causal factors and determine the true loss potential of the occurrence and develop practical recommendations to prevent recurrence. A major weakness of many accident investigations is the failure to establish and consider all factors—human, situational, and environmental—that have contributed to the accident. Reasons for this failure are:

- inexperienced or uninformed investigator
- reluctance of the investigator to accept responsibility
- narrow interpretation of environmental factors
- erroneous emphasis on a single cause
- judging the effect of the accident to be the cause
- arriving at conclusions too rapidly
- poor interviewing techniques
- delay in investigating the accident

An experienced investigator must be ready to acknowledge as contributing causes any and all factors that may have contributed to the accident. What may at first appear to be a simple, uninvolved accident may, in fact, have numerous contributing factors which become more complex as analyses are completed.

Investigators must safeguard themselves during an investigation and protect the collected evidence. The investigator must remember:

- The scene of an accident may be more dangerous than it was prior to the accident. For example, electrical equipment may have been damaged in an accident. The investigator must make certain that the equipment is disconnected and if necessary locked out before proceeding.
- To use personnel protective equipment, if there is the possibility that the investigator may be exposed to toxic materials.
- To establish and delegate clear responsibilities to safety committee members.
- Physical evidence is sometimes mishandled, rendering such evidence useless and making it more difficult to find the cause(s) of the accident. Thus, if this evidence were needed in a legal case, the fact that it was lost or impaired would destroy its value.

5.4 What to Look For

During the accident investigation many questions must be asked. Because of the infinite number of accident-producing situations, contributing factors, and causes, it is impossible to provide a complete list of questions that will apply to all accident investigations. The following questions are generally applicable and considered most often by accident investigators:

1. What was the injured person doing at the time of the accident? Performing an assignment? Shop maintenance? Working on a personal project? Assisting another worker?
2. Was the injured working on a task he was authorized to do? Was the worker qualified to perform the task? Was the victim familiar with the process, equipment, and machinery?
3. What were other workers doing at the time of the accident?
4. Was the proper equipment being used for the task at hand (screwdriver instead of can opener to open a paint can, file instead of a grinder to remove burr on a bolt after it was cut)?
5. Was the injured person following approved procedures?
6. Is the process, operation, or task new to the shop?
7. Was the injured person being supervised? What was the proximity and adequacy of supervision?

8. Did the injured receive hazard recognition training prior to the accident?
9. What was the location of the accident? What was the physical condition of the area when the accident occurred?
10. What immediate or temporary action(s) could have prevented the accident or minimized its effect?
11. What long term or permanent action(s) could have prevented the accident or minimized its effect?
12. Had corrective action been recommended in the past but not adopted?

During the course of the investigation, the above questions should be answered to the satisfaction of the investigator(s). If other questions come to mind as the investigation continues, they should also be answered and recorded.

5.5 Conducting Interviews

Interviewing accident or injury victims and witnesses can be very difficult if the interview is not handled properly. The individual being interviewed often is fearful and reluctant to provide the interviewer with accurate facts about the accident. The accident victim may be embarrassed, afraid of disciplinary action, or hesitant to talk for any number of reasons. A witness may not want to provide information that might place blame on friends, fellow workers, or possibly himself. To obtain the necessary facts during an interview, the interviewer must first eliminate or reduce fear and anxiety by developing rapport with the individual being interviewed. It is essential that the interviewer create a feeling of trust and establish lines of communication before beginning the actual interview.

Once rapport has been developed, the following five-step method should be used during the actual interview:

1. Discuss the purpose of the investigation and the interview. Emphasize that the purpose of the investigation it to find facts relevant to the accident not to find fault.
2. Have the individual relate his version of the accident with minimal interruptions. If the individual being interviewed was the one injured, ask him to explain where he was, what he was doing, how he was doing it, and what happened. If practical, have the injured

person or eyewitness explain the sequence of events, which occurred at the time of the accident. When someone is at the scene of the accident, they will be able to relate facts that might otherwise be very difficult to explain.

3. Ask questions in order to clarify facts or fill in any gaps.
4. The interviewer should relate his understanding of the accident to the injured person or eyewitness. Through this review process there will be ample opportunity to correct any misunderstanding that may have occurred and clarify, if necessary, any details of the accident.
5. Discuss methods of preventing recurrence. Ask the individual for suggestions of how to eliminate or reduce the hazard which caused the accident. By asking the individual for his ideas and discussing those ideas, the interviewer shows sincerity and places emphasis on the fact-finding purpose of the investigation.

The following is an example of how not to conduct an interview:

Investigator: O.K., Bill, tell me how you cut your hand. Start as far back as you can remember. I have to write it all down.

Bill: Well, it was like this. I had to rip a 4' x 8' (1.2 m. x 2.4 m.) sheet of plywood on the table saw, and I knew the guard would be in the way. So I...

Investigator: You took it off!

Bill: Well, yeah, I did but...

Investigator: You know you were told never to remove the guard. If you listened, you wouldn't have been hurt.

Bill: Yes, I know, but I didn't have much choice—

Investigator: That's what they all say. Haven't you heard enough from your supervisor about never using the saw without the guard in position?

Bill (pensive look—doesn't answer)

Investigator: See what happens when you don't listen! Well, all right, be more careful in the future and follow directions. Safety is important. Do you understand that?

Bill (nods grimly)

Investigator: O.K., go back to work now and remember to be more careful and listen to your supervisor or you will be in trouble again soon.

How do you think that Bill felt at the end of this interrogation? Do you think that Bill will report a minor injury again? The investigator's first shortcoming was that he acted and sounded as if he was disgusted because he had to take the time to make out a report. He put the worker on the defensive from the start and interrupted the worker before he could finish his statements. The interviewer was impersonal and abrupt and did not express sympathy or concern for the worker's injury. He even ended the interview on a sour note and never acquired a complete understanding of the accident.

The following illustrates the correct method of conducting an interview:

Investigator: How's your hand, Bill? Does it still hurt? Did you get proper care from the health nurse?

Bill: Yes, sir, thank you. She did a good job. The cut wasn't too deep, and it should heal O.K.

Investigator: Well, Bill, I would like to take a little of your time to go over what happened. Before you tell me, I'd like to tell you why I think it is important to check out every injury occurring in our business. Quite simply, by going over accidents carefully, often a lot can be learned to prevent similar accidents in the future. Please don't take the questions I'm about to ask personally, and don't worry about admitting that you did something wrong. I'm not trying to blame anyone. What I learn from you may prevent another worker from being injured in the future.

Bill: I understand and I'll do my best to help.

Investigator: O.K., let's go over to where you were when the accident occurred. (They arrive at the table saw area in the shop.) Bill, will you explain what you were doing and how you were doing it when the accident happened? Take your time, and try to remember as many things as you can which occurred just before the accident.

Bill: I had to rip a 4' x 8' (1.2 m. x 2.4 m.) sheet of 1/2" (1.3 cm.) plywood. Since the portable circular saw was broken, I had to either cut it by hand or use the table saw. Of course I knew that the saw guard would be in the way so I removed it and made the cut with no problem.

Investigator: Well, then, how did you cut your hand?

Bill: I know it sounds dumb, but when I was placing the guard back on the saw, I leaned over to make sure it was aligned correctly. As I did, I slipped on something and pushed my hand onto the blade.

Investigator: Was there sawdust, oil or anything else on the floor in the area of the saw? — like this sawdust (pointing to the floor)?

Bill: There probably was. Some guys never clean up after finishing their work.

Investigator: Well, Bill, let me see if I have a clear picture of this. You didn't cut your hand while the saw was running. You did it while installing the guard that you removed to get your job done. While you were installing it, you slipped and jammed your hand against the blade.

Bill: Well, not when I was putting it on-when I was making the final alignment.

Investigator: O.K. Bill, how do you think this accident could have been prevented?

Bill: If the portable saw had been available, I probably wouldn't have got hurt.

Investigator: We'll get that saw fixed right away. What about the table saw?

Bill: I should have told my supervisor and got some help. Also, if we had one of those swing-type guards I've seen, I wouldn't have had to remove the guard.

Investigator: That's a good idea. I'll recommend that we look into purchasing one. Any suggestions about the sawdust on the floor?

Bill (smiling): I'll make sure that I sweep the area around shop equipment before I go to work. It's no big deal.

Investigator: O.K., Bill. I appreciate your cooperation and your suggestions. One thing more: the next time you need to change a standard procedure in order to get your work done, let your supervisor know about it before you start.

Bill (grinning): All I can say is I will. Thanks.

The preceding example suggests ways to conduct an interview. The investigator was friendly but at the same time created an image of competence. He took the time to tell Bill what he was doing and why. He conducted his interview at the scene, not in an office far removed from where the accident occurred. He listened without interrupting, was not sarcastic and didn't appear to blame Bill for the accident. In this way he found that the apparent cause of the accident was not the real cause. He carefully and expertly guided the worker into making practical suggestions for correcting the hazards. In conclusion, he was able to encourage the worker to seek help without putting him on the defensive.

All accident investigations do not go as smoothly as the one in the above example. However, by using these techniques an investigator can conduct effective accident investigations, acquire the data needed to pinpoint the cause(s) of the accident under investigation, and encourage employee participation in the investigation process.

5.5.1. Key Points for Conducting Interviews

Interviewers should remember and follow these important guidelines:

- conduct interviews as soon after the accident as practical
- delay interviews with the injured until he has received medical treatment no matter how minor his injuries
- interview one person at a time
- avoid making witnesses feel that they are informers
- be diplomatic
- put witnesses at ease
- explain the purpose of the investigation
- keep questions as simple as possible
- avoid the implied answer or leading question
- never ridicule a witness
- give the person being interviewed the opportunity to present his version in its entirety and without interruption
- review the details of acquired information
- discuss methods to prevent recurrence

5.6 Accident Report Forms

Many good report forms exist for recording the facts surrounding an accident situation. At a minimum, accident reports should contain the following information:

- identification of persons involved (name, address, age)
- time of accident (hour, day, month)
- place of accident
- type of injury
- identification of all witnesses
- severity of injury (amount of lost time, cost of injury, name of attending doctor or first aid attendant, record of treatment)
- description of property and material damage

- exact description of the accident
- a full description of the accident stating, for example, whether the person fell or was struck, and all the factors contributing to the accident
- identification of the machine, tool, appliance, gas, liquid, or other agent which was most closely associated with the accident
- if a machine or vehicle was involved, identification of the specific part involved (e.g., gears, pulley or motor)
- a judgment about the way in which the machine, or tool, was involved
- description of mechanical guards or other PPE which were in use
- statement about whether the person or persons used the safeguards provided
- description of the unsafe action which resulted in the accident
- recommendations relative to preventing or controlling similar hazards

5.7 Recording and Classifying Accident Data

Ultimately, the data on each accident report form must be recorded and classified in such a manner that important relationships may be drawn and decisions concerning accident reduction can be made by management and/ or members of the safety and health committee. Before each meeting of the committee, the chairman should present a summary of the accident(s) to be reviewed by the entire committee. Copies of the complete report should also be available to the committee. Among the facts important in a summary report are the:

- case number—the number assigned to each report for future identification and recall (e.g., 79-1)
- name of injured person
- date of injury
- where accident happened—specific place/area where accident occurred
- nature of injury—type of physical injury (e.g., cut, abrasion, chemical burn)

- body part—the part of body injured
- source of injury—the object, substance, exposure or bodily motion which directly produced the injury (e.g., saw blade, abrasive wheel)
- tools, equipment used—the tools, equipment or machines used when the accident occurred (e.g., metal lathe)
- time lost—the actual number of days and hours lost as a result of the accident
- hazardous condition—the condition which directly caused the accident (e.g., improperly guarded saw, oil spot on floor)
- human errors—the act of commission or omission which directly caused the accident (e.g., operating without authority, horseplay, operating at unsafe speed, misreading instruments, failing to follow instructions)
- costs—medical (doctor and hospital costs associated with the injury) and other (noninsured costs such as administrative time, investigation time, additional training, and compensation costs).

An important duty of the safety and health committee is to review the accident record forms, to recommend corrective action and improvements to prevent similar accidents from occurring in the future, and to determine the presence of any previously undetected hazards or weaknesses in the health and safety program.

Summary

All accidents in the workplace should be thoroughly and immediately investigated in order to determine both the direct and indirect causes, ways to prevent similar accidents, and to document the facts of the accident in the event of future legal actions.

In order to gain the full cooperation of those involved in the accident and in order to uncover hidden causes, the investigation should be a fact-finding process rather than a faultfinding process. The investigation should be initiated by the supervisor and include personal interviews with those involved. The investigation is concluded by a submitting a detailed report to Safety and Health Committee.

Bibliography

Lee, Jeffrey S., D.R. Lillquist, F.J. Sullivan (1997) "Risk Communication in the Workplace" Chapter 41 in The Occupational Environment - Its Evaluation and Control, AIHA, Fairfax, VA.

Bochnak, Peter M., (1999) "How to Establish Industrial Loss Prevention and Fire Protection" Chapter 12, page 436, in the Handbook of Occupational Safety and Health, John Wiley, N.Y., N.Y.

Dougherty, Thomas M., (1999) "How to Conduct an Accident Investigation" Chapter 5, in the Handbook of Occupational Safety and Health, John Wiley, N.Y., N.Y.

6

Safety and Health Inspection Techniques

A SAFETY and health inspection is a monitoring activity conducted in the workplace to locate and report existing and potential hazards which have the potential to cause injury or illness. Safety and health inspections are an essential element of a hazard control program. Every work place and each of its processes and operations contain some existing or potential hazards which arise through normal working or production procedures or through changes or modifications. One way of keeping abreast of hazards is by conducting continuous and periodic inspections of the workplace. Such inspections should be mandatory.

There are two important reasons for conducting safety and health inspections. The primary reason is to detect potential hazards so that they can be corrected before an accident occurs. A secondary purpose is to utilize the information collected as a basis for improving efficiency and effectiveness within the organization.

The inspection process can be approached from either a negative or a positive viewpoint. The negative or fault-finding viewpoint has an emphasis on criticism and is non-productive. The positive or fact-finding viewpoint places emphasis on the location of potential hazards and should guide the inspection process.

6.1 Types of Inspections

There are two types of safety and health inspections, periodic and continuous. A periodic safety and health inspection is, by design, deliberate, thorough, and systematic. Such inspections are often conducted by safety and health committee on a monthly or bi-monthly basis. The advantage of this type of inspection is that it provides for the detection of changes in operations or equipment in time for effective countermeasures to be provided.

Continuous or ongoing inspections should be conducted by supervisors, workers, and maintenance personnel as part of their instructional, supervisory, or maintenance responsibilities. Continuous inspections involve noting an apparent or potential hazardous condition or unsafe act and either correcting it immediately or making a report to initiate a corrective action. The ultimate goal of the safety and health inspection program is to promote vigilance on the part of each supervisor and worker to examine, correct, and report any condition which has accident potential. This continuous monitoring function is critical to a successful program.

6.2 Who Should Make Inspections

Safety and health inspections can and should be conducted by several persons in the organization. The team leader or supervisor should make continuous inspections and be alert to changing conditions, operations, and work methods. Some operations may require the supervisor to make several inspections to be certain that all safety precautions are being taken. To be sure that there are no unsafe conditions to which workers are exposed, the supervisor should make inspections at the beginning of each day or shift.

Supervisors play a key role in the inspection program. Supervisors should make weekly rounds of the facility recording unsafe conditions and practices and forwarding the information gathered to the management and, if conditions warrant, to the Safety and Health Committee. When a supervisor makes a regular inspection, it reinforces the organizations interest in the safety and health program and inspires interest and cooperation in others.

Generally, the Safety and Health Committee is responsible for all formal inspections. This committee includes supervisors, workers, shop foremen, maintenance personnel, and an employee representative. The diverse knowledge of committee members will reveal hazards which otherwise might be overlooked by persons less familiar with specific operations.

Schedules should be established to permit the committee to accomplish a thorough inspection within the constraints of members' time and availability. The length and frequency of inspections depend on the size and layout of the facility. Where there is new construction or installations or where there are changes in processes, operations, or materials, the committee may need to make special, unscheduled inspections to be certain that safety requirements are being met.

It is important that those chosen to conduct inspections should have special safety training and have their skills continually updated to keep them abreast of the new standards, regulations, and control strategies.

6.3 Inspection Procedures

An inspection program implies the regular, systematic, and continuous comparison of safety and health standards to deviations in the work environment. A safety and health inspection program requires:

- sound knowledge of the production or source process
- knowledge of applicable standards, regulations, codes and their application
- a series of systematic inspection steps
- a method of gathering, reporting, evaluating, and using the information

Many different types of checklists are available for use in safety and health inspections, varying in length from thousands of items to just a few. See Appendix B for a sample check list. Each type has its place and when properly used can be very useful. Generally, the longer checklist refers to OSHA standards and can be used during an initial survey to determine which standards apply.

Regardless of how complete the inspection checklist, the results of the inspection are only as good as the individuals conducting it. Inspectors must be realistic and use their ability, experience, and intuition to their fullest. A hazard observed during an inspection must be recorded even though it was not identified on the checklist. The inspection checklist must be used as an aid to the inspection process, not as an end in itself.

6.4 What Should be Inspected

When planning the inspection program, inspectors should consider the following:

1. Materials and substances used in the business should be viewed with respect to their potential to cause an injury, occupational illness, fire, property damage, or other hazard.
2. Machinery, equipment, and tools must be free from material defects

and other hazards. Particular attention should be given to machinery and the points of operation, including all moving parts as well as accessories (flywheels, gears, shafts, pulleys, equipment, belts, couplings, sprockets, chains, controls, lighting, brakes, exhaust systems, and grounding).

3. Personal protective equipment (PPE) and safety equipment must be available and used where there is a reasonable probability that an injury can be prevented. All PPE must be suitable to protect against specific hazards and be stored in a sanitary method.

4. Working and walking surfaces (stairs, ladders, scaffolds, and ramps) must be functionally safe, meet existing safety standards, and be properly maintained.

5. Environmental factors such as illumination, ventilation, and noise control devices must be in compliance with current standards.

6. Housekeeping, sanitation, waste disposal, material storage, and related items.

7. Medical services, first aid facilities, and a means for transporting the injured must be available at all times.

8. Electrical equipment, including switches, breakers, fuses, special fixtures, insulation, extension cords, tools, motors, and grounding must be determined to be in compliance with the regulations.

9. Chemical storage, handling, use, and transportation must be viewed with respect to adequate exhaust systems, warning signs, protective clothing, and other allied equipment.

10. Fire protection and extinguishing systems such as alarms, sprinklers, fire doors, exit signs, and extinguishers deserve particular scrutiny including review of egress plans for emergency exits.

11. The scheduling and effectiveness of the preventive maintenance program should be assessed.

Should the inspection identify a hazardous machine, piece of equipment, or operation the inspector should report the condition to the supervisor or person in charge and if necessary, shut down the machine or equipment and tag it immediately. In addition to the tag, a lockout is recommended.

6.5 Assessing Health Hazards

A thorough inspection requires that emphasis be given to both safety and health hazards. In order to be able to recognize health hazards during an inspection process, the inspector should be knowledgeable of the environmental factors in the workplace which have the capacity to cause sickness or illness to workers and supervisors. The inspector must consider:

- the nature of the product being produced
- the raw materials being used
- materials and substances being added in the process
- by-products produced
- the equipment involved
- the cycle of operations
- operational procedures being used
- health and safety controls utilized
- number and level of exposures to harmful chemical, biological, and physical agents

In addition, inspectors should consider the following guidelines:

1. List all hazardous chemical, physical, and biological agents on site.
2. Determine where the hazardous health agents are and the state in which they exist (dust, fumes, mists, vapors, smokes, gases).
3. Determine the threshold limit value for all chemical, physical, and biological agents, and compare them against actual levels in the workplace.
4. Determine which processes and equipment are capable of producing unsafe levels during operations.

6.6 Recording Hazard Facts and Using the Data

During the inspection, precautions should be taken to locate and describe each hazard found. As hazards are uncovered, a clear description of the hazard should be recorded. It is also important to determine which hazards present the most serious consequences and are most likely to occur. Classifying hazards properly places them in the proper perspective, assigns priorities, and aids in correcting the hazardous condition.

How the information acquired from an inspection activity is used is as important as the inspection process itself. It is necessary that inspectors bring problems and recommendations for corrective action to the attention of management. General recommendations could include:

- changing the procedure or improving the process, operation, or work task
- limiting the exposure by relocating the present process in such a way as to make it less hazardous while at the same time providing better results
- redesigning a tool or fixture or changing the workers' work pattern to reduce the hazard potential
- providing more training to personnel engaged in a particular operation
- providing personal protective equipment

Based on the hazards uncovered and the recommendations made, management must decide what course of action to take. Usually these actions will be based on the cost effectiveness of the recommendations. For example, it may be cost effective and practical to substitute a less toxic material that works as well as the highly toxic material presently in use. On the other hand, a particular hazardous machine may be too costly to replace. In this case, the choice may be the less expensive alternative of installing machine guards to correct the problem.

Information from inspections should not be used for punitive action. The information should be viewed as the basis for establishing priorities and implementing programs that will reduce accidents and illness, improve working conditions, raise morale, and increase the overall effectiveness of the organization by reducing losses.

Summary

Safety and health inspections are conducted for two reasons: first, to detect potential hazards so that they can be corrected before an accident occurs, and second, to utilize the information collected as a basis for improving efficiency and effectiveness within the organization. Safety and health inspections are performed either periodically or continuously. Depending upon the type, inspections are usually conducted by supervisors and members of the safety and health committee. The inspection should cover the appropriate areas, items, and processes of the workplace and include an assessment of health hazards. Each hazard identified by the inspection should be recorded and classified.

Bibliography

Firenze, Robert J., (1978) *The Process of Hazard Control,* Kendall/Hunt Publishing Company, Dubuque, Iowa 52001.

Gordon, A.W., R.W. Michaud (1995) "Principles of Environmental Health and Safety Management" Government Institutes, Rockville, MD.

Horowitz, M.R., M.F. Hallock (1999) "Recognition of Health Hazards in the Workplace" Chapter 4, in the Handbook of Occupational Safety and Health, John Wiley Publisher, N.Y., N.Y.

U.S. Department of Labor, Occupational Safety and Health Administration, *Safety and Health Inspections for an Effective Safety and Health Program,* February 1977.

7

Principles of Good Shop Planning

ONE OF THE primary resources used in planning the layout and arrangement of a work area should be the supervisor or group leader. Their knowledge and experience are of exceptional value to administrators and architects in planning new work areas or altering or expanding old ones. The supervisor's knowledge of how operations take place and what is needed complements the architect's knowledge of the feasibility of the various design and construction alternatives. Together, they can reduce, if not eliminate, many of the problems encountered in building renovation, construction, and expansion.

This marriage of knowledge, experience, and imagination will prove most beneficial if the supervisor and architect give high priority to safety and health needs while planning the work area layout and arrangement. Numerous accidents, occupational illnesses, explosions, and fires can be prevented or minimized if suitable measures are taken early in the planning stage.

7.1 Shop Layout

In order to avoid difficulties of organization and supervision, it is generally recommended that the facility length should be no more than twice its width. The space allotted to shop work should be not less than 55 square feet (5.1 square meters) per worker. All parts of the facility should be visible to the supervisor. Window space should be not less than one-fifth of the floor space and ceiling height should be between 10.5 and 14 feet (3.2 and 4.3 meters). An open assembly area should be provided where large projects can be assembled. These general space guidelines vary greatly depending on the type of operation. Water, air, hydraulic, electrical and other piping or conduit systems should be contained in the walls to eliminate the need for constant cleaning.

7.1.1 Working Spaces

Floors, ramps, stairs, ladders, and scaffolds deserve special consideration in shop design and are discussed in Chapter 8, Safe Working Surfaces. Floor materials should be easy to clean and as slip-resistant as possible. Safe load limits must be considered along with the total weight burden. Because floors begin to sag when overloaded, it is important to know their weight capacity before installing equipment or machinery or storing heavy materials. To be safe, floors should have weight capacities four times the static load or six times the moving load.

7.1.2 Aisles

Main aisles should parallel the flow of materials in process. Main aisles should be four feet (1.2 meters) wide. Aisle spaces should be kept clear of material and equipment at all times. Painting a clearly visible white or yellow line on the floor should identify the edges of all aisles. Tool rooms and emergency equipment should be located off the main aisles.

7.1.3 Exits

Exits should be sufficient both in number and size and located so that, in case of fire or other emergency, the workers can be quickly and safely evacuated. Plans should be adequate and conform to OSHA standards, National Fire Protection Association (NFPA) Building Exits Code, and to state and local requirements. Changing or adding exits after a building has been constructed is very costly.

To a large extent, the number and width of exits are determined by the building occupancy. In high hazard occupancy, no person should be more than 75 feet (22.9 meters) from an exit. For medium and low hazard, 100 to 150 feet (30.5 to 45.8 meters) is permissible. NFPA Standard No. 101 specifies that access to exits provided by aisles, passageways, or corridors should be convenient to every occupant and that the aggregate width of passageways and aisles should be equal to or greater than the required width of the exit. All shops should have at least two exits, one of which should be wide enough to permit the passage of large equipment.

Exit doors should be clearly visible, illuminated, labeled, and open in the direction of exit travel. They should not be locked or blocked by machinery or equipment. Exits should be located where they will not be eliminated by any future additions to the building.

7.2 Designing for Specific Safety Needs

7.2.1 Storage of Materials

Planners must realize that efficient shop management depends on the availability of and access to tools and materials. In addition, sufficient project storage areas are necessary to avoid confusion, theft, damaged projects, and general discontent among the employee population. Consideration should be given as to what type of storage area is most appropriate. Storage area types include closed, completely open, or a combination of both.

The size of storage rooms and areas will depend upon the amount and size of stock to be kept on hand. Storage areas should be equipped with both vertical and horizontal racks. Racks should be so designed and constructed that heavy stock can be pulled out without tipping the racks. In the lumber storage area, vertical stacking has several distinct advantages:

- reduced warpage
- less dust collects and the stock remains cleaner
- reduced chances of stock falling on employees
- stock can be removed without requiring the assistance or the use of ladders
- smaller pieces of stock can be removed without first having to move larger pieces
- increased ease in taking inventory

Where space permits, separate racks should be used for each variety of stock. Pieces of stock, which are of irregular size, do not lend themselves readily to storage in a rack. They should be stored on fiat shelves beneath workbenches.

In the typical small business tools, supplies, and small repair parts such as nuts, bolts, washers, and rivets are best stored in bins, drawers, cabinets with drawers, or similar storage spaces.

Worker lockers should be perforated for ventilation and large enough to accommodate shop clothing. It is a good idea to specify that the tops of lockers be sloped to prevent excessive build-up of dirt and accumulation of materials on top. Lockers should be fastened to the shop floor to prevent them from being overturned.

7.2.2 Personal Service Facilities

An area often overlooked in the design and layout of shops is the procurement and placement of personal service facilities such as drinking fountains, wash basins and soap dispensers. These items play an important role in maintaining worker health.

A general rule is to install drinking fountains at convenient locations throughout the shop, provided it is located away from machinery or operations such as welding heat-treating metals. Planners will he assisted by specifications from the American National Standards Institute when purchasing and installing drinking fountains (ANSI C33.82, 1972).

Each shop should be equipped with a minimum of one wash basin, a two-foot (60 cm) trough, or a circular or semi-circular wash fountain with hot and cold running water for every twenty workers. A good quality soap distributed from a dispenser provides for ordinary hygiene and protects against dermatitis. Paper towels in covered dispensers should be available along with a closed disposal receptacle in close proximity to the washing facility.

7.2.3 Quick Drenching Facilities and First Aid Supplies

Safety showers and eyewash fountains for quick drenching and/or flushing of the eyes and body must be provided when a potential for exposure to injurious corrosive materials exists. First aid supplies, approved by the American Red Cross or other authoritative source, must be readily accessible. These supplies should be in sanitary containers with individually sealed packages of sterile gauze, bandages, and dressings. Other items often needed are adhesive tape, triangular bandages (to be used as slings), inflatable plastic splints, scissors, and mild soap for the cleansing of wounds or cuts.

7.2.4 Exhaust and Ventilation Systems

Industrial shop practices create various dusts, gases, smoke, vapors, and mists that, unless intercepted, will enter the shop atmosphere. These contaminants can produce a variety of occupational related illnesses, such as diseases of the lung dermatitis.

Planners must pay particular attention to the design and installation of both general and local exhaust ventilation systems. General dilution systems are designed to remove air throughout the shop atmosphere at pre-determined intervals and replace it with air that is free of contaminants. Local exhaust systems are designed to prevent a contaminant from reaching the operator by capturing it near its source and carrying it away to special collectors where the general ventilation. system can dispose of it or carry it outside the building.

Exhaust and ventilation systems are most critical in places where solvents are used, fumes accumulate, and where dust is produced. Planners should consider the installation of vacuum systems and special openings designed to remove or store waste to be collected and disposed of by other means. Exhaust and ventilation systems are more fully discussed in Chapter 12, Health Hazards.

7.2.5 Fire Protection

Although good shop plans will specify the provision of fire detection, alarm, and extinguishing systems, planners must supplement these systems with portable fire extinguishers which are readily accessible and easy to use. Safety containers for flammable liquids, fire blankets, and approved receptacles for oily rags and waste materials should be placed at critical locations. The subject of fire protection will be discussed in Chapter 11, Fire Protection.

7.2.6 Electrical Requirements

More fires are caused by electrical malfunction than by any other cause. OSHA has adopted the National Electrical Code, NFPA 70-1971, ANSI C1-1971, as a national consensus standard. The purpose of the code is the practical safeguarding of persons and buildings and their contents from hazards arising from the use of electricity. The code contains basic minimum provisions considered necessary for safety. Planners, maintenance personnel, and supervisors should be familiar with these requirements and should regularly inspect the following items for compliance:

1. Each means of disconnection (circuit breaker or fuse box) must be legibly marked to indicate its purpose unless its purpose is evident. See Figure 7.1.
2. Frames of electrical motors, regardless of voltage, must be grounded.
3. Exposed noncurrent-carrying metal parts of fixed equipment that may become energized under abnormal conditions must be grounded under any of the following circumstances:
   a. in wet or damp locations
   b. if in electrical contact with metal
   c. if operated in excess of 150 volts to ground
   d. in a hazardous location
4. Exposed noncurrent-carrying metal parts of the following plug connected equipment, which may become energized, must be grounded or double-insulated and distinctly marked:
   a. portable, hand-held, motor-operated tools
   b. appliances
   c. any equipment operated in excess of 150 volts to ground
5. Outlets, switches, and junction boxes, must be covered.
6. Flexible cords may not be:
   a. Used as a substitute for fixed wiring
   b. Run through holes in walls, ceilings or floors
   c. Run through doors, windows, etc.
   d. Attached to building surfaces.
7. Flexible cord must be fastened so that there is no pull on joints or terminal screws and must be replaced when frayed or when the insulation has deteriorated. See Figure 7.2.
8. All splices in electrical cord must be brazed, welded, or soldered or join the conductors with suitable splicing devices. Any splices, joints, or free ends of conductors must be properly insulated.

7.2.7 Illumination

The shop planner must consider the quantity and quality of illumination required for various tasks, the problem of glare, and the placement of specialized lighting equipment in hazardous areas. The topic of illumination will be examined in greater detail in Chapter 10, Illumination and Color Safety.

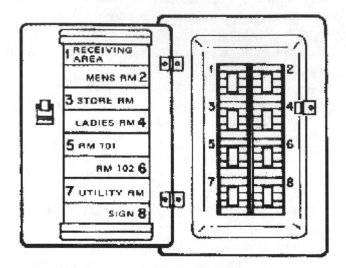

Fig. 7.1 Proper Labeling of Circuit Breakers.

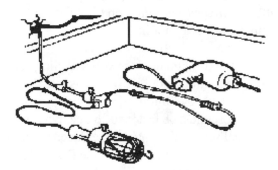

Fig. 7.2 Misuse of Flexible Cords.

7.3 Placement of Machinery

Regardless of the type of shop planned, the design and arrangement of equipment, machinery, tools, and materials should be engineered for the most effective and efficient hazard controlled operations. Consideration should be given to:

- the flow of materials
- placing machines adjacent to ones needed for successive operations

- providing sufficient space so that workers do not interfere with each other
- preventing interference between operations and operator
- determining the maximum amount of space needed for machines used with large pieces of stock
- placing machines near materials
- allowing space for hand trucks
- allowing space for cleaning and maintenance

7.3.1 Flow of Materials

Because the manner by which materials are brought into and are handled in the shop can produce hazards, materials flow is a major concern during planning. The planner should trace the route the materials travel in the shop, from the time they are received through the various stages of transportation and storage through the final stage of completion and disposal. Tracing the route of materials and analyzing each operation and movement of materials from the standpoint of hazards is consistent with the theory of Hazard Analysis discussed in Chapter 3. After the analysis is complete, suggestions can be made to control or eliminate the hazards discussed.

7.3.2 Flow of Operations

If operations require the worker to move from one machine to another for successive operations, the machines should be located adjacent to or as close to each other as possible to shorten the distance the worker must travel. Reducing the need for crisscrossing and backtracking helps reduce the danger of collision and exposure to potential hazards from other equipment.

Enough space should be provided between machines to prevent the workers from getting in each other's way, while permitting the supervisor to provide the necessary training and supervision. In specialized shops, the space requirements range from as little as 30 square feet (9.1 m.) per worker for drafting to as much as 100 square feet (30.5 m.) per worker for the machine or auto mechanics shop.

7.3.3. Locating Equipment and Machinery

Special care must be taken to locate equipment to eliminate interference between the operations and the operators. When feasible, machines should be placed at a 45° angle to window walls in order to ensure that the shop receives the maximum amount of natural light. Placing the machines at a 45° angle also ensures that operators will be out of alignment with the moving parts of adjacent machines, thereby reducing the danger from machine accessories or materials which may be thrown from neighboring equipment.

Since additional space may be needed, the largest piece of material each machine can handle should be determined. For example, a lathe to be used for machining long bars fed through the headstock needs more space to the left of the machine than one which is to be used only for chuck work. Certain machines, such as the metal working planer and shaper, need to be placed so that sufficient clear space remains when tables or rams are operating at them maximum distances. Other machines, such as cutoff saws and shears, should be placed near the materials storage areas in order to reduce hazards from handling large pieces of stock.

Heavy duty machinery and equipment should be placed as close as possible to the entrance through which heavy material is received. Consideration also should be given to the feasibility of installing electric hoisting devices. All heavy equipment should be leveled and securely fastened to floors. The placement of felt, cork, rubber, or other shock-absorbing materials under machines is recommended in order to reduce vibration and noise.

When deciding where to place machinery, planners should allow space for hand trucks to be brought as close as possible to unload jigs, boxes of materials, and so forth. Machinery should be located so that there is sufficient room for cleaning, maintenance, and repair work.

Methods used in placement of equipment usually involve scaled drawings of floor plans indicating fixed obstructions, such as supporting pillars in walls, windows and door openings, and the relationship of the room to other service areas. The "single dimension method" involves showing the relative location of all equipment and facilities on a drawing. The "two-dimensional method" is more frequently used and involves arranging flat patterns on the floor plan drawing. The patterns are to scale and in the shape of the floor area required for each item of equipment. A more revealing technique is to set up three-dimensional scale models of equipment on a drawn-to-scale plan. Some equipment manufacturers furnish models upon request, and other models can be carved from softwood or made from cardboard.

The planner should make several layouts before deciding which will be implemented. The introduction of a single item of equipment may demand the rearrangement of the entire floor plan.

7.4 Criteria for Purchasing Machinery

The time spent establishing the criteria for the purchase of machinery for the shop will be well spent, and many problems will be eliminated or reduced. Among the criteria for selecting shop machines include the following.

1. Provisions must exist for the automatic lubrication of critical parts and for effective collection systems.
2. Machine parts subject to wear and/or needing periodic adjustments or lubrication should be easily accessible.
3. Automatic feeds and systems for waste removal should be present. Dust collectors on machinery reduce the amount of airborne particulate matter.
4. Provisions must be made for the continual removal of metal particles, fumes, mists, gases, and vapors during the operation of the machinery.
5. Provisions should be made for reducing noise and vibration through enclosures, shock mountings, and other attenuation and dampening techniques.
6. Electrical on-off switches should be located within easy reach of operators.
7. Emergency stop buttons and main power disconnect should be provided.
8. Operating controls should be color-coded according to standards.
9. Operating levers should be protected to prevent accidental starts.
10. Machine controls should be located in a manner that students will not be required to be in close proximity to the point of operation while activating the controls.
11. Guards should be provided at all points of operation as specified by the manufacturer and OSHA regulations.
12. It should be difficult to start the machine unless guards are in place and access doors are closed and latched.
13. Power transmission components such as belts and pulleys should be protected to prevent contact with moving parts.
14. Overload devices should be built into the machine.
15. All electrical equipment, especially hand-held equipment, should have an effective grounding system.
16. Adequate illumination should be provided for all points of operation.
17. Wherever possible, all sharp corners and edges should be rounded.

Because of highly competitive marketing, some manufacturers of machine tools find it advantageous to list safety devices designed for the protection of operators as auxiliary equipment. Shop personnel must be familiar with such items and make sure that they are included in the original purchase order.

Safeguarding the worker from dangerous parts of the machinery and equipment is a primary concern in writing specifications for shop equipment. A well-guarded machine, in addition to being safe to work on, is valuable from a psychological standpoint. When a worker's fear of a machine is alleviated, his concentration can be devoted to his work. Machine guarding is discussed in detail in Chapter 14.

Summary

The well planned shop design and layout coupled with the proper types and placement of equipment and machinery are the basis of safe workplace. Considerations for planning the shop include the design requirements for specific safety needs, placement of machinery, and the purchasing of machinery with adequate safety controls.

Bibliography

Dougherty, Thomas M. (1999) "Risk Assessment Techniques" Chapter 16 in the Handbook of Occupational Safety and Health, John Wiley & Sons, N.Y., N.Y.

U.S. Department of Health and Human Services, National Institute for Occupational Safety and Health, Occupational Safety and Health in Vocational Education February 1979.

U.S. Department of Labor, Bureau of Labor Standards (1967) Planning Layout and Arrangement for Safety, Bulletin 289.

8

Safe Working Surfaces

SAFE WALKING and working surfaces are an important element in the safety program. The five major working and walking surfaces discussed in this chapter are floors, ramps, stairs, ladders, and scaffolds. In the workplace, falls on or from these surfaces are a leading cause of injuries.

There are two broad classifications of falls, those from the same level and those from different levels. As would be expected, analyses of injuries indicate that falls from higher levels usually result in severe injuries. However, falls on the same level occur more frequently and can also result in serious injuries. In order to reduce slips and falls in the workplace, the management must ensure the proper:

- selection and placement of flooring materials
- selection of the most efficient systems for transporting people and material from one level to another (ramps, ladders, and stairs)
- maintenance of working and walking surfaces

8.1 Floors

Unsafe floors are a primary source of accidents in the workplace. Floors are hazardous to employees when they are not properly maintained, when they are not kept free from materials and other obstructions, or when they are uneven or slippery.

Slippery floors are usually the result of water or other liquids such as grease accumulating on the floor following a spill, or the result of improper water drainage. Repairing leaks, providing spill pans, and cleaning up spilled substances immediately following a spill can reduce the likelihood of accidents. Improper drainage can be corrected by providing an adequate number of drains at the low points of floors. In some situations, the floor may need to be resurfaced to provide the slope necessary for proper drainage.

Materials, tools, and work in process can obstruct the flow of traffic in the shop. Workers should be instructed not to use the floor as storage areas for equipment, materials, or projects. Adequate storage areas should be provided for these items.

The most common causes of uneven floors in the shop include:

- warping from water or moisture
- loose boards
- excessively worn surface planks
- settling of the building
- improper alignment of floors during the construction of additions to rooms
- depressions, cracks, or ruts caused by dragging heavy items across floors or from the wheels of materials-handling equipment (rubber tires on equipment help to compensate for this problem)

8.2 Ramps

Ramps, usually constructed of timbers, concrete, metal, or asphalt, are the simplest means of getting from one elevation to another but they can be a source of accidents. Workers can slip on, slide down, or fall off ramps. They can also trip on uneven ramp surfaces and lose control of wheeled vehicles while transporting them down a ramp. Ramps can also collapse if they are subject to excessive strain.

The recommended maximum rise for ramps is one foot for every ten feet of distance. For example, a ramp rising eight feet would be 80 feet long. This rise requirement may prohibit the use of ramps in areas due to the length needed for the ramp. Ramps exceeding this recommended rise are too steep to be used safely, especially with wheeled vehicles. If ramps are to be used for wheeled traffic between levels, they should have a solid curb on open sides and, to prevent bottlenecks, should be as wide as the aisles or road they service.

Ramps should be built with the lowest degree of slope practical, where the recommended slope is six degrees. Ramps should not exceed a slope of 20 degrees. Figure 8.1 illustrates the recommended slopes.

According to the Building Exits Code (Vol. 4, National Fire Codes-NFPA), ramps used in connection with exits shall be of substantial construction and adequately designed for use as exits. Ramps in assembly areas, may, depending on the building's capacity, have the following maximum slopes:

- Class A (capacity, 1000 people or more) - one inch per foot
- Class B (capacity, 200—1000) - 1-3/16 inches per foot)
- Class C (under 200) - two inches per foot

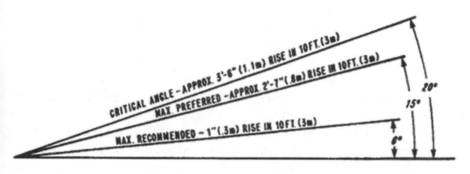

FIG. 8.1 Recommended ramp slopes.

Steep ramps should be surfaced with an anti-slip material such as abrasive metal plates, non-slip compounds, or abrasive paints. When necessary for hand-trucking operations, cleats should be evenly spaced eleven to sixteen inches apart and should not interfere with the operation of trucks. Planks should not overlap and should run the long way of the ramp.

8.2.1 Standard Railings and Toeboards

Standard guardrail systems are recommended for ramps, especially for sections of the ramp which are more than four above the floor. Four inch toeboards should be provided on the open sides if workers can pass near or under the ramp, if the ramp is over or near machinery, or if material in transport would create a hazard if dropped. A standard guardrail system consists of a top rail, an intermediate or midrail, and posts. The height of the rail should be 42 inches (plus or minus two inches) as measured from the upper surface of the top rail to the floor, runway, or platform. The requirements for wood railings, pipe railings, and structural steel railings contained in OSHA 1910.23(e)(3) are:

1. Wood railings - the posts shall be made of at least two inch by four

inch stock spaced less than six feet apart and the top and intermediate rails shall be of at least two inch by four inch stock. If the top rail is made of two right angle pieces of one inch by four inch stock, the posts may be spaced on eight foot centers with a two inch by four inch intermediate rail.

2. Pipe railings – the posts, top, and intermediate railings shall be at least 1.5 inches in diameter with posts spaced not more than eight feet on centers.

3. Structural steel railings - the posts, top, and intermediate rails shall be of 2x2x? inch angles or other metal shapes of equivalent bending strength with posts spaced not more than eight feet on centers.

Rail frames must be anchored to platforms to withstand a thrust of 200 pounds applied in any direction at any point of the top rail. The intermediate or midrail should be approximately halfway between the top rail and the floor, platform, runway or ramp. A toeboard must be at least four inches high from its top edge to the level of the floor, platform, runway, or ramp. The toeboard should be made of a substantial material and be securely fastened with less than a ¼ inch clearance above floor level.

In addition to their use on ramps, standard railings are required by General Industry Safety and Health Standards (OSHA 1910.23) to be placed at every:

1. open-sided floor or platform four feet or more above the adjacent floor or ground level. These areas must be guarded on all open sides, except where there is an entrance to a ramp, stairway, or fixed ladder.

2. stairway opening (on all open sides) except the entrance to the stairway.

3. ladderway floor opening. A standard guardrail system and toeboard on all sides must guard these openings. To prevent a person from walking directly into the opening, a passage through the railing should be constructed.

4. runway or catwalk four feet or more above ground level. These openings must have protection on all open sides.

5. scaffolds or platforms ten feet or more above the ground.

As a general condition, a standard toeboard is required wherever people walk beneath the open sides of a platform or under similar structures or where things could fall from the structure onto machinery or workers below.

8.3 Fixed Stairs

Accidents involving stairs can arise from many sources including irregular tread or risers, loose coverings, worn surfaces, stepping on objects, bumping head on ceiling, and tripping. Flights have from four to seventeen treads between landings. Twelve treads are recommended as an average for comfort and uniformity. Fixed stairs in the shop must conform to OSHA requirements for safety (OSHA 1910.24 and 23) which include the following standards:

1. Stair Strength. Fixed stairways must be designed and constructed to carry a load five times the anticipated normal live load. At a minimum, they must be able to safely carry a moving concentrated load of 1,000 pounds.
2. Risers. Riser heights must be uniform throughout the stairs and they should be greater than 6.5 inches and less than 9.5 inches high.
3. Tread Widths. Tread widths, like riser heights, must be uniform throughout the stairs. The tread width must be greater than 9.5 inches. In addition, a nosing is recommended. All treads must be reasonably slip-resistant.
4. Nosings. Stair treads and the top landing of a stairway, where risers are used, should have a nose which extends ½ to one inch beyond the face of the lower riser. Nosings should have an even leading edge and must have a non-slip finish.
5. Stair Width. The minimum width is 22 inches.
6. Stairway Railings and Guards. If the flight of stairs has four or more risers, the following guardrail system is required:
    a. on stairways less than 44 inches wide having both sides enclosed, at least one handrail, preferably on the right hand descending
    b. on stairways less than 44 inches wide having one side open, at least one stair railing on the open side
    c. on stairways less than 44 inches wide having both sides open, one stair railing on each side
    d. on stairways more than 44 but less than 88 inches wide, one handrail on each enclosed side and one stair railing on each side
    e. on stairways 88 or more inches wide, one handrail on each enclosed side, one stair railing on each open side, and one intermediate stair railing located approximately midway of the width

7. Vertical Clearance. Vertical clearance above any stair tread to an overhead obstruction must be at least seven feet, measured from the leading edge of the tread.
8. Lighting. All stairs should be adequately lighted.
9. Handrails. Stairs must have handrails 30 to 36 inches high as measured from the tread at the upper face of the riser.
10. Angle of Stairway Rise. The angle to the horizontal made by the stairs must be between 30° and 50°.

8.4 Fixed Ladders

Ladders must be purchased wisely, maintained properly, and used carefully. Ladders can fall or slip, workers can fall or slip from them, and objects can drop or fall from them. In addition, metal ladders can cause electrical shock.

At various locations in the shop, fixed ladders may be installed for a variety of general and maintenance uses. These ladders should meet the following requirements as set forth in OSHA Safety and Health Standards 1910.27:

1. A fixed ladder must be permanently fastened to an upright surface. Fixed ladders are usually constructed of metal.
2. A ladder must have cages or wells if it rises more than twenty feet from the floor. Cages must extend a minimum of 42 inches above the top of a landing unless other acceptable protection is provided. Cages must extend down the ladder to a point seven or eight feet above the base of the ladder. A platform is required every thirty feet for caged ladders and every twenty feet for unprotected ladders, that is, when no ladder safety device is used.
3. Fixed ladders must be designed to withstand a single concentrated load of at least 200 pounds.
4. Rungs of metal ladders must have a minimum diameter of ¾ inch. Rungs of wood ladders must have a minimum diameter of 1? inches.
5. Rungs must be at least sixteen inches wide, be spaced no more than twelve inches apart, and be uniform throughout the length of the ladder. Rungs, cleats, and steps must be free of splinters and burrs.
6. If their construction and location requires, ladders must be treated with a preservative to resist deterioration.
7. The preferred pitch for safe descent is 75° to 90°. Unless caged or

equipped with a ladder safety device, ladders with a 90° pitch must
have a 2.5 foot clearance on the climbing side.
8. There must be at least a seven inch clearance in back of the ladder
to provide adequate toe space.

8.5 Portable Ladders

Portable ladders are used throughout the workplace for a variety of
tasks. Among the most common types of portable ladders are those made
of wood, steel, aluminum, magnesium alloy, and fiberglass. A straight ladder
consists of two beams (side rails) and rungs (cross members). The top end
of the ladder is called the tip and the bottom the base.

A standard ladder has rungs spaced twelve inches on center and at least
two metal cross braces set at a maximum distance of ten feet apart. Single
ladders must be no more than thirty feet long. Two-section extension ladders
must not exceed sixty feet. Defective ladders must not be used. Ladders
are defective if they have defects in the wood, crossgrain knots, cracks,
checks, shades, decay, loose or broken rungs, or pitch pockets.

An extension ladder consists of a bottom section called the bed and a
movable top section called the fly. The fly is extended from the bed by a
lanyard and pulley arrangement and locks on the rungs of the bed ladder
with automatic locks called dogs. Extension ladders should be examined
to be sure that the automatic locks are operable, free from defects, and
work efficiently by gravity alone. In addition, the size and condition of the
rope and the condition of the hoisting pulley should be monitored. Worn or
under-sized rope should be replaced as well as any rusted parts and pulleys
which are not anchored properly or do not turn freely.

Lapping sections are those ladder sections that overlap with another to
add strength to the overlapping section. On two-section extension ladders,
the minimum overlaps for the two sections (specified in OSHA 1910.25)
are as follows:

- for ladders up to and including 36 feet - a three-foot overlap
- for ladders between 36 and 48 feet - a four-foot overlap
- for ladders from 48 to 60 feet - a five-foot overlap

To provide adequate traction, all portable ladders should be equipped
with non-slip bases. Four common types of non-slip ladder bases are
universal, rubber suction, spike, and toothed. Each type of base is illustrated
in Figure 8.2.

1. The universal ladder base can be used on solid surfaces with the corrugated surface down or on soft surfaces (such as the ground) with the spike turned down. These bases are available with corrugated surfaces made of cork, rubber, cord, or abrasive materials.
2. The rubber suction surface base is available in rubber, neoprene, or cord and is excellent for wet, smooth surfaces.
3. The spike base has spikes made of steel or bronze and are used outdoors.
4. Toothed bases are used on concrete floors, sidewalks, or asphalt surfaces.

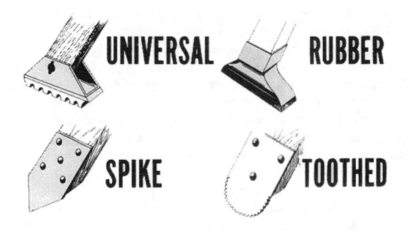

FIG. 8.2 Non-slip ladder bases.

Because of the adverse effect of the solvents on rubber, rubber bases should not be used in areas where oil or solvents may come into contact with the ladder. Neoprene and certain other plastics are better suited for these situations. Ladder bases must be properly maintained. When the bases no longer serve the intended purpose they should be discarded and replaced.

8.6 Ladder Safety

No matter how well constructed and maintained, ladders are safe only if used properly. The following practices should be observed when placing ladders:

1. The horizontal distance from the base to the vertical plane of the support should be approximately one-fourth the ladder length between supports. For example, a twelve-foot (3.7 m.) ladder must be placed so that the bottom is three feet (0.9 m.) away from the object against which the top is leaning.
2. Horizontally placed ladders should not be used as a runway or scaffold. Single and extension ladders are designed for use in a nearly vertical position and cannot be used safely in a horizontal position or in a position in which the base is placed at a greater distance from the support than indicated in number 1 above.
3. A ladder should not be placed in front of a door that opens toward the ladder unless the door is locked, blocked, or guarded.
4. Portable ladders should be placed so that both side rails have secure footing. Solid footing should be ensured to prevent the ladder from sinking.
5. The ladder feet should be placed on a substantial and level base, and should never be placed on movable objects.
6. The ladder should never lean against an unsafe backing such as loose boxes or barrels.
7. When using a ladder to access high places, the ladder should be securely lashed or otherwise fastened to prevent it from slipping.
8. When using a ladder to access a scaffold, both the bottom and top of the ladder should be secured to prevent displacement.
9. The ladder side rails should extend at least three feet above the top landing.
10. The ladder should not be placed near live electric wiring or against any operational piping.
11. Metal ladders should not be used where they could come in contact with electric circuits. If such a situation cannot be avoided, proper safety measures should be taken to prevent short circuits or electrical shock.

The following practices should be observed when ascending or descending ladders:

1. Hold on with both hands when going up or down. If material must be handled while ascending or descending, the material should be raised or lowered with a rope either before or after climbing to the desired level.
2. Always face the ladder when ascending or descending.

3. Never slide down a ladder.
4. Shoes should be free from grease, mud, and other slippery substances.
5. Do not climb higher than the third rung from the top on straight or extension ladders or the second tread from the top on stepladders.

As part of the safety and health inspection program, ladders should be thoroughly inspected every three months. The routine inspection form in Table 8.1 lists the items to be inspected. An accurate record of each inspection should be kept.

The Safety Code for Portable Wood Ladders, ANSI A14.1, states that "ladders should be kept coated with a suitable protective material. The painting of ladders is satisfactory provided the ladders are carefully inspected by competent and experienced inspectors prior to painting and provided the ladders are not for resale." A clear wood preservative such as linseed oil is preferred over paint since it does not cover up defects and it provides better overall traction. If a ladder is to be coated, it must be coated completely, otherwise the uncoated portions attract moisture which can become trapped in the wood and promote decay.

8.7 Scaffolds

Scaffolds are temporary elevated working platforms designed to support both workers and materials. The major hazards associated with scaffolds include:

- people or equipment falling from the scaffold
- accidents when getting down from the scaffold
- objects falling onto the scaffolds
- mobile scaffolds rolling while in use
- the setting, shifting, breaking loose, or collapse of the scaffold

The three major scaffold types are wood pole, tube and coupler, and tubular welded frame. Wooden pole scaffolds may be classified according to their use, that is, either light duty (used in plastering and lathing) or heavy duty (used in stone-masonry and bricklaying). Tube and coupler scaffolds consist of tubing (which serves as posts, bearers, braces, ties, and runners), a base supporting the posts, and special couplers which serve to connect the uprights and join the various members. Tubular welded frame scaffolds are built of prefabricated welding sections consisting of posts and bearers with

TABLE 8.1
Ladder Inspection Checklist

Item to be Checked	Needs Repair	Condition O.K.
General:		
Loose steps or rungs		
Loose nails, screws, bolts, or other metal parts		
Cracked, split, or broken uprights, braces, steps, or rungs		
Silvers on uprights, rungs, or steps		
Damaged or worn non-slip bases		
Stepladders:		
Wobbly (from side strain)		
Loose or bent hinge spreaders		
Stop on hinge spreaders broken		
Loose hinges		
Extension Ladders:		
Loose, broken, or missing extension locks		
Defective locks that do not seat properly when the ladder is extended		
Deterioration of rope from exposure to acid or other destructive agents		
Trolley Ladders:		
Worn or missing tires		
Wheels that bind		
Floor wheel brackets broken or loose		
Floor wheels and brackets missing		
Ladders binding in guides		
Ladder and rail stops broken, loose, or missing		
Rail supports broken or section of rail missing		
Trolley wheels out of adjustment		
Trestle Ladders:		
Loose hinges		
Wobbly		
Loose or bent hinge spreaders		
Stop on hinge spreader broken		
Center section guide for extension out of alignment		
Defective locks for extension		
Sectional Ladders:		
Worn or loose metal parts		
Wobbly		
Fixed Ladders:		
Loose, worn, or damaged rungs or side rails		
Damaged or corroded parts of cage		
Corroded bolts and rivet heads on inside of metal stacks		
Damaged or corroded handrails or brackets on platforms		
Weakened or damaged rungs on brick or concrete slabs		
Base of ladder obstructed		
Fire Ladders:		
Markings illegible		
Improperly stored		
Storage obstructed		

intermediate connecting members braced with diagonal or cross braces. Tubular welded frame scaffolds are quicker and easier to set up than wooden pole scaffolds.

Tubular metal scaffolds are the most widely used type because they are readily available, economical, versatile, and adaptable to all scaffolding needs. Most manufacturers and suppliers of tubular metal scaffolding provide engineering services to ensure the proper design of the scaffolding. In addition, many manufactures will provide erection and dismantling services.

Mobile or rolling scaffolding are made of either caster-mounted sections of tubular metal scaffolds or other components designed specifically for the purpose. When mobile or rolling scaffolding is used, additional precautions must be taken to ensure safety. These precautions are provided in OSHA Safety and Health Standards 29 CFR 1910.29.

The following general requirements for all types of scaffolding are contained in OSHA 1910.28:

1. All scaffolding must have sound, rigid footing and be capable of holding the intended load without settling or shifting. Unstable objects such as barrels, boxes, loose bricks, or concrete blocks must not be used to support scaffolds or planks.
2. Guardrails and toeboards must be used on all open sides and ends of platforms which are more than ten feet above the ground or floor (except needle beam scaffolds and floats in use by structural iron workers). Scaffolds four to ten feet high, which are less than 45 inches wide must also have guardrails.
3. An access ladder or equivalent safe access must be provided.
4. Scaffolds and their components must be able to support at least four times the maximum intended load. Scaffolds must not be in excess of the working load for which they are intended. Wire or fiber rope used for scaffold suspension must be capable of supporting at least six times the intended load.
5. All planking or platforms must be overlapped at least twelve inches or secured from movement.
6. Planks must extend over the end supports not less than six inches or more than eighteen inches (not more than twelve inches on construction sites), and should be secured from falling off the platform.
7. Scaffolds must be secured when in use and must not be moved when in use or occupied.

8. All scaffolds must be maintained in a safe condition at all times. Unsafe scaffolds should be removed from the site for disposal or the defective parts immediately replaced or repaired.
9. The poles, legs, or uprights of scaffolds must be plumb and securely and rigidly braced to prevent swaying and displacement.
10. Planking must be scaffold grade for the species of wood used if the scaffold is of the wood pole type. The maximum permissible spans for two-by-nine inch or wider planks are provided in Table 8.2.
11. Where there are overhead hazards, overhead protection must be provided. No one should work on scaffolds during storms or high winds.

TABLE 8.2
Maximum Permissible Spans for Two-by-Nine inch or Wider Planks

	Material				
	Full Thickness Undressed Lumber			Nominal Thickness Lumber	
Working Load (p.s.f.)	25	50	75	25	50
Permissible Span (ft.)	10	8	6	8	6

Note: The maximum permissible span for 1¼ x 9 inch or wider plank of full thickness is four feet with a medium loading of 50 p.s.f

Summary

Unsafe walking and working surfaces are a leading cause of injuries in the workplace. The five major working and walking surfaces include floors, ramps, stairs, ladders, and scaffolds.

Floors must be properly maintained, be free from obstructions and spills, and not exceed the recommended slope. Ramps must be equipped with the appropriate guardrail systems and toeboards. Ladders must be maintained properly, used carefully, and inspected periodically. In addition, ladders should meet all OSHA requirements to ensure their safe placement and use. Three types of scaffolding are commonly used. Each type of scaffolding must meet the general OSHA requirements.

Bibliography

Safety Requirements for Portable Metal Ladders, A14.2-1972, American

National Standards Institute, New York, New York 10018. 1972.

U.S. Department of Labor, Bureau of Labor Standards, Safe Working Surfaces, Bulletin 292, 1967.

9

Maintaining the Small Business Work Area

Maintenance is an important aspect of the safety and health program. Insufficient or improper maintenance can result in accidents, property damage, and equipment breakdown. The difference between an average maintenance program and a superior one is that an average program is aimed at maintaining conditions while a superior program is aimed at improving them.

Even where safety and health are prime considerations in the planning and implementation of a safety and health program, improperly maintained hazards can surface and production can be adversely affected. Preventive maintenance is the orderly, scheduled activity by the maintenance organization to prevent equipment breakdown; prolong the useful life of equipment and buildings; and reduce the potential for injury, illness, and property damage. Preventive maintenance is a program of mutual support creating safer conditions, eliminating costly delays and breakdowns, and prolonging equipment life.

9.1 Preventative Maintenance: A Shared Responsibility

Although many activities are usually done by maintenance personnel, preventive maintenance must be understood as a shared responsibility by management and employees. During their regular inspections, supervisors and employees must ensure that hazardous conditions are not present because of poor housekeeping. The Safety and Health Committee calls attention to deficiencies, helps maintenance personnel set up schedules for servicing or replacing machinery and equipment parts, and ensures that preventive maintenance functions are performed correctly.

From the time the workers enters the shop, they should be taught to correct the potential hazards detected rather than to ask himself whether he created the poor housekeeping condition or whether he is responsible for cleaning up. Because the safety and health of everyone in the facility depends upon it, properly maintaining the work area is everyone's job.

9.2 Functions of a Maintenance

Maintenance has four main functions which affect safety and health:

1. *Installing, constructing, and maintaining buildings, facilities, equipment, and machinery.* Ensuring that the facilities design, layout, construction, and installation work conforms to good engineering practices, is an important first step toward the reduction of hazards and potential accidents.
2. *Providing utility services (heat, light, power, and compressed air) for production operations.* During these activities, maintenance ensures that acceptable standards are met and that hazards, which have the capacity to cause accidents, are quickly identified and removed or controlled.
3. *Providing for the cleaning of facilities and the disposal of scrap materials and waste.* Windows and lighting fixtures must be cleaned regularly to provide the necessary illumination for a safe work environment. Certain materials must not be drained into a public sewage system. Chemical wastes and acids should be placed in prescribed containers for removal.
4. *Providing planned preventive maintenance on all buildings, electrical systems, machinery, and equipment.*

9.3 Advantages of Preventative Maintenance

Advantages of preventive maintenance include safer working conditions, more training time on the equipment, decreased "down time" of equipment due to breakdown, an increased lifespan of the equipment.

Poor maintenance can cause accidents. If a guard is not replaced after it has been removed for lubrication or repair, then the maintenance process has created a new hazard. Maintenance should eliminate or control hazards not create them. For example, a local exhaust system might, during a routine maintenance activity, be found with a substantial amount of dust around the exhaust hood. This situation prohibits the exhaust system from functioning properly and creates a safety hazard.

Satisfactory operations are contingent upon the buildings, equipment, machinery, portable tools, and safety devices being in good operating condition and being maintained in such a manner that production activities are not interrupted while repairs are being made or equipment is replaced. When safe and properly maintained tools are issued, employees have an added incentive to provide proper care for them and maintain their safe operating condition.

9.4 Principles of Maintenance Management

The basic principles of maintenance management are organization, motivation, and control. Organizing means establishing policies and procedures for operating the program, designating and assigning staff to supervise and carry out the maintenance activities, providing them with the means to get the job done, and rewarding outstanding performance. Maintenance personnel must have the proper tools, materials, and equipment to do the job when it needs to be done, without undue delay and without requiring the machines or equipment to be partly torn down or out of order any longer than necessary.

Motivation involves instilling a strong sense of responsibility in all maintenance personnel to assure that they will perform their duties conscientiously and carefully. This involves training in safe procedures during maintenance and in keeping good records.

Control is concerned with the actual activity of preventive or corrective maintenance. The more efficient preventive maintenance is, the fewer times more expensive corrective actions will be necessary. Overall control is a supervisory function. Maintenance management requires that one person in the organization be responsible for seeing that all phases of the program are operating in accordance with company policy and regulatory procedures.

9.5 Components of a Preventative Maintenance Program

A preventive maintenance program has four main components:

1. scheduling and performing periodic maintenance
2. keeping records of service and repairs

3. repairing and replacing equipment and equipment parts
4. providing spare parts control

9.5.1 Scheduling and Performing Periodic Maintenance

Maintenance can be scheduled on either a time or use basis. Factors to be considered include the age of the machine, the number of hours per day the machine is used, past experience, and the manufacturer's recommendations. Manufacture's specifications provide procedures to be followed for safe and economical use of the equipment. Examples of activities which need to be scheduled include:

- lubrication of equipment
- replacing belts, pulleys, fans, and similar parts
- checking and adjusting brakes, tool rest, cables, and hoists

9.5.2 Record Keeping

Two types of records should be kept. The first is a maintenance service schedule for each piece of equipment used. The schedule should indicate the date the equipment was purchased or placed in operation, its cost (if known), the shop in which it is used, each part to be serviced, the kind of service required, and the frequency of service and the person assigned to do the servicing.

The second type of record to be kept for each piece of equipment is a repair record. A repair record includes an itemized list of parts replaced or repaired and the name of the person who did the work.

9.5.3 Repair and Replacement

In addition to scheduled adjustments and replacements, maintenance personnel must look for malfunctioning or broken equipment. Repairs should be made in accordance with the manufacturer's specifications. Equipment may need to be sent back to the manufacturer or to their representative for repair. Maintenance personnel should be aware of the manufacture's limitations and recognize that the manufacture's experience and expertise may not be sufficient for all types of repairs.

Maintenance personnel, along with supervisors and Safety and Health Committee members, have a responsibility to tag and/or lock out defective equipment. Tags are not to be considered a complete warning method but should be used until the hazard can be eliminated. For example, a "Do Not Start" tag on power equipment should be used only until the switch or source of entry on the system can be locked out.

The following four types of accident prevention tags are required for use (see OSHA 1910.145):

1. *Caution* tags warn against potential hazards or caution against unsafe practices.
2. *Danger* tags indicate that an immediate hazard exists and that special precautions are necessary.
3. *Do Not Start* tags must be conspicuously located and used when energizing the equipment would cause a hazardous condition.
4. *Out of Order* tags indicate that the equipment or machinery is out of order and should not be used.

FIG. 9.1 Four types of accident prevention tags.

9.5.4 Spare Parts Control

Another element of the total preventive maintenance program is the survey of spare parts requirements. In order to keep needed parts on hand, it is necessary to periodically review the material required for repair orders and the delivery schedule of such parts. If maintenance personnel keep purchasing agents informed of their anticipated stock needs, it can prevent lengthy "down time" while waiting for parts to arrive.

9.5.5 Examples of Effective Preventative Maintenance

Specific examples of the ways in which preventive maintenance creates a safe work environment include:

1. Ensuring electrical wires have adequate insulation.
2. Proper storage of compressed gas cylinders.
3. Tagging out unsafe equipment.
4. Maintaining brakes on materials-handling equipment.
5. Replacing guards on machinery.
6. Proper labeling of high and low pressure steam lines, compressed air, and sanitary lines.
7. Maintaining boilers and pressure vessels.
8. Preventing oil from being thrown onto the insulation of electrical windings and the adjacent floor by not over-oiling motor bearings.

9.6 Program Evaluation and Support

As part of its inspection function, the Safety and Health Committee or management representative should will periodically check to make sure that preventive maintenance functions are being adequately defined and satisfactorily performed. Supervisors and employees also will be exercising a control function by seeing that equipment is maintained correctly. Maintenance personnel should understand that inspections are aimed at fact-finding, not faultfinding, and that they share with other staff an obligation for maintaining a safe facility.

The supervisor and worker can support the preventive maintenance program by contributing their own insights. The following examples

illustrate ways in which the effectiveness of a program can be measured
and improved:

1. *Hand tools.* Employees know the condition of the tools in use at
   the machine and on the benches. Is there a system of replacing or
   repairing defective tools?
2. *Electric wires, operating switches, and control boards.* Are these
   kept in good condition? Are "temporary" repairs, alterations, and
   additions eliminated? Temporary jobs tend to become permanent
   unless carefully limited to those jobs that are absolutely necessary
   and made standard immediately after the emergency has passed.
3. *Sound of operating equipment.* Employees soon become accus-
   tomed to the tune of equipment, each of, which has a characteristic operating sound. They can tell when a machine is overloaded
   by the noise, the grunt, or the squeal. They can tell that maintenance is required because of the rattle and vibration.
4. *Servicing and repair records.* Employees will know when
   servicing schedules are being kept. These records must be rigidly
   adhered to, particularly as they affect electrical hoisting equipment, pressure vessels, guards, cranes, slings, chains and tackle,
   extension cords, portable motor-operated tools, and personal
   protective equipment.

9.7 Role of Housekeeping in Safety and Health Programs

Housekeeping plays an important role in the safety and health
program. Good housekeeping reduces accidents, improves morale, and
increases shop efficiency and effectiveness. In a clean and orderly work
environment, the workers enjoy working and can accomplish their tasks
without interference and interruption.

A production area, by nature, contains tools, which must be kept
organized, and machinery, which must be kept clean. Flammable substances and materials, which require special storage and removal, may
also be present. These materials and substances include dust, scrap metal
filings and chips, waste liquids, and scrap lumber.

Housekeeping is a continuous process, both throughout and at the
end of each shift. Both the supervisor and the workers have housekeeping responsibilities. This function represents one of the single most
important ingredients in reducing losses and injuries in the workplace.

9.7.1 Benefits of Good Housekeeping

A good housekeeping program incorporates the housekeeping function into all processes, operations, and tasks. The ultimate goal is for each worker to view housckeeping as an integral part of his performance, not as a supplement to the job. A well-administered housekeeping program also results in other immediate and long-range benefits.

When the work area is clean and orderly and the housekeeping program becomes a standard part of operations, less time and effort will be spent keeping it clean, making needless repairs, and replacing equipment, and fixtures.

When workers can concentrate on their required tasks without excess scrap material, tools, and equipment interfering with their work, they can create higher quality products acquire higher degrees of skill.

When everything has an assigned place, the chance of materials and tools being moved or misplaced is reduced. The shop foreman and supervisor are able to quickly determine what is missing. Different colors of paint can be used on the tools to identify the department to which they belong. Tool racks or holders should be painted a contrasting color as a reminder to employees to return the tools to their proper places. The space directly behind each tool should be painted or outlined in color to call attention to a missing tool.

Losses are reduced and efficiency is increased when employees and supervisors handle supplies and materials with proper care.

When aisle and floor space is uncluttered, movement within the work area as well as the maintenance of machinery and equipment is easier and safer.

When the facility has adequate workspace and when oil, grease, water, and dust are removed from floors and machinery, employees are less likely to slip, trip, fall, or inadvertently come into contact with dangerous parts of machinery.

When an area is clean and orderly, the morale of employees and supervisors is improved. Workers understand and respect orderliness and cleanliness. When an area is kept free from accumulations of combustible materials which may burn upon ignition or, in the case of certain material relationships, spontaneously ignite without the aid of an external source of ignition, the chances of fires are minimized. Furthermore, an orderly work area permits easy exits during an emergency by keeping exits and aisles free from obstructions. A neat and orderly area also makes it easier to locate and obtain fire emergency and extinguishment equipment.

9.8 Conditions Which Indicate Poor Housekeeping

Examples of poor housekeeping include:

1. *Objects on the floor and in aisles.* This category includes articles which fall from machinery or are dropped in transit, material left over following a repair job, objects which are stored in aisles, and materials which are stored improperly.
2. *Improperly stored tools and equipment.* When tools and equipment which are not in immediate use are left on workbenches or on the floor, storage control breaks down and housekeeping problems result.
3. *Improperly stored or stacked stock.*
4. *Inadequate waste disposal.* Signs of inadequate waste disposal include overflowing waste receptacles, lack of disposal of chips and oil, and the lack of a system for the safe reclamation of cutting oils.
5. *Dirt, grime, and general disorder.* Dust accumulations on windows, skylights, and lighting fixtures reduce illumination and increase eyestrain. Other indicators of poor housekeeping include:
   - scrap materials on benches, floors, or equipment
   - mops and brooms stored in areas other than those assigned
   - sinks and areas around them with accumulations of dirt and grime
   - rubbish on floors and in corners
   - oil, grease, cleaning compounds, and solvents spilled and not cleaned up
6. *Improper storage and disposal of combustible materials.* Good housekeeping requires that oil and solvent-soaked rags are not allowed to accumulate and that combustible materials are stored in suitable receptacles.

Summary

   Maintenance is an important aspect of the safety and health program. Properly maintained machinery and equipment leads to safer working conditions, decreased "down time" of equipment, and longer equipment life. The principles of maintenance management include organization, motivation, and control. The components of a preventative maintenance program include scheduling and performing periodic maintenance, keeping proper service and repair records, repairing and replacing equipment and equipment parts, and providing spare parts control.

Like preventive maintenance, housekeeping is a shared responsibility. Good housekeeping requires planning, supervision and constant attention. During their regular inspections, supervisors, and employees must make sure that hazardous conditions are not present because of poor house-keeping. Maintenance personnel and the Safety and Health Committee must call attention to deficiencies they note during their inspections.

Bibliography

National Safety Council (1993) "Accident Prevention Manual for Industrial Operations", Itasca, IL.

U.S. Department of Health and Human Services, National Institute for Occupational Safety and Health, *Maintaining Facilities and Operations,* September 1979.

U.S. Department of Labor, Bureau of Labor Standards, *Housekeeping for Safety,* Bulletin 295, 1967.

U.S. Department of Labor, Bureau of Labor Standards, *Preventive Maintenance for Safety,* Bulletin 290, 1967.

U.S. Department of Labor, Occupational Safety and Health Administration, *General Industry OSHA Safety and Health Standards (29 CFR 1910).*

10

Illumination and Color Safety

IN CHAPTER 7, adequate illumination was mentioned as essential to a safe and healthy shop environment. Along with such factors as machinery and equipment placement, materials storage, ventilation, aisle arrangement, space for work, materials flow, first aid, and electrical considerations, illumination and color play an important role in a effective safety program.

The principal reasons for work area lighting are to protect workers from eyestrain, reduce losses in visual performance, and enable employees to more readily detect hazards in the work environment. The benefits of adequate illumination include:

- the minimization of errors - The quality of work is thereby increased and the number and severity of accidents is decreased.
- the improved ability to detect defects - The project's quality is thereby improved.
- the reduction in the time necessary to determine fine details and to make fine measurements
- improved housekeeping - A well-illuminated shop will make rubbish and other waste products more visible, thus encouraging prompt removal.
- an improved social climate - Better lighting promotes a more cheerful work environment and enables colors to be more visible.

10.1 Factors in Adequate Illumination

The factors associated with illumination can be grouped into two broad categories- quantity and quality. Quantity is the amount of light that produces sufficient brightness to illuminate the task and its surroundings. The quality of illumination pertains to the distribution of brightness in a visual environment and includes its color, direction, diffusion, and degree of glare.

10.1.1 Quantity of Illumination

The desired quantity of light for operations in the workplace depends primarily upon the task being done. As the quantity of illumination is increased, so is the ease, speed, and accuracy in accomplishing the task.

The quantity of illumination is measured in units called foot-candles, an index of the ability of a light source to produce illumination. An illumination meter gives a direct reading of the number of foot-candles of light reaching the working plane. A foot-candle is the amount of illumination on a surface of one square foot, all parts of which are at a distance of one foot from a standard candle.

Generally, those tasks requiring fine detail, low contrasts, and prolonged work periods require higher illumination levels than more casual or intermittent tasks involving high contrast. The current minimum levels of illumination for industrial areas as recommended by the Illuminating Engineering Society (IES) are given in ANSI/IES RP-7-1979. Table 10.1 lists the recommended foot-candles required for various tasks.

Generally, calculations of foot-candles are based on the illumination of horizontal surfaces. But most work is done on slanting and vertical surfaces where illumination may be only one-half to one-third of that on horizontal surfaces. Compensation must be made and additional illumination provided for such surfaces. The shop's lighting system should also provide proper levels of illumination on working areas located along the sides of the shop where the illumination is considerably less than the average calculated level for the central areas of the shop.

10.1.2 Quality of Illumination

Quality of illumination pertains to the distribution of brightness in the visual environment. Glare, diffusion, direction, uniformity, color, brightness, and brightness ratios all have a significant effect on visibility and the ability to see easily, accurately, and quickly. Tasks performed over long periods of time and demanding discernment of fine details require illumination of a high quality.

The ability to see detail depends upon a difference in brightness between the detail and its background, but the eyes function most comfortably and efficiently when the difference is kept within a certain range. The ratios between areas within the field of vision should not exceed those shown in Table 10.2, which lists the maximum brightness ratios recommended by the Illuminating Engineering Society in ANSI/IES RP-7-1979. The goal in shop illumination should be to have the task brighter than the surroundings.

TABLE 10.1
Recommended Foot-Candles

The following recommendations represent the minimum on the task at any time for adults with normal and better than 20/30 corrected vision.

Task	Foot-candles
Forging	50
Foundry work	
Annealing (furnaces)	30
Core making	
Fine	100
Medium	50
Pouring	50
Garages (auto shop)	
Repairs	100
Active traffic areas	20
Machine Shops	
Rough bench and machine work	50
Medium bench and machine work, ordinary automatic machines, rough grinding,	
medium buffing and polishing	100
Fine bench and machine work, fine automatic machines, medium grinding, fine	
buffing and polishing	500*
Extra-fine bench and machine work, grinding, fine work	1000*
Paint shops	
Dipping, simple spraying, firing	50
Rubbing, ordinary hand painting and finishing art, stencil and special spraying	50
Fine hand painting and finishing	100
Extra-fine hand painting and finishing	300*
Printing and photoengraving	
Printing	
Color inspection and appraisal	200*
Machine composition	100
Composing	100
Presses	70
Proofreading	150
Photoengraving	
Etching, staging, blocking	50
Routing, finishing, proofing	100
Tint laying, masking	100
Sheet metal shops	
Miscellaneous machines, ordinary bench work	50
Presses, shears, stamps, medium bench work, spinning	50
Punches	50
Tin plate inspection (galvanized)	200**
Scribing	200**
Welding	
General illumination	50
Precision manual arc welding	1000*

TABLE 10.1 Continued

Task	Foot-candles
Woodworking	
Rough sawing and bench work	30
Sizing, planing, rough sanding, medium quality machine and bench work, gluing,	
veneering, coopering	50
Fine bench and machine work, fine sanding and finishing	100

*The recommended foot-candles may be obtained by a combination of general lighting plus special supplementary lighting. Care must be taken to design and install a system which not only provides sufficient light but directs and diffuses the light and protects the eyes. Insofar as it is possible, glare (both direct and reflected) and objectionable shadows should be eliminated.

**In such tasks, the mirror-like surface of the material means that special care must be taken in selecting and placing lighting equipment and/or orienting the work to reduce glare.

TABLE 10.2
Recommend Maximum Brightness Ratios

	Classification (see below)		
	A	B	C
Between tasks and adjacent darker surroundings	3 to 1	3 to 1	5 to 1
Between tasks and adjacent lighter surroundings	1 to 3	1 to 3	1 to 5
Between tasks and more remote darker surfaces	10 to 1	20 to 1	*
Between tasks and more remote lighter surfaces	1 to 10	1 to 20	*
Between lighting units (or windows, skylights) and surfaces adjacent to them	20 to 1	*	*
Anywhere within normal field of view	40 to 1	*	*

*Brightness ratio control not practical.
A - Interior areas where reflectances of entire space can be controlled.
B - Areas where reflectances of immediate work area can be controlled but not those of remote surroundings.
C - Areas where reflectances cannot be controlled and environmental conditions can be altered only with difficulty.

10.2 Factors Associated with Poor Illumination

In addition to the factors of insufficient quality and quantity of illumination, other less tangible factors contribute to accidents. These factors include direct glare, reflected glare, dark shadows, and excessive visual fatigue. When a worker moves from brighter to darker surroundings or from darker to brighter surroundings, his eyes must adapt to the new light levels. During the time it takes the eyes to adapt, the worker's vision is compromised, creating a situation which could cause accidents.

Glare is defined as any brightness within the field of vision which causes reduced visibility and discomfort, annoyance, or eye fatigue. The two types of glare are direct and reflected. Direct glare is the result of a source of illumination within the field of view, whether that source is artificial or natural. To reduce direct glare:

- Decrease the brightness of the light source. For example, use a shade.
- Reduce the area of high brightness. For example, install window coverings.
- Increase the angle between the glare source and the line of vision. For example, position the light source so that it falls outside the normal field of vision.
- Reduce the contrast by increasing the brightness of the area surrounding the glare source. For example, make certain that surface reflectances are at recommended levels.

Indirect or reflected glare is caused by very bright objects or surfaces or by differences in brightness reflected from shiny surfaces such as ceilings, walls, machinery. This type of glare is often more annoying and fatiguing to the eyes than direct glare because it is so close to the line of vision that the eye cannot gain relief by moving away from it. To reduce reflected glare:

- Decrease the brightness of the light source while maintaining the limits prescribed in the standard. For example, replace an exposed incandescent lamp with a fluorescent light.
- Shield and/or diffuse the light source. For example, use lighting units with "batwing" distribution.
- Change the position of either the work or the light source. For example, raise the position of a wall light fixture.
- Physically reduce the reflecting source. For example, use a dull, flat paint.
- Reduce the contrast by increasing surrounding brightness. For example, bring surface reflectances up to recommended levels.

10.3 Types of Lighting

Interior lighting systems are divided into five types: direct, semi-direct, general diffuse or direct-indirect, semi-indirect, and indirect.

Direct lighting aims practically all of the light downward, directly toward the usual working area. Since direct units may cause disturbing shadows, it is best used to light areas where tasks are vertical or near vertical.

Semi-direct lighting units direct 60 to 90 percent of their light downward. The increased ceiling illumination from the semi-direct distribution of light reduces the brightness ratio between the lighting fixture and the ceiling, softens shadows, and increases diffusion.

General diffuse or direct-indirect lighting refers to lighting units in which the downward and upward components are approximately the same (each 40 to 60 percent of the total light output). General diffuse lighting units emit light about equally in all directions. Direct-indirect lighting units emit very little light at angles near the horizontal and are generally preferred since they are lower in brightness in the direct-glare zone. This is the type most commonly used type for general shop lighting.

Semi-indirect lighting units direct most of their light (60 to 90 percent of the total light output) upward. The major portion of the light reaching the normal horizontal work plane is reflected from the ceiling and upper walls. When this type of lighting is used, it is important to ensure high reflectance by keeping the ceiling and upper walls as clean as possible. This use is generally limited to areas where reflected glare from mirror-like work surfaces must be reduced.

Indirect lighting units emit from 90 to 100 percent of their light upward and are seldom used in industrial operations. While this lighting is generally the most comfortable, it is the least utilized and is often difficult to maintain. Figure 10.1 illustrates each type of lighting.

10.4 Factors in Selecting Lighting Units

10.4.1 Lighting Types and Locations

Incandescent or fluorescent lighting units may be used in the shop. Low-bay improved-color mercury lighting units may also be used. The lighting should follow the best industrial lighting practices, with special emphasis on assisting the accuracy and safety of manual operations.

In printing shops, parabolic specular metal reflectors or prismatic closure plates, both of which spread the brightness of the lamps out over the width of the reflector, should be used over make-up racks and composing stone. Multiple presses need light, preferably from continuous fluorescent lighting units, directed from the side into the machines. The lighting of the printing shop should have an upward component to light ceiling and upper walls.

In machine shops and metal working areas, a grid system of semi-direct fluorescent lighting units will yield good results. Localized lighting should supplement general lighting near high-speed tools and saws.

Many areas in the shop may be classified as hazardous. These locations require the use of specialized lighting equipment (vapor-proof, explosion-proof and dust-proof lighting units) which provides required illumination without introducing hazards to life and property.

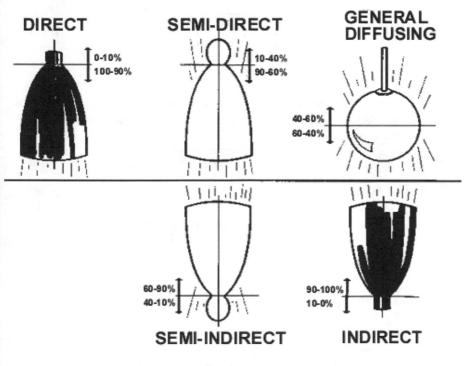

FIG. 10.1 Types of Illumination.

10.4.2 Lighting Maintenance

Important factors to consider when selecting suitable lighting units for the work area include the rate at which the lighting unit will collect dirt and the ease with which the lighting unit can be cleaned and lamps replaced. For example, open-bottom or louvered types offer the advantage of having no bottom surface to collect dust and dirt. Closed-bottom dust-tight units often are selected to prevent dirt and dust accumulation on the lamps and reflecting surfaces.

In order to maintain the proper quantity and quality of illumination, it is necessary to set up a maintenance schedule. The schedule should include:

- cleaning fixtures
- cleaning room surfaces and windows. The collection of dirt on room surfaces may decrease their reflectance, thereby reducing the amount of visible light on the working plane. Rough surfaces, which encourage the accumulation of dirt, should be avoided.
- replacing worn-out globes and tubes. Care also should be exercised during relamping to ensure adequate sealing against dust. Even a small space left in the gasket seal will allow the lighting unit to take-in dirt.
- Repainting fixtures and room surfaces. The maintenance schedule should be considered while planning shop illumination to allow for safe access to the globes, tubes, fixtures, windows, and room surfaces.

10.5 The Role of Color

Illumination and color are partners in affecting tasks which involve the eyes. There are several advantages to using color in the shop. Color can:

- provide brightness and contrast which can positively affect visual performance
- make the shop environment more pleasant, raising morale
- result in better housekeeping and improved repair and maintenance
- be used as a coding technique to spotlight hazards

The ease and efficiency with which workers see can be aided by the judicious selection of colors for the work area and its contents. Conversely, when eyes are exposed for long periods of time to extreme monotony or high contrast in color and brightness, they will function less efficiently.

Using color to increase brightness and contrast is a device easily adapted to the shop setting. For example, a white object will appear much brighter than a dark one under the same illumination. Contrast tends to increase visibility by making the task stand out against the background. The speed of visibility is a direct function of brightness and contrast. Increasing either brightness or contrast will shorten visibility time, increase production, decrease errors and mistakes, and possibly give the workers those extra fractions of a second needed to avoid injury.

Besides the effect on visual ability, color can be used to make the working environment more interesting and pleasant. For example, the colors cream, ivory, and buff are "warm" physiologically. Light gray finishes are excellent for either the background or for equipment and machinery. Green and blue tints give a cool effect.

Unpainted metal parts of machines, which are usually black or gray, allow dust, dirt, and grease to accumulate unnoticed, covering up defects in the machine. Painting encourages workers to keep machines clean and enables defects to be spotted before they become a hazard.

The use of color as a technique for coding various controls has been met with great success. For instance, piping systems which carry dangerous contents would be painted yellow and fire protection systems would be painted red. Such color-coding is of value from both a safety and maintenance standpoint.

10.6 Color for Accident Prevention Signs

Color is widely used in accident prevention signs. In these cases, uniformity in the color and design of signs is essential. ANSI Z35.1-1972, "Specifications for Accident Prevention Signs," specifies the color combinations for industrial accident prevention signs. The specifications are as follows:

- Danger - immediate and grave danger or peril. White lettering within red oval on black rectangular background in upper panel; black or red lettering on white background in lower panel.
- Caution - lesser hazards. Yellow lettering on black background in upper panel; black lettering on yellow background in lower panel.
- General Safety - white lettering on green background on upper panel black or green lettering on white background on lower panel.
- Fire and Emergency - white letters on red background in upper panel; red on white background optional for lower panel.

Codes for other common uses include:

- Information - blue is used on informational signs and bulletin boards which do not contain safety information
- In-Plant Vehicle Traffic - standard highway signs (ANSI D6.1-1971)
- Exit Marking - red letters on white background

10.7 Recommended Color Standard for Marking Hazards

Seven colors have been designated by the American National Standards Institute Standard Z53.1-1971, "Safety Color Code for Marking Physical Hazards," for spotlighting hazards and assisting in the visual identification of equipment and other items in the shop. The colors are red, orange, yellow, blue, green, white and black, and purple.

10.7.1 Red

Red identifies danger, emergency stops on machines, and fire protection equipment. OSHA regulations require red lights at barricades and at temporary obstructions. Red must also be used for:

- danger signs
- stop buttons or electrical switches used for emergency stopping of machinery
- portable containers (including safety cans) holding flammable liquids
- fire hydrants, pumps and sirens
- location of fire extinguishers
- fire exit signs
- fire buckets or pails
- fire alarm boxes
- fire hose locations
- post indicator valves for sprinkler systems
- sprinkler piping

10.7.2 Orange

Orange is used for calling attention to dangerous parts of machines or energized equipment which may cut, crush, shock, or otherwise injure. Orange is also used to emphasize instances where enclosure doors are open or when guards around moving equipment are open or removed, exposing unguarded hazards. Examples of how the orange color standard is used include:

- The inside of movable guards (those which can be opened or removed) are painted orange to attract the attention of the machine operator.
- Safety starting buttons and boxes are painted orange as a warning of the potential hazard involved and as a "hands-off" notice to unauthorized persons.
- The edges of exposed parts of gears, pulleys, rollers, cutting devices, power jaws, and similar devices are painted orange to warn against contact.
- The insides (as a minimum) of the box door or cover of open fuse, power, and electrical switch boxes are painted orange to warn of exposed live wires and electrical equipment.

10.7.3 Yellow

According to OSHA regulations (29 CFR 1910.144), "yellow shall be the basic color for designating caution and for marking physical hazards such as striking against, stumbling, falling, tripping, and 'caught in between.' Solid yellow, yellow and black stripes, yellow and black checkers (or yellow with suitable contrasting background) should be used interchangeably, using the combination which will attract the most attention in the particular environment." Examples of how the yellow color standard is used include:

- exposed and unguarded edges of platforms, pits, and walls
- fixtures suspended from ceilings or walls which extend into normal operating areas
- handrails, guardrails, or top and bottom treads of stairways where caution is needed
- lower pulley blocks and cranes
- markings for projections, doorways, traveling conveyors, low beams and pipes, the frames of elevator ways, and elevator gates
- materials-handling equipment (or areas thereon), such as forklifts and hand trucks

- pillars, posts or columns which might be struck
- vertical edge of horizontally sliding pairs of fire doors

10.7.4 Blue

Blue, in the railroad and other industries, is the standard color for warnings against starting, using, or moving equipment which is under repair. In the construction industry, yellow is used for these same purposes. Blue is also used for warning markers such as painted barriers and flags. These markers should be located at the power source or starting point of machinery and be displayed on:

- ovens and vats
- tanks
- kilns
- boilers
- electrical controls
- scaffolding
- ladders

10.7.5 Green

Green is used as the standard color for designating safety and for showing the location of first aid equipment, that is, equipment other than fire-fighting equipment. It is recommended that the use of "safety green" be restricted for protection of the worker, so that the color will have value in acquainting him with the location of safety and allied devices and so that their location will come to mind readily in an emergency. Examples of safety and allied devices include:

- stretchers and stretcher cabinets
- first aid cabinets
- first aid kits
- respiratory containers
- emergency showers
- safety bulletin boards

10.7.6 Purple

Purple is the standard color for designating radiation hazards.

10.7.7 White and Black

White, black, or a combination of these are traffic and housekeeping markings. Solid white, solid black, single color striping, alternate stripes of black and white, or black and white checkers should be used in accordance with local conditions. Examples of places in which white and black markings could be used include:

- dead ends of aisles or passageways
- location and width of aisle ways
- stairways (risers, direction, and border limit lines)
- location of refuse cans
- white corners for rooms or passageways
- drinking fountains and food dispensing equipment locations
- clear floor areas around first aid, fire fighting, or other emergency equipment

10.8 Advantages and Limitations of the Safety Color Code

Introducing workers to the safety color code has two advantages. First, the workers who know the code are, by working around the colors each day, constantly absorbing the safety information it provides. This may allow workers to react spontaneously in time of emergency. Second, because the safety color code has been adopted widely by industry and construction, workers are using an information system which is easily applied to other workplaces.

There are two limitations of the standard safety color code. Too many color identifications in the environment can be confusing and fatiguing. For maximum emphasis, the number of markings should be kept to a minimum. It is better to eliminate a hazard than to mark it with a color warning. The second limitation is the fact that the code is only a supplement to a program aimed at properly guarding or, wherever possible, eliminating hazardous conditions.

10.9 How We See

In general, one perceives objects in three ways: reflection, transmission and silhouette. Perception by silhouette is the detection of an object and its contour via its darker outline contrasting against lighter surroundings. Silhouette lighting is used as low-level safety lighting, where hazardous objects or obstacles are seen against lighted surroundings (protective and emergency lighting, outdoor passageways and roadways).

Transmission involves the revealing of details through the variation and transmission of white light or the changing of color through materials that are susceptible to penetration. Transmission involves the inspection of translucent materials.

By far the most common method of seeing is by reflected light, where light and dark areas or details are revealed by a difference in reflection. Highlights and shadows are actually the result of the amount of light reflected by an object which in turn helps us to perceive the world in three dimensions.

10.10 Factors Affecting Vision

The visibility of an object is determined by its size, contrast, time of viewing, and brightness. Each of these factors are interrelated so that a deficiency in one may be compensated by augmenting one or more of the others. As size increases, visibility increases, and up to a certain point seeing becomes easier. If the size of an object is small, a person should use more light, hold the object closer to his eye or use a magnifier.

To be readily visible, each detail of the perceived object must differ in brightness (or color) from the surrounding background. If discrimination is dependent solely on brightness differences, visibility is at a maximum when the contrast of details within the background is the greatest. Therefore, within the available limits, the contrast between the object and its immediate background should be made as high as possible. Where it is impractical to provide good contrast conditions, higher levels of illumination help compensate for poor contrast.

The speed with which a visual task can be performed is related to illumination, size, and contrast. By increasing illumination, the time required for seeing will be shortened. Tasks of high contrast and large size generally require less time than tasks of low contrast and small size.

Size, contrast, and to some extent time, are factors inherent in the task itself. Brightness is the visibility factor that is most controllable. Within wide limits, brightness can be controlled by varying the amount and distribution of light and can, to some extent, compensate for deficiencies in other factors.

10.11 The Theory of Color

Some substances absorb certain wave lengths and reflect others. This property gives the object its color. An object appears green because it absorbs most wave lengths but reflects green. Black is not a color, but black objects appear so because most of the light is absorbed and little is reflected. Yellow occupies the point of highest visibility in the color spectrum. Most authorities agree that black lettering on a yellow background has the greatest legibility.

"Bright" colors appear larger than "deep" colors. Experiments have proven that yellow is seen as the "largest" of hues, followed by white, red, green, blue, and black. Therefore, signs, objects, and rooms can be made to appear to expand or contract in accordance with arrangements of color brightness within the field of vision. Ceilings can be made to appear higher or lower; walls, nearer or farther away; and selected forms or shapes, larger or smaller.

The following definitions are useful in discussing color:

- Color (chrome, hue, shade, tint, tinge) is a property of an object visible only in light.
- Hue is the attribute of a color which distinguishes it from other colors; for example, red from yellow, or blue from green. It also applies to variations in a color when mixed with another color.
- Tone is the attribute in which a color holds a position in a dark-to-light scale, such as navy blue which differs in tone from light blue. Tone is also known as saturation.
- Tint, a light or delicate touching with color, is also sometimes used to mean the slight alteration of a color.
- Tinge is an infusion or trace of color.
- Shade has come to mean simply any color.

Summary

The proper quantity and quality of illumination is essential to a safe work environment. The quantity of light is measured in units of foot-candles.

The quality of light pertains to its distribution of brightness and its effect on visibility.

There are five types of lighting, each with unique illumination patterns and uses. A maintenance schedule should be provided for each fixture to ensure they maintain the proper quantity and quality of illumination. The five types of lighting include direct, semi-direct, general diffusing, semi-indirect, and indirect.

The color of light plays an important role in the workplace. Color can provide brightness and contrast which positively affects visual performance, make the shop environment more pleasant, result in better housekeeping and improved repair and maintenance, and be used as a coding technique to spotlight hazards. Specific colors are used in accident prevention signs to signal danger, caution, general safety precautions, and fire and emergency equipment. The recommended color standards for marking hazards include seven colors: red, orange, yellow, blue, green, white and black, and purple.

Objects are perceived via reflection, transmission, or silhouette. The visibility of an object is determined by four interrelated factors: size, contrast, time of viewing, and brightness. Objects and walls can be made to appear larger or smaller and ceilings appear nearer or farther away, depending on their color.

Bibliography

National Safety Council (1974) *Accident Prevention Manual for Industrial Operations*, Chicago.

National Safety Council (1979) *Fundamentals of Industrial Hygiene*, Chicago.

McCormick, Ernest J. (1970) *Human Factors Engineering*, 3rd ed., McGraw-Hill, New York, New York 10020.

11

Fire Protection

PREVIOUS CHAPTERS have been concerned with principles of good shop planning including the construction of safe working surfaces, maintenance of facilities, and the provisions for adequate illumination. Two other important considerations in planning the shop are fire protection and safe storage and handling of hazardous materials.

This chapter discusses the methods whereby fires in the small business facility can be prevented and controlled. Topics covered include the fundamentals of fire; common ignition sources and ways to control them; the classes of fires and precautions for controlling each; the requirements for storing flammable and combustible materials; the requirements for an adequate fire system, including detection devices, sprinkler systems, and portable fire extinguishers; and emergency procedures.

Common causes of fires and the precautions necessary to prevent their occurrence are rather straightforward. These causes and precautions include the:

- safe storage and handling of flammable and combustible materials
- installation of detection and alarm systems
- use of automatic sprinkler systems
- use of special systems to control specific hazards
- use of portable fire extinguishers
- use of proper training

11.1 Fundamentals of Fire

Everything necessary to start a fire is present in shops. Fire is the combination of oxygen and fuel at the temperature necessary to sustain combustion. Combustion is the chemical process in which fuel and oxygen unite at a rapid rate, producing light and heat. Fuel, oxygen (air), heat, and an uninhibited chain reaction are required to produce combustion. Figure 11.1. the fire pyramid, illustrates these four items.

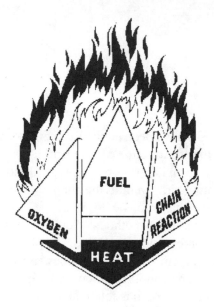

FIG. 11.1 Fire Pyramid.

Fuel is usually in abundant supply from sources such as gasoline, paint, solvents, plastics, oily rags, paper, wood, hydrogen generated during battery charging, cleaning materials, sawdust, paper, and scrap. Common chemical elements such as carbon, hydrogen, and sulfur also serve as fuel sources.

The oxygen necessary for a fire is in plentiful supply from the open air. The air we breathe is, in fact, 21 percent oxygen. The shop also contains oxidizing agents which support combustion. These agents include iron oxides, nitric acids, and nitrogen oxides.

Fuel will not burn, or the union of oxygen and fuel will not take place, until a certain temperature is reached. Various fuels ignite at various temperatures. Two terms are used in defining the temperature at which a fuel ignites are flash point and auto-ignition temperature.

The flash point is the minimum temperature at which a liquid gives off vapor in sufficient concentration to form an ignitable mixture with air near the surface of the liquid. Those liquids with the lowest flash points are the most hazardous. Flash point temperatures cannot be considered constants, for they are related to other conditions. For example, the amount of vapor that will accumulate in the air above a volatile liquid depends upon the vapor pressure or the relative saturation of air with the vapor and the surface area of the volatile liquid. Flammable liquids are those with flash points below 100°F, combustible liquids are those with flash points at or above 100°F.

The auto-ignition temperature is the lowest temperature at which a flammable gas- or vapor-air mixture will ignite from its own heat source or from a contacted heat surface without a spark or flame. Vapors and gases will ignite spontaneously at a lower temperature in oxygen than in air. The majority of vapors and gases will not self-ignite in air until they reach temperatures of about 500°F to 900°F.

Heat may be produced by friction, an electric spark, chemical action, the rays of the sun or the heat from other burning materials. The source of ignition may also come from open flames, sparks, heating systems, welding, hot metal, and electrical equipment.

The fourth requirement for fire is an uninhibited chain reaction. This concept emphasizes the need for chemical reactions between the fuel and oxidizer to progress without interference. The most common way of preventing fires is by eliminating either the fuel or the heat (source of ignition). To extinguish a fire, it is necessary to remove one of the four requirements, for example:

1. Fuel – Mechanically remove or seal off fuel or divert or shut off the flow of burning liquids or gases.
2. Oxygen - Exclude the air supply by smothering or dilution.
3. Heat – Using a suitable cooling agent, cool the burning material until it's temperate is below its ignition point.
4. Chain Reaction - Interrupt the chemical chain reaction of the fire by using dry chemical or Halon extinguishing agents.

Methods of extinguishing fires will be discussed in more detail later in this chapter.

11.2 Heat Transfer

Before discussing fire protection, it is necessary to understand the methods of heat transmission. The most common means of spreading or propagating a fire is the direct contact of flammable or combustible material with a flame. Conduction is the transfer of heat from one body or object to another through an intervening conducting medium, usually a solid. Convection is heat transmission by a circulating medium, either gas or liquid.

Since heated media expand and rise, the hot gases and smoke convey heat and toxic gases to the upper level floors. The more intense the fire, the greater the speed with which the smoke and heat rise. Convection of smoke, heat, and toxic gases through vertical openings presents grave dangers to persons on the upper floors of buildings. Some toxic gases from fire are carbon monoxide (CO) and sulfur dioxide (SO_2).

Heat waves or rays given off by a heated body travel in all directions in straight lines until they are absorbed or reflected by another object. The amount of heat radiated from the source increases rapidly as the temperature of the source rises. The amount of radiant heat reaching an exposed object depends upon the temperature difference between the heat source and the object and the distance between them.

Explosions of flammable dust usually occur in a series, with the later explosions causing the most serious damage. Fire, which may or may not cause another explosion, occurs after the initial explosion.

11.3 Ignition Sources

A study of approximately 25,000 industrial fires traced the origins of these fires to the following general causes:

- 23% electrical
- 18% smoking
- 10% friction
- 8% overheated materials
- 7% hot surfaces
- 7% burner flames
- 5% combustion sparks
- 4% spontaneous ignition
- 4% cutting and welding
- 3% exposure
- 3% incendiarism
- 2% mechanical sparks
- 2% molten substances
- 1% chemical action
- 1% lightning

- 1% static sparks
- 1% miscellaneous

Electric failures and misuse of electrical equipment are the principal causes of industrial fires. Some specific causes of such fires are electrical arcing, short circuits, overloaded circuits and equipment, and substandard wiring. Proper maintenance and periodic inspection of electrical equipment and proper job training will reduce the hazards of this ignition source. Haphazard wiring, poor connections, and "temporary" repairs must be brought up to standard. Fuses should be of the proper type and size. Circuit breakers should be checked to ensure that they have not been blocked in the closed position (to prevent overloading) and to see that moving parts do not stick.

Smoking and matches are the second most common cause of fire. This ignition source can be controlled if workers are taught to follow smoking regulations and to smoke only in non-hazardous, permitted areas. Because the careless disposal of ashes and matches can present problems in permitted smoking areas, proper receptacles should be provided to prevent fire in these areas. Prohibition of smoking without the proper attention to education has caused fires through clandestine smoking. Where regulations permit, smoking areas should be provided and workers should understand why it is hazardous to smoke in prohibited areas.

Fires caused by friction usually result from hot bearings, misaligned or broken machine parts, choking or jamming of materials, and poorly adjusted power drives and conveyors. Friction can be eliminated by proper maintenance, lubrication, and regular inspection of all mechanical equipment.

Another cause of fire is overheated materials. Overheated materials can result from any processes or operations that require the heating of flammable materials and liquids and ordinary combustibles. Fires caused by hot surfaces are usually the result of conduction, convection, and radiation of heat from boilers, hot ducts or flues, steam pipes or electric lamps, which ignite flammable liquids and combustibles. Control of such fires requires:

- proper clearances of combustible materials around boilers and steam pipes
- adequate insulation and air circulation between hot surfaces and combustibles
- proper job training and careful supervision
- well-maintained automatic temperature control devices

Fires caused by burner flames are usually the result of improper use or poor maintenance of portable torches, boilers, driers, and portable heating equipment. The sources of fires started by sparks can usually be traced to burning rubbish, engine stacks, foundry cupolas, furnaces, and welding stations. Burners, stoves, and furnaces should be properly adjusted and maintained with adequate clearances from combustibles and with spark arrestors on exhaust flues or pipes. Other preventive measures include keeping open flames away from combustible materials, employing adequate ventilation and combustion safeguards, and proper design, operation, and maintenance.

Spontaneous ignition occurs when combustibles and oxygen in the air are heated sufficiently to begin a reaction which continues until the combustible materials reaches a temperature at which the reaction becomes self-sustaining. This temperature level is known as the ignition point or ignition temperature. Spontaneous ignition is usually the result of improper disposal of oily waste and rubbish and of deposits in ducts and flues. Methods to control these types of spontaneous ignitions include:

- providing safe containers for all substances subject to spontaneous heating
- providing for prompt and regular disposal of the contents of such containers
- using nonflammable cleaning solvents
- providing for regular cleaning of ducts and flues
- isolating storages subject to spontaneous heating

Fires that occur as the result of cutting and welding operations are usually caused by sparks and hot metal landing near welding surfaces, or by defective gauges or deteriorated gas lines on the welding apparatus. Proper training of workers in the recognition of hazards during cutting and welding will help reduce these ignition sources.

Fires that occur through exposure are usually caused by heat from adjoining or nearby buildings. Incendiary fires are started maliciously by employees, workers, or intruders. Fires caused by mechanical sparks are usually generated by metal in machines or during grinding and crushing operations. Proper care in cleaning and keeping stock free of foreign metallic pieces will prevent sparks from causing fires.

Fires caused by molten substances are generally the result of molten metal being released from a ruptured furnace or spilled during handling. Fires from these sources can be prevented by the proper operation and maintenance of equipment. Chemical fires start when chemicals react with other chemicals or materials, or when decomposition causes chemicals to be unstable. The best preventions are the proper operation and careful handling and storage of chemicals.

Fires from static are the result of contact and separation of materials which ignite flammable vapors, dusts, and fibers. These static charges are generated upon agitation and mixing equipment, belts, and splash-filling of tanks. These types of fires can be prevented by proper grounding, bonding, ionization, and humidification.

11.4 Static Electricity

Static electricity is generated by the repeated contact and separation of dissimilar materials. For example, static electricity is generated when a fluid flows through a pipe or from an orifice into a tank.

The principle hazards created by static electricity are fire and explosion caused by spark discharges containing sufficient energy to ignite flammable or explosive vapors, gases, or dusts which may be present. Workers may also be shocked by static electricity, causing an involuntary reaction such as falling which may lead to injury.

Static sparks are most dangerous where flammable vapors are present, such as at the outlet of a flammable liquid fill pipe, a delivery hose nozzle, near an open flammable liquid container, and around a tank truck fill opening or barrel bunghole. A spark between two bodies occurs in the presence of an insufficient electrical conductive path between them. Grounding and bonding of flammable liquid containers prevent static electricity from causing a spark. Recommended Practice on Static Electricity, NFPA Standard No. 77, should be consulted for details.

11.5 Bonding and Grounding

The terms "bonding" and "grounding" are often used interchangeably due to a lack of understanding of their meanings. The purpose of bonding is to eliminate a difference in potential between objects. The purpose of grounding is to eliminate a difference in potential between an object and ground. Bonding and grounding are effective only when the bonded objects are conductive.

Although bonding will eliminate a difference in electrical potential between the objects that are bonded, it will not eliminate a difference in potential between those objects and the earth unless one of the objects possesses a ground to earth. Therefore, bonding will not eliminate the static charge but will equalize the potential between the objects bonded so that a spark will not occur between them.

Flammable liquids are capable of building up electrical charges when they flow through piping, when they are agitated in a tank or a container, or when they are subjected to vigorous mechanical movement such as spraying or splashing. Proper bonding and grounding of the transfer system usually drains off this static charge to ground as fast as it is generated. However, rapid flow rates in transfer lines can cause very high electrical potentials on the surface of liquids, regardless of vessel grounding.

Furthermore, some petroleum liquids are poor conductors of electricity, particularly the pure refined products, and even though the transfer system is properly grounded, a static charge may build up on the surface of the liquid in the receiving container. The charge accumulates because static cannot flow through the liquid to grounded metal as fast as it is being generated. The accumulated static charge can result in a static spark with sufficient energy to ignite a flammable air-vapor mixture. This high static charge is usually controlled by reducing the flow rates, avoiding violent splashing with side-flow rates, and using relaxation time.

When flammable liquids are transferred from one container to another, they must be effectively bonded and grounded. This practice prevents electrical discharge or sparks from accumulating a static charge during the transfer process.

Flat moving belts are sources of static electricity unless they are made of a conductive material or coated with a conductive belt-dressing compound designed to prevent the accumulation of static charges. V-belts, however, do not usually create sufficient static charge to cause concern. Nonconductive materials passing through or over rolls also create charges of static electricity. Bonding and grounding systems should be checked regularly for electrical continuity.

11.5.1 Ionization and Humidity Control

In addition to bonding and grounding, there are two other ways that separated charges of static electricity can be recombined before sparking potentials build up. First, the surrounding environment can be ionized. If the air in the immediate area is ionized with a charged body, a conducting path will be provided. Air can be ionized by using high voltage or radioactive static eliminators. Second, the humidity of the air can be controlled. Keeping the relative humidity in the area above 60 percent will condense enough moisture on the surface of the objects to allow static charges to dissipate.

11.6 Classes of Fires

The discussion thus far has been related to the causes of fires. This section examines the classes of fires. The National Fire Protection Association (NFPA) has defined four general fire classifications, class A, B, C, and D.

11.6.1 Class A Fires: Ordinary Combustible Materials

Class A fires, sometimes called surface burning fires, occur from ordinary combustible materials such as wood, cellulose, paper, cloth, excelsior, and rubber. Class A fires are extinguished by lowering the ignition temperature of the materials with water. Under certain circumstances, these fires may be extinguished by the blanketing and smothering effects of dry chemical and carbon dioxide fire extinguishers.

Because ordinary combustible materials serve as fuel for fire, the following precautions are necessary:

1. Provide a program of adequate disposal of all combustible wastes and rubbish designed specifically for the operations or processes involved.
2. Provide for regular inspection of the waste storage area.
3. Provide a program of internal housekeeping which will prevent any accumulation of waste and which will result in safe, clean work areas.
4. Provide a program of external housekeeping to prevent accumulation of waste, brush, or high grass around buildings.

Combustible substances such as dusts, plastics, textiles, waste materials, and cleaning supplies present potential fire hazards. Sawdust and coal dust are examples of combustible dusts. Deposits of combustible dust on floors, beams, and machines are subject to flash fires and dust suspended in air can explode violently. Prevention or removal of dust accumulation on structural members, walls, and ceilings is necessary. Precautions against the hazards presented by dust include:

1. Where possible, provide local exhaust systems for dust collection.
2. Provide a program for frequent vacuum cleaning of structural members, ceilings, and walls. The vacuum cleaning equipment should be explosion proof.
3. Remove or control the accumulation of hazardous dust at all ignition sources.

Burning plastics such as nitrocellulose, polystyrene, cellulose, rayon and polyvinyl chloride are particularly dangerous because they produce large amounts of smoke and because their fumes are extremely toxic. Special care should be taken to keep plastics away from ignition sources.

Some textiles such as rayon, cellulose fibers, and cotton textiles are highly flammable and have a high rate of flame spread. Though flame retardant treatments help to slow down flame spread, such treatments may be washed out after a few launderings. Workers must be cautioned to keep clothing away from ignition sources and, where necessary, to wear aprons. Combustible waste materials, such as oily shop rags or paint rags, must be stored in covered metal containers and disposed of daily.

The materials used for cleaning can also create fire hazards. All oily mops must be stored in closed approved meal containers. Combustible sweeping compounds, such as oil-treated sawdust, can be a fire hazard. Floor coatings containing solvents with low flash points can be dangerous, especially near sources of ignition.

11.6.2 Class B Fires: Flammable Liquids, Gases, and Greases

Class B fires are those that occur in flammable liquids, gases, and greases. Class B fires are extinguished by limiting the amount of air available to support combustion. Dry chemical, carbon dioxide, foam and halogenated hydrocarbon agent type fire extinguishers are recommended for this class of fire. Solid streams of water are likely to spread the fire, but under certain circumstances water fog nozzles can be effective.

Flammable liquids are those with flash points below 100°F. Flammable liquids are hazardous because of their ease of ignition (low flash point), flammable range, and burning intensity. The degree of danger is determined largely by the flash point of the liquid, the concentration of vapors in the air and the possibility of a source of ignition at or above a temperature sufficient to cause the mixture to burst into flame.

Flammable liquids vaporize and form flammable mixtures when in open containers, when leaks or spills occur, or when heated. Examples of flammable liquids include gasoline, ethyl alcohol, benzene, turpentine, and naphtha. Fluid commodities containing liquids such as paints, varnishes, and cleaning solutions should be considered flammable liquids and classed according to the flash point of the mixture.

Combustible liquids are those having flash points at or above 100°F. Combustible liquids are usually safe to handle at normal temperatures, but these liquids should not be heated. Lube oils, kerosene, cresols, benzyl alcohol, cooking oil, mineral spirits, and palm oil are examples of combustible liquids. Table 11.1 classifies some common products according to their flash points.

TABLE 11.1
Classification of Typical Flammable and Combustible Products

Class		Product	Flash Point	Boiling Point
I	A	Gasoline (some) Pentane	Lower than 73°F	Lower than 100°F
	B	Acetone Denatured alcohol Gasoline (some) Naphtha, VM and P Toluene	Lower than 73°F	At or above 100°F
	C	Xylene	At or above 73°F	
II		Kerosene Mineral spirits Naphtha Stoddard Solvent	At or above 100°F	
III		Asphalt Break fluid Fuel oil no. 4 Fuel oil no. 5 Fuel oil no. 6	At or above 140°F	

Because of their low flash point and low ignition temperatures, flammable liquids are dangerous and require the following precautions:

1. When possible, avoid the use of highly flammable liquids by using nontoxic and nonflammable (or less flammable) liquids instead. For example, benzene can be replaced by toluene in most lacquers, synthetic-rubber solutions, and paint removers.
2. Keep flammable liquids in approved fire resistant safety containers, never in glass containers. All flammable liquids should be kept in

closed containers when not in use.

3. Limit the amount of flammable liquids in the shop area to that needed for one day.
4. Provide safe operating procedures, including local exhaust systems, for all processes.
5. Remove or control all ignition sources such as static electricity, smoking, and open flames.
6. Provide for adequate clearances between flammable liquid containers or safety cans and any heat sources.
7. Provide adequate ventilation for all operations involving the use or storage of flammable liquids.
8. Anticipate flammable liquid spills and provide means to control and limit spillage and suitable absorptive materials for use in cleaning up spills. Promptly clean all spills of flammable or combustible liquids.
9. Ensure that the connections on all drums containing flammable and combustible liquids are vapor and heat tight.
10. Always use and handle flammable liquids with extreme caution, no matter how familiar they are.
11. Store large amounts of flammable liquids in a separate fire resistant building or vault, which conforms to the recognized standards. Storage tanks should be properly vented and supported by masonry or, in diked areas, by poured concrete supports.

The hazards of flammable gases are generally the same as those of flammable liquids, except that commercial gases are usually contained in compressed gas cylinders. Flammable compressed gases usually burn with a greater intensity than exposed flammable liquids. Acetylene, propane, hydrogen, natural gas, and butane are examples of flammable gases.

11.6.3 Class C Fires: Electrical Equipment

Class C fires occur in electrical equipment and require the use of non-conducting extinguishing agents such as dry chemical, carbon dioxide, and compressed gas. Details of fire prevention and protection for Class C fires are discussed in Chapter 16, Electrical Safety.

11.6.4 Class D Fires: Combustible Metals

Class D fires occur in combustible metals such as magnesium, potassium, zirconium, lithium, sodium, powdered aluminum, zinc, and titanium. Graphite base type extinguishing agents and sand are used for class D fires. Precautions against fires in combustible metals include:

1. Providing for frequent collection of combustible metal chips and shavings.
2. Preventing any presence of oil or grease near or in finely divided particles of combustible metals.
3. Providing covered, plainly labeled, clean, dry steel containers for combustible metal particles which are to be salvaged and storing them in an isolated storage yard or at a safe distance from all buildings.

11.7 Fire Protection Requirements for Storing Hazardous Materials

All four classes of fires can and do occur. Class D fires are a particular concern because flammable liquids and hazardous substances are necessary to many shop operations. Therefore, special attention must be paid to fire protection requirements for storing flammable and combustible materials. In this section, the requirements for storing such materials in containers, cabinets, rooms, and the outdoors are discussed.

11.7.1 Storage Containers

The National Fire Protection Association (NFPA) publishes standards which govern fire protection in the workplace. The code establishes the following storage provisions for flammable liquids:

1. No container for Class I or Class II liquids shall exceed a capacity of one gallon, except that safety cans can be of two gallons capacity.
2. Not more than ten gallons of Class I and Class II liquids combined shall be stored outside of a storage cabinet or storage room, except in safety cans.
3. Not more than 25 gallons of Class I and Class II liquids combined

shall be stored in safety cans outside of a storage room or storage cabinet.

4. Not more than 60 gallons of Class III-A liquids shall be stored outside of a storage room or storage cabinet.

5. Quantities of flammable and combustible liquids in excess of those set forth in this section shall be stored in an inside storage room or storage cabinet.

11.7.2 Storage Cabinets

Special cabinets are available for storing flammable and combustible liquids in the shop. All such cabinets must be clearly marked with the words "Flammable—Keep Fire Away." According to the NFPA, the cabinets must be designed so that the internal temperature does not exceed 325°F (162.8⁰C) when subjected to a ten-minute fire test using the standard time-temperature curve specified by the NFPA.

For a metal cabinet, the code requires that the bottom, top, door, and sides be at least No. 18 gauge sheet iron and double-walled with at least 1½ inches of air space between the walls. Joints must be made tight by riveting or welding. Doors must have three-point locks and the doorsill must be raised at least two inches above the bottom of the cabinet.

The NFPA code also allows the use of wooden cabinets if they are properly constructed. Tests indicate that wooden cabinets can be at least as effective, and in many cases are more effective, than metal cabinets. The NFPA specifies that a wood cabinet used for storing flammable liquids must have the bottom, sides, and top constructed of an approved grade of plywood at least one inch thick. Joints must be rebated and fastened in two directions with flathead wood screws.

The NFPA code specifies that not more than 60 gallons of flammable or 120 gallons of combustible liquids may be stored in a single storage cabinet and not more than three such cabinets may be located in a single fire area.

When central storage facilities are not available, storage cabinets provide a convenient method for storing flammable and toxic chemicals. Storage cabinets are equipped with locks and provide excellent security. They also are equipped with pipe connections to facilitate the connection of the cabinets to a mechanical exhaust system. The cabinets should be exhausted to prevent the accumulation of toxic or explosive chemical vapors. One manufacturer of storage cabinets for flammable liquids recommends that they be exhausted at a rate of twenty cubic feet per minute.

11.7.3 Storage Rooms

Because both flammable and non-flammable materials are often stored in the same area, storage rooms should be built to conform to NFPA codes for storage of flammable liquids. Storage rooms in the industrial shops should not have an opening that communicates with the public portion of the building. The NFPA code requires that:

1. The floor in the storage room must be at least four inches lower than the surrounding floors or that there be a noncombustible, liquid-tight raised sill or ramp at least four inches high between the storage area and adjacent rooms or buildings.
2. All doors must be approved, self-closing fire doors.
3. The room must be liquid-tight where the walls join the floor.

Table 11.2 lists the quantities of flammable liquids which may be stored in rooms of various sizes. Note that rooms with separate fire protection systems are allowed far greater quantities of flammable liquids.

TABLE 11.2
Storage in Inside Rooms

Fire Protection Provided*	Fire Resistance	Maximum Size	Total Allowable Quantities (gals./sq. ft. floor area)
Yes	2 hours	500 sq. ft.	10
No	2 hours	500 sq. ft.	4
Yes	1 hour	150 sq. ft.	5
No	1 hour	150 sq. ft.	2

*Fire protection systems shall be sprinkler, water spray, carbon dioxide, dry chemical, halon, or other approved systems.

Containers with capacities of more than 30 gallons (113.6 liters) of flammable liquid should not be stored on top of other such containers. Large containers should be stored on or near the floor. The upper shelves should not above the comfortable reach of an average-sized person. Smaller containers may be stored on the higher shelves. Table 11.3 lists the maximum number of gallons that may be stored on various indoor storage levels. Note that those liquids classed as A, B, and C are flammable, and Class II and III liquids are combustible.

TABLE 11.3
Indoor Storage Containers

Class Liquid	Storage Level	Protected Storage Maximum per Pile (gallons*)	Unprotected Storage Maximum per Pile (gallons*)
A	Ground and upper floors	2,750 (50)	660 (12)*
	Basement	Not permitted	Not permitted
B	Ground and upper floors	5,500 (100)	1,375 (25)
	Basement	Not permitted	Not permitted
C	Ground and upper floors	16,500 (300)	4,125 (75)
	Basement	Not permitted	Not permitted
II	Ground and upper floors	16,500 (300)	4,125 (75)
	Basement	5,500 (100)	Not permitted
III	Ground and upper floors	55,000 (1,000)	1,375 (250)
	Basement	8,250 (450)	Not permitted

Note 1: When two or more classes of materials are stored in a single pile, the maximum gallonage permitted in that pile shall be the smallest of the two or more separate maximum gallonages.

Note 2: Aisles shall be provided so that no container is more than 12 feet from an aisle. Main aisles shall be at least 3 feet wide and side aisles at least 4 feet wide.

Note 3: Each pile shall be separated form each other by at least 4 feet.

*Numbers in parentheses indicate corresponding number of 55-gallon drums.

OSHA regulations require that there be one clear aisle at least three feet (9 m) wide and that lights in the storage area be explosion-proof.

Storage room floors should be constructed out of material that is resistant to chemicals and readily cleaned. All electrical outlets and equipment must be well grounded. The room should be kept cool, but not cold enough to freeze the reagents stored in it.

All inside storage rooms must be equipped with either a gravity or a mechanical exhaust system to remove hazardous vapors. If flammable liquids are stored, then a mechanical exhaust ventilation should be used.

The exhaust duct should be located near the floor level (one foot above). Both the exhaust and inlet air openings should be arranged to allow air to move across all portions of the floor to prevent accumulation of flammable vapors. The NFPA code requires that the exhaust system be capable of completely changing the air within the storage room at least six times each hour. As a rule of thumb, the ventilation system should be capable of removing 10,000 cubic feet of air for every gallon of liquid vaporized.

At least one fire extinguisher having a rating of at least 12-B units must be located outside of, but not more than ten feet from, the door opening into any room used for storage. Where flammable and combustible liquids are stored outside of a storage room but still inside a building, at least one portable fire extinguisher with a rating of at least 12-B units must be located not less than ten feet (3 m) or more than 25 feet from the storage area. Open flames and smoking must not be permitted in flammable or combustible liquid storage areas.

Both storage cabinets and storage rooms must provide for two other considerations related directly to fire protection, security and chemical exposure protection. In addition to locking cabinets and storage rooms, the development of a security system that accounts carefully for hazardous materials is required. The security system must include an inventory of stored items and a record of all materials removed. This record should include the name of the person who removed each item and the purpose and time of the removal. Careful record-keeping is mandatory, both for fire protection and as part of the broader safety and health program. Controlling what chemicals are stored and how they are stored protects against chemical exposure.

Some chemicals are incompatible and may form a violent reaction if they come into contact with another. For example, strong oxidizing materials should not be stored next to organic materials and flammable solvents should not be stored next to acids. This does not mean that incompatible materials have to be stored in another room or cabinet. They can be stored on another shelf or the other side of the room. Ventilation is an important part of chemical exposure protection and, as stated above, is required in storage rooms and cabinets. Table 11.4 provides examples of incompatible chemicals.

TABLE 11.4
Incompatible Chemicals

Chemical	Keep Out of Contact With:
Acetic acid	Chromic acid, nitric acid, hydroxyl compounds, ethylene glycol, perchloric acid, peroxides, permanganates
Acetylene	Chlorine, bromine, copper, fluorine, silver, mercury
Alkaline metals, such as powdered aluminum or magnesium, sodium, potassium	Water, carbon tetrachloride or other clorinated hydrocarbon, carbon dioxide and the halogens
Ammonia, anhydrous	Mercury (in manometers, for instance), chlorine, calcium hypochlorite, iodine, bromine, hydrofluric acid (anhydrous)
Ammonium nitrate	Acids, metal powders, flammable liquids, chlorates, nitrites, sulfur, finely divided organic or combustible materials
Aniline	Nitric acid, hydrogen peroxide
Bromine	Same as for chlorine
Carbon, activated	Calcium hypochlorite and all oxidizing agents
Chlorates	Ammonium salts, acids, metal powders, sulfur, finely divided organic or combustible materials
Chlorine	Ammoniz, acetylene, butadiene, butane, methane, propane (or other petroleum gases), hydrogen, sodium carbide, turpentine, benzene, finely divided metals
Chlorine dioxide	Ammonia, methane, phosphine, hydrogen sulfide

TABLE 11.4 (Continued)

Chemical	Keep Out of Contact With:
Chromic acid	Acetic acid, naphthalene, camphor, glycerine, turpentine, alcohol and flammable liquids in general
Copper	Acetylene, hydrogen peroxide
Cumene hydroperoxide	Acids-organic and inorganic
Flammable liquids	Ammonium nitrate, chromic acid, hydrogen peroxide, nitric acid, sodium peroxide and the halogens
Fluorine	Isolate from everything
Hydrocarbons (butane, propane, benzene, gasoline, turpentine, etc.)	Fluorine, chlorine, bromine, chromic acid, sodium peroxide
Hydrocyanic acid	Nitric acid, alkalis
Hydrofluric acid, anhydrous	Ammonia, aqueous or anhydrous
Hydrogen peroxide	Copper, chromium, iron, most metals or their salts, alcohols, acetone, organic materials, aniline, nitromethane, flammable liquids, combustible materials
Hydrogen sulfide	Fuming nitric acid, oxidizing gases
Iodine	Acetylene, ammonia (aqueous or anhydrous), hydrogen
Mercury	Acetylene, fulminic acid, ammonia
Nitric acid (concentrated)	Acetic acid, aniline, chromic acid, hydrocyanic acid, hydrogen sulfide, flammable liquids, flammable gases
Oxalic acid	Silver, mercury
Perchloric acid	Acetic anhydride, bismuth and its alloys, alcohol, paper, wood
Potassium	Carbon tetrachloride, carbon dioxide, water
Potassium chlorate	Sulfuric and other acids
Potassium Perchlorate (see also "Chlorates")	Sulfuric and other acids
Potassium permanganate	Glycerine, ethylene glycol, benzaidehyde, sulphuric acid
Silver	Acetylene, oxalic acid, tartaric acid, fulminic acid, ammonium compounds
Sodium	Carbon tetrachloride, carbond dioxide, water
Sodium peroxide	Ethyl or methyl alcohol, glacial acetic acid, acetic anhydridge, bezaidehyde, carbon disulfide, glycerine ethylene glycol, ethyl acetate, methyl acetate, furfural
Sulfuric acid	Potassium chlorate, potassium Perchlorate, potassium permanganate (or such compounds with similar light metals, such as sodium lithium, etc.)

11.7.4 Outdoor Storage

If flammable and combustible liquids are stored outside, the area should be graded to divert possible spills away from buildings. The storage area should be kept free of debris and weeds and be protected from tampering or trespassing. Smoking must be prohibited. Table 11.5 lists the limits on outdoor container storage.

Special regulations apply to outdoor gasoline storage. Instructors will need to contact appropriate local and state authorities to see under what conditions, if any, storage in excess of 60 gallons is permitted.

TABLE 11.5
Outdoor Container Storage

Class	Maximum per pile (gal.) See note 1	Distance between piles (ft.) See note 2	Distance to property line that can be built upon (ft.) See notes 3 and 4	Distance to street, alley public way (ft.) See note 4
IA	1,000	5	20	10
IB	2,200	5	20	10
IC	4,400	5	20	10
II	8,800	5	10	5
III	22,000	5	10	5

Note 1: When 2 or more classes of materials are stored in a single pile, the maxim gallonage in that pile shall be the smallest of the 2 or more separate gallonages.
Note 2: Within 200 feet of each container, there shall be a 12 foot wide access way to permit approach of fire control apparatus.
Note 3: The distances listed apply to properties that have protection for exposures as defined. If there are exposures, and such protection for exposures does not exist, the distances in column 4 shall be doubled.
Note 4: When total quantity stored does not exceed 40 percent of maximum per pile, the distances in columns 4 and 5 may be reduced 50 percent, but not less than 3 feet.

11.8 Limiting Fire Spread

Should a fire occur, its spread must be limited if its potential for damage is to be reduced. Several structural devices help retard the spread of fire. A fire wall, made of noncombustible materials such as brick or block, will prevent a fire from burning through it. A fire wall will also prevent heat conduction through the wall thereby preventing a fire on the other side. The length of time a fire wall will resist fire spread varies, depending on the type of construction, from one to four hours. Because it is usually self-supporting, a fire wall will remain standing even after the building collapses. Properly used fire walls divide a building into distinct fire areas.

Although they are necessary openings in a fire wall, fire doors do not offer the protection of a solid fire wall and should be used only where necessary. Standard fire doors may be clad in steel or sheet metal with wooden or noncombustible cores, or they may be hollow metal doors. The length of time each door can protect an opening varies from 45 minutes to three hours. All fire doors should be hung in noncombustible frames. The three most common types of fire doors are swinging, horizontal sliding, and vertical sliding. Fire doors should not be hooked in the open position since it compromises the protective capability of the door.

Air ducts can convey hot fire gases and smoke throughout a building. To prevent this hazard, fire dampers should be installed in all ducts that penetrate fire walls and in branch lines to other fire areas. A detection system should be used to stop all fans operating in the duct work when smoke or high temperatures are detected. Fire dampers and detection systems should be installed in accordance with recognized NFPA standards.

An adequate fire detector and/or suppression system is one of the best investments a business can make. Once installed, it must be tested and inspected regularly. A good fire detection or suppression system will provide:

- detection devices in all hazardous and concealed areas
- manual trip boxes at all exits and other appropriate locations
- horns or bells throughout the building to sound the alarm
- alarm signals giving the location of the fire and at the same time, notifying the fire department
- a water-flow alarm which sounds when the automatic sprinkler system is activated
- an automatic sprinkler system supplemented where necessary by special systems to control specific hazards
- portable fire extinguishers to act as a first line of defense
- a regular inspection and maintenance program to keep the system in proper operating condition

11.9 Fire Detection Devices

There are four main types of fire detection devices, each detects fire at a distinct stage. The four stages of fire are:

1. ionization-incipient stage
2. photoelectric-smoldering stage
3. infrared-flame stage
4. thermal-heat stage

Ionization detectors respond to the combustion particles produced in the incipient stage. These particles, too small to be visible, are created by chemical decomposition and are generated before smoke or flame is visible and before significant heat develops. When the combustion particles rise to the ceiling, the ionization detector sounds an alarm.

When the fire progresses to the smoldering stage, the combustion particles increase in number and become visible as smoke. Photoelectric detectors, commonly called smoke detectors, detect this smoke and sound the alarm and shut off air flow in ducts.

As the smoldering stage continues, the fire reaches the point of ignition and flames appear. The fire in now in the flame stage where the smoke levels decrease and the heat levels increase. Infrared detectors detect the infrared energy produced by the fire.

At the final heat stage, large amounts of heat, flame, smoke, and toxic gases are produced. Thermal detectors respond to this heat energy. There are two general types of thermal detectors: fixed temperature devices and rate of rise detectors. Thermostats are the most frequently used fixed temperature detectors. The bimetallic thermostat utilizes the difference in coefficients of thermal expansion of two metals, which laminate into a single strip that bends when heated, closing the electrical contacts.

In the rate of rise system, a rapid rise in temperature heats the air in a tubing system or air chamber. This rise in pressure trips the device and sounds the alarm. If the temperature rises slowly, the pressure bleeds off through a compensating port.

11.10 Automatic Sprinkler Systems

After a fire is detected, it must be extinguished. Automatic sprinklers are the most versatile and dependable form of fire protection available. Since their initial use in industrial plants around 1850, they have been refined and improved so that most fire protection engineers consider them to be essential fire fighting equipment.

A sprinkler system is an integrated system of underground and overhead piping which includes a suitable water supply, a controlling valve, and a device for actuating an alarm when the system is in operation. It is usually activated by heat from a fire and discharges water over the fire area. In Standard 13 of the National Fire Code, the NFPA defines five systems, of which only two, the wet-pipe and dry-pipe systems, are likely to be used in small businesses.

According to the NFPA, wet-pipe is a system employing automatic sprinklers attached to a piping system containing water and connected to a water supply so that water discharges immediately from sprinklers opened by a fire. Antifreeze is necessary if portions of the system are subjected to freezing temperatures.

The dry-pipe system is defined as one employing sprinklers attached to a piping system containing air or nitrogen under pressure, the release of which as from the opening of sprinklers permits the water pressure to open a valve known as the dry-pipe valve. The water then flows into the piping system and out the opened sprinklers.

Because it takes longer for the dry-pipe system to put water on the fire, it should be installed only where a wet-pipe system is not practical (for example, in rooms or buildings which cannot be properly heated). However, an approved dry-pipe system is preferred over shutting off the water supply entirely during cold weather.

11.10.1 Temperature Ratings

Sprinklers should be selected on the basis of occupancy and temperature rating. Ratings are based on standardized tests in which a sprinkler head is immersed in a liquid the temperature of which is raised until the head operates. The proper sprinkler head rating is determined by the maximum room temperature at ceiling level under normal working conditions. Table 11.6 lists the operating temperature, color, and ceiling temperatures for a variety of sprinkler heads.

TABLE 11.6
Sprinkler Head Rating and Color

Rating	Operating Temp. (°F)	Color	Maximum Ceiling Temp. (°F)
Ordinary	135-170	Uncolored*	100
Intermediate	175-225	White*	150
High	250-300	Blue	225
Extra High	325-375	Red	300
Very Extra High	400-475	Green	375
Ultra High	500-575	orange	475

*The 135°F sprinklers of some manufactures are half black and half uncolored. The 175°F sprinklers of the same manufacturers are yellow.

11.10.2 Maintenance

Dependable sprinkler protection systems require systematic maintenance and inspection. NFPA 13A, Care and Maintenance of Sprinkler Systems, provides specific maintenance requirements, including inspection of control valves, testing water flow, and reading water and air pressure gauges. Because 35 percent of all sprinkler system failures are caused by closed valves, it is imperative that controlling valves be kept open and water supplies maintained.

The local fire department should be familiar with the automatic sprinkler equipment including its location, the arrangement of control valves, connections for fire department use, and the extent of protection offered by the system. Such knowledge can save precious minutes and prevent faulty use in the event of a fire.

11.11 Special Systems

In addition to, and as a supplement to automatic sprinkler systems, special fire control systems can be installed where water is not an acceptable extinguishing agent. These systems include foam, dry chemical, carbon dioxide, and water spray.

NFPA II, Standard for Foam Extinguishing Systems, defines foam as an "aggregate of tiny gas-filled or air-filled bubbles, lighter than the lightest oils." When applied, it forms "a coherent floating blanket on flammable and combustible liquids lighter than water and prevents or extinguishes fire by excluding air and cooling the fuel." It is used primarily for the protection of flammable liquid storage areas and tanks. The air (mechanical) foam has replaced chemical foam, which is now considered obsolete. Air foam consists of bubbles of air produced when air and water are mechanically agitated with a foam-making agent.

According to NFPA 17, Dry Chemical Extinguishing Systems, dry chemical is a "finely divided powder, usually sodium bicarbonate, with additives to prevent caking and to increase flow-ability." Dry chemical systems are widely used to extinguish a rapidly spreading surface of fire typical of flammable liquids. Because it is electrically non-conducting, it is often used on fires involving electrically energized equipment. It can also be used on ordinary combustibles when the fire is of a surface nature and where rapid flame knock-down is beneficial.

Carbon dioxide (CO_2) is a colorless, odorless, electrically non-conductive inert gas which "extinguishes fire by reducing the concentration of oxygen and/or the gaseous phase of the fuel in the air to the point where combustion stops" (NFPA 12, Carbon Dioxide Extinguishing Systems). CO_2 systems are used to protect gaseous and liquid flammable processes and materials, engines using gasoline and other flammable fuels, electrical equipment, ordinary combustibles, and hazardous solids. Because the CO_2 concentration dilutes the oxygen in the air, this system may create an atmosphere that will not sustain life. The area must be thoroughly ventilated after the fire is extinguished.

Water spray systems use what is called wet water, that is, water to which an approved wetting agent has been added. The wet water is discharged from a device capable of separating the water into spray. Adding a wetting agent reduces the surface tension of the water and increases its penetrating, spreading, and/or emulsifying ability. According to NFPA 15, Water Spray Systems, such systems are particularly effective on fires involving flammable liquids, combustible solids, and electrical equipment.

11.12 Portable Fire Extinguishers

Portable fire extinguishers are the first line of defense against smaller fires. They are necessary even if the shop is equipped with automatic sprinklers or other fixed protection devices. The National Safety Council lists six requirements for effective portable extinguishers. The portable extinguishers must be:

1. a reliable type
2. the right type for each class of fire that may occur in the area
3. in sufficient quantity to protect against the exposure in the area
4. located where they are readily accessible for immediate use
5. maintained in perfect operating condition, inspected frequently, checked against tampering, and recharged as required
6. operable by the persons who are in the area, who can find them, and who are trained to use them effectively and promptly

11.13 Types of Portable Fire Extinguishers

Portable fire extinguishers can be divided into three basic categories: water, gaseous, and dry chemical. A dry powder type is available for Class D fires. Units are classified with a letter designation, that is, A, B, C, or D or a combination of letters to indicate the classes of fire for which they are designed. Class A and B units also have a numerical rating indicating their approximate extinguishing potential. Table 11.7 contains a complete list of extinguisher characteristics.

TABLE 11.7
Extinguisher Characteristics

Extinguishing Agent	Method of Operation	Capacity	Horizontal Range of Stream (ft.)	Approximate Time of Discharge	Protection Required Below 40 deg. F (4 deg. C)	UL or ULC Classifications*
Water	Stored Pressure	2 1/2 gal.	30-40 ft.	1 min.	Yes	2-A
Water	Pump	1 1/2 gal.	30-40 ft.	45 sec.	Yes	1-A
	Pump	2 1/2 gal.	30-40 ft.	1 min.	Yes	2-A
	Pump	4 gal.	30-40 ft.	2 min.	Yes	3-A
	Pump	5 gal.	30-40 ft.	2-3 min.	Yes	4-A
Water (Anti-freeze Calcuim Chloride)	Cartridge & Stored Pressure	1 1/4, 1 /12 gal.	30-40 ft.	30 sec.	No	1-A
	Cartridge & Stored Pressure	2 1/2 gal.	30-40 ft.	1 min.	No	2-A
	Cartridge & Stored Pressure	23 gal. (wheeled)	50 ft.	3 min.	No	20-A
Water (Wetting Agent)	Cartridge & Stored Pressure	25 gal. (wheeled)	35 ft.	1 1/2 min.	Yes	10-A
	Cartridge & Stored Pressure	45 gal.	35 ft.	2 min.	Yes	30-A
	Chemically generated expellent	1 1/4, 1 1/2 gal.	30-40 ft.	30 sec.	Yes	1-A
Water (Soda Acid)	Chemically generated expellent	2 1/2 gal.	30-40 ft.	1 min.	Yes	2-A
	Chemically generated expellent	17 gal. (wheeled)	50 ft.	3 min.	Yes	10-A
	Chemically generated expellent	33 gal. (wheeled)	50 ft.	3 min.	Yes	20-A

TABLE 11.7 (Continued)

Extinguishing Agent	Method of Operation	Capacity	Horizontal Range of Stream (ft.)	Approximate Time of Discharge	Protection Required Below 40 deg. F (4 deg. C)	UL or ULC Classifications*
Loaded Stream	Stored Pressure	2 1/2 gal.	30-40 ft.	1 min.	No	2 to 3-A and 1-B
Loaded Stream	Cartridge & Stored Pressure	33 gal. (wheeled)	50 ft.	3 min.	No	20-A
Foam	Pressurized	21 oz.	4-6 ft.	24 sec.	Yes	1-B
	Chemically generated expellent	1 1/4., 1 1/2 gal.	30-40 ft.	40 sec.	Yes	1-A, 2-B
	Chemically generated expellent	2 1/2 gal.	30-40 ft.	1 1/2 min.	Yes	2-A:4-B to 2-A:6-B
Foam	Chemically generated expellent	5 gal.	30-40 ft.	2 min.	Yes	4-A:6-B
	Chemically generated expellent	17 gal. (wheeled)	50 ft.	3 min.	Yes	10-A: 10-B to 10-A:12-B
	Chemically generated expellent	33 gal. (wheeled)	50 ft.	3 min.	Yes	20-A:20-B to 20-A-40-B
AFFF	Stored Pressure	2 1/2 gal.	20-25 ft.	50 sec.	Yes	3-A-20-B
		2 1/2 to 5 lb.	3-8 ft.	8 to 30 sec.	No	1 to 5-B:C
		10 to 15 lb.	3-8 ft.	8 to 30 sec.	No	2 to 10-B:C
Carbon Dioxide	Self expellent	20 lb.	3-8 ft.	1 0 to 30 sec.	No	10-B:C
		50 to 100 lb. (wheeled)	3-1 0 ft.	1 0 to 30 sec.	No	10 to 20-B: C

TABLE 11.7 (Continued)

Extinguishing Agent	Method of Operation	Capacity	Horizontal Range of Stream (ft.)	Approximate Time of Discharge	Protection Required Below 40 deg. F (4 deg. C)	UL or ULC Classifications*
Dry Chemical (Sodium Bicarbonate)	Stored Pressure	1 lb.	5-8 ft.	8 to 10 sec.	No	1 to 2-B:C
	Stored Pressure	1 1/2 to 2 1/2 lb.	5-8 ft.	8 to 12 sec.	No	2 to 5-B:C
	Cartridge and Stored Pressure	2 3/4 to 5 1/2 lb.	5-20 ft.	8 to 20 sec.	No	5 to 10-B:C
	Cartridge and Stored Pressure	7Y2 to 30 lb.	5-20 ft.	10 to 25 sec.	No	10 to 120-B: C
	Nitrogen cylinder or Stored Pressure	75 to 350 lb. (wheeled)	15 to 45 ft.	20 to 105 sec.	No	40 to 240-B:C
Dry Chemical (Potassium Bicarbonate)	Stored Pressure	1 to 2 lb.	5-8 ft.	8 to 10 sec.	No	1 to 5-B:C
	Stored Pressure	2 1/4 to 5 lb.	5-12 ft.	8 to 10 sec.	No	5 to 20-B:C
	Cartridge and Stored Pressure	5 1/2 to 10 lbs.	5-20 ft.	8 to 20 sec.	No	1 0 to 130-B: C
	Cartridge and Stored Pressure	16 to 30 lbs.	10-20 ft.	8 to 25 sec.	No	40 to 120-B: C
	Nitrogen cylinder or Stored Pressure	125 to 300 lbs. (wheeled)	15-45 ft.	30 to 60 sec.	No	80 to 480-B:C
Dry Chemical (Potassium Chloride)	Stored Pressure	2 to 2 1/2 lbs.	5-8 ft.	8 to 10 sec.	No	5 to 10-B:C
	Cartridge or Stored Pressure	5 to 10 lbs.	5-20 ft.	8 to 25 sec.	No	20 to 40-B:C
	Cartridge or Stored Pressure	19 1/2 to 30 lbs.	15-45 ft.	8 to 25 sec.	No	60 to 80-B:C
	Nitrogen cylinder or Stored Pressure	50 to 160 lbs. (wheeled)	15-45 ft.	30 to 60 sec.	No	120 to 160-B:C
Dry Chemical (Ammonium Phosphate)	Stored Pressure***	1 to 5 lbs.	5-12 ft.	8 to 1 0 sec.	No	1 to 2-A and 2 to 10-B:C
Dry Chemical (Ammonium Phosphate)	Stored Pressure or Cartridge	2 1/2 to 8 1/2 lbs.	5-12 ft.	8 to 12 sec.	No	1 to 3-A and 1 0 to 40-B: C
	Stored Pressure or Cartridge	9 to 17 lbs.	5-20 ft.	1 0 to 25 sec.	No	2 to 10-A and 10 to 60-B: C
	Stored Pressure or Cartridge	17 to 30 lbs.	5-20 ft.	1 0 to 25 sec.		3 to 20-A and 30 to 80-B:C
	Nitrogen Cylinder or Stored Pressure	50 to 315 lbs. (wheeled)	15-45 ft.	30 to 60 sec.		20 to 40-A and 60 to 320-B:C

TABLE 11.7 (Continued)

Extinguishing Agent	Method of Operation	Capacity	Horizontal Range of Stream (ft.)	Approximate Time of Discharge	Protection Required Below 40 deg. F (4 deg. C)	UL or ULC Classifications*
Dry Chemical (Foam Compatible)	Cartridge and Stored Pressure	4 3/4 to 9 lbs.	5-20 ft.	8 to 10 sec.	No	10 to 20-3:C
	Cartridge and Stored Pressure	9 to 27 lbs.	5-20 ft.	10 to 25 sec.	No	20 to 30-B:C
	Cartridge and Stored Pressure	18 to 30 lbs.	5-20 ft.	10 to 25 sec.	No	40 to 60-B:C
	Nitrogen cylinder and Stored Pressure	150 to 350 lbs. (wheeled)	15-45 ft.	20 to 150 sec.	No	80 to 240-B: C
Dry Chemical (Foam Compatible)	Cartridge and Stored Pressure	2 1/2 to 5 lbs.	5-12 ft.	8 to 10 sec.	No	10 to 20-B:C
	Cartridge and Stored Pressure	9 1/2 to 20 lbs.	5-20 ft.	8 to 25 sec.	No	40 to 60-B:C
	Cartridge and Stored Pressure	19 1/2 to 30 lbs.	5-20 ft.	10 to 25 sec.	No	60 to 80-B:C
(Potassium Chloride)	Nitrogen cylinder and Stored Pressure	50 lbs. (wheeled)	15-45 ft.	30 sec.	No	1 20-B: C
Dry Chemical (Foam Compatible)	Stored Pressure	5 to 11 lbs.	11-22 ft.	13 to 18 sec.	No	40 to 80-B:C
		9 to 23 lbs. 175 lbs. (wheeled)	1 5 -30 ft.	17 to 33 sec.	No	60 to 160-B: C
(Potassium Bicarbonate Urea)	Stored Pressure	175 lbs. (wheeled)	70 ft.	62 sec.	No	480-B:C
Bromotrifluoro-methane	Self Expellent	2 1/2 lbs.	4-6 ft.	8 to 1 0 sec.	No	2-B:C
	Stored Pressure	4 1/2 lbs.	6-1 0 ft.	8 to 10 sec.	No	5-B:C
Bromochlorodi-fluoromethane		2 to 4 lbs.	8-12 ft.	8 to 12 sec.	No	2 to 5-B:C
	Stored Pressure	5 1/2 to 9 lbs.	9-15 ft.	8 to 15 sec.	No	1-A and 10-B:C
		16 to 22 lbs.	14-16 ft.	10 to 18 sec.	No	1 to 2-A and 20 to 80-B:C

Note: 1 oz.=29.6 mi; 1 lb. = 0.454 kg; 1 ft. = 0.305 m; 1 gal. = 3.785liters.

*UL and ULC ratings checked as of December 27, 1974. Readers concerned with subsequent ratings should review the pertinent "Lists" and "Supplements" Issued by these Laboratories: (Write Underwriters Laboratories Inc., 207 East Ohio St., Chicago, Illinois 6061 1, or Underwriters' Laboratories of Canada, 7 Crouse Road, Scarborough, Ont., Canada MIR 349).

**Carbon–dioxide extinguishers with metallic horns do not carry a "C" classification.

***Some small extinguishers containing ammonium phosphate dry chemical do not carry on "A" classification.

Vaporizing liquid extinguishers (Carbon tetrachloride or chlorobromomethane base) are not recognized in this standard.

11.13.1 Water Type Extinguishers

Perhaps the most common portable fire extinguisher is the 2½ gallon pressurized water unit which is recommended for Class A hazards. This unit requires little skill to operate and has a discharge time of about one minute enabling the inexperienced user to have more time to fight a fire. Unlike gaseous or dry chemical units, it makes little noise and creates no clouds. Water units are available in either stored-pressure or pump-tank models (see Figure 11.2).

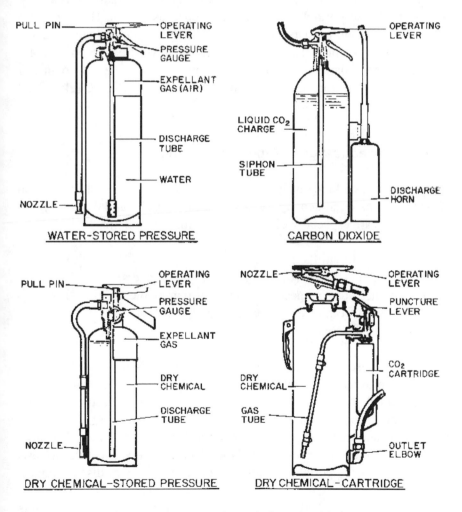

FIG. 11.2 Major types of portable fire extinguishers.

Stored-pressure units have a pressure gauge at the top. The entire main tank is pressurized with dry air or nitrogen. Although anti-freeze can be added to water units, only the manufacturer's recommended solution should be used. As a rule, water units must be protected from temperatures below 40°F. Therefore, they are limited to indoor heated locations.

Although pump-tank water units are uncommon they are effective. These units have a double-action pump that pushes water out as the operator pumps a plunger. They are not internally pressurized.

A third water-based portable unit is available. It is filled with aqueous film-forming foam (AFFF). These 2½ gallon stored-pressure units are the only water-based portable units effective on Class A and B (flammable liquids) fires. Their primary ingredient is water, to which a foaming agent is added to form a premixed solution. The agent is effective on ordinary combustibles and flammable liquids but the water base makes it a conductor of electricity. Therefore, these units are not suitable for Class C (electrical) fires.

11.13.2 Gaseous-Agent Type Extinguishers

There are two kinds of gaseous-agent portable extinguishers: carbon dioxide and haloganated hydrocarbon (Halon 1211 and 1301). Such extinguishers are especially effective on the following items:

- gas and liquid flammable materials
- electrical hazards (transformers, circuit breakers)
- engines utilizing gasoline and other flammable fluids
- ordinary combustibles (paper, wood, textiles)
- hazardous solids

Because gaseous agents leave no residue and dissipate readily, these extinguishers are used to protect more costly equipment. They are ideal for use on small electrical fires and on fires in switchgear or electrical motors.

Portable carbon dioxide extinguishers are available in five- to twenty-pound sizes and may be recognized by their large, cone-shaped discharge horns. CO_2 gas is self-expelling, and these units are very effective on Class B and C fires. CO_2 portable extinguishers have two major disadvantages:

1. Wind and draft adversely affect the discharge requiring these units to be used only indoors and at close range.
2. The dilution of the oxygen in the air by the CO_2 concentration can

create atmospheres that will not sustain life. Thorough ventilation is necessary after their use.

Halogenated extinguishers are among the newest type of portable units. They are stored-pressure models and are most effective on Class B and C hazards. Larger units, those with more than nine pounds of agent, are rated for Class A fires as well. The major drawback of Halon portable extinguishers is their cost. Although they are extremely effective and clean to use, they are perhaps the most expensive units on the market.

11.13.3 Dry Chemical Type Extinguishers

Dry chemical units are designed for use on Class B and C fires. The notable exception is the multipurpose dry chemical (monoammonium phosphate), which is effective on all three major classes of fires. These extinguishers are available in either stored-pressure or cartridge-operated units.

In the stored-pressure unit, an expellant (usually dry nitrogen, an inert gas) and extinguishing agent are stored in one chamber, discharge is controlled by a valve. A pressure gauge at the top of the unit indicates readiness for use. Normally, pressure ranges from 100 to 195 psi.

In cartridge-operated units, the expellant gas is stored in an auxiliary cartridge adjacent to the agent-containing shell. The unit is not pressurized until the cartridge is punctured. These units have no pressure gauge and must be weighed to determine their readiness for use. It is easier to recharge a cartridge unit than a stored-pressure unit, but the stored-pressure unit is more widely used. It is best suited for locations where infrequent use is anticipated and where skilled personnel with professional recharge equipment are available. Though both regular and multipurpose dry chemical units are very effective fire fighters, duration time is short and the units are among the most sensitive to operator error.

The major disadvantage to the multipurpose dry chemical extinguisher is that it clings to hot surfaces and forms a sticky film making cleanup difficult. Therefore, this type of unit should not be located near delicate equipment or machinery. If metal surfaces do become coated with monoammonium phosphate, they should be wiped with a wet cloth and dried as soon as possible to prevent rust or corrosion. Although regular dry chemical does not adhere to hot surfaces, it does leave a residue and should be brushed away as soon as possible after use.

Cartridge-operated portable units containing dry powder for use on Class D fires are available, but combustible metal fires present special problems. Each metal has distinctive burning characteristics that must be known before the fire can be combated effectively. Dry powder is frequently stored in a bucket with a shovel in the vicinity of the combustible metal hazard. Should a fire occur, the burning metal must be covered with a layer of powder at least two inches deep. The National Safety Council's Data Sheet 1-567-79, Fire Protection for Combustible Metals, is an excellent source for further information.

11.14 Obsolete Fire Extinguishers

Some older fire extinguishers still in use today are considered unacceptable and unsafe (see Table 11.8). Obsolete models should be removed. All inverting-type extinguishers should be replaced.

Table 11.8
Non-Acceptable and Obsolete Types of Extinguishers

Type	Description
Non-Acceptable	Stores pressure water and/or antifreeze--Brass or fiberglass shells
	Dry chemical (over 2½ lb. capacity)--Brass shells
	Soda-acid--Brass or copper shells
	Foam--Brass or copper shells
	Cartridge-operated water--Brass or copper shells
	Cartridge-operated loaded stream--Brass or copper shells
Obsolete Models	Soda-acid, foam, and cartridge-operated water types (including antifreeze and loaded stream with stainless steel shells are recommended to be replaced because: (1) parts are no longer available; substitute parts should never be used as they may create a serious danger, (2) method of operation is very difficult; does not have control valve for on-and-off operation, and (3) listing approval has been withdrawn by Underwriters laboratories, Inc. and by Factory Mutual.

Because these units are no longer manufactured, suitable replacement parts are not available. These 2½ gallon units, including soda-acid, foam, cartridge-operated water, and cartridge-operated loaded stream types, are no longer listed by Underwriters Laboratory or approved by Factory Mutual.

Of these units, those with brass or copper shells are considered dangerous. The NFPA Committee on Portable Fire Extinguishers has determined that the reliability and safety of extinguishers with copper and brass shells cannot be ascertained by standard hydrostatic test methods. Stored-pressure water types with brass shells are subject to "creep." The bottom of the unit is soft soldered, and may blow out when the unit is used.

Dry chemical units with brass or fiberglass shells are also considered unsafe. The shells, which are almost identical to those of the discontinued inverting types, are susceptible to extensive corrosion and have an unacceptably high failure rate during hydrostatic testing. Some units have exploded.

Stored-pressure units with fiberglass shells tend to rupture upon recharge. The manufacturer has recalled all of these units. Any that remain should be taken out of service immediately. Withstanding a hydrotest is no assurance that the unit is safe. Furthermore, the test may weaken the fiber structure, causing the unit to explode during use. If there is doubt about the safety or reliability of a unit, it should be removed from service and the distributor, manufacturer or NFPA should be consulted.

11.15 Unit Placement

Becoming familiar with the types of units available is only the first step in understanding portable fire extinguishers. Placing the proper units in areas where they will be the most useful is also critical. Some of the factors to be considered are accessibility, visibility, ease of handling, and appropriateness to the location.

11.15.1 Accessibility

Extinguishers should be located close to potential hazards but not so close that they would be damaged by fire or become inaccessible if a fire occurred. For example, extinguishers should be located outside the door of a storage room rather than within the room itself where they might become inaccessible. If the hazard is a dip tank, the fire extinguisher should not be mounted on the side of the tank. If the tank catches fire, the extinguisher is likely to be inaccessible. A better place for the unit is away from the tank on a wall or support column along normal paths of travel.

Table 11.9 lists the minimum extinguisher rating and maximum travel distance to extinguisher for Class A hazards as required by NFPA. Table 11.10 lists the maximum area to be protected per extinguisher. This table can be used to determine the number of extinguishers required. Table 11.11 provides NFPA's minimum extinguisher ratings and maximum travel distances to extinguisher for Class B hazards.

These NFPA standards form the basis for most mandatory rules adopted by federal, state, and local agencies. Local fire authorities will help apply these requirements to the individual shop. They also can determine which shop areas fall under light, ordinary, and extra hazard occupancy.

TABLE 11.9
Size and Placement for Class A Hazards

	Light (Low) Hazard Occupancy	Ordinary (Moderate) Hazard Occupancy	Extra (High) Hazard Occupancy
Minimum extinguisher rating	1A	2A	3A
Maximum floor area per unit of A	3,000 sq. ft.	1,500 sq. ft.	1,000 sq. ft.
Maximum floor area per extinguisher	11,250 sq. ft.*	11,250 sq. ft.*	11,250 sq. ft.*
Maximum travel distance to extinguisher	75 ft.	75 ft.	75 ft.

*11,250 sq. ft. is considered a practical limit.

TABLE 11.10
Maximum Area to be Protected per Extinguisher

Class A Rating Shows on Extinguisher Nameplate	Light Hazard Occupancy (ft.)	Ordinary Hazard Occupancy (sq. ft.)	Extra Hazard Occupancy (sq. ft.)
1A	3,000	-	-
2A	6,000	3,000	2,000
3A	9,000	4,500	3,000
4A	11,250	6,000	4,000
6A	11,250	9,000	6,000
10A	11,250	11,250	10,000
20A	11,250	11,250	11,250
40A	11,250	11,250	11,250

TABLE 11.11
Extinguisher Ratings and Travel Distance

Type of Hazard	Basic Minimum Extinguisher Rating	Maximum Travel Distance to Extinguishers	
		feet	meters
Light (low)	5B	30	9.15
	10B	50	15.25
	10B	30	9.15
Ordinary (moderate)	20B	50	15.25
	40B	30	9.15
Extra (high)	80B	50	15.25

Note: The specified ratings do not imply that fires of the magnitudes indicated by these rating will occur, but rather to five the operators more time and agent to handle difficult spill fires that may occur.

11.15.2 Visibility

To ensure proper accessibility and visibility, extinguishers should be placed at exits where they can be used to fight fires while still allowing the firefighter to escape. In this location workers will see the extinguisher each time they enter or leave and will know where to find it if it is needed. Care must be taken not to obstruct the exit.

A fire extinguisher must not be blocked or hidden by stock, equipment, or machines. It must be located where it will not be damaged by equipment, corroded by chemical processes, or exposed to the elements. Its location should be made conspicuous. For example, if it is hung on a column or post, a red band can be painted around the post. The extinguisher must be kept clean and should not be painted in a way that will camouflage it or obscure labels or markings.

11.15.3 Ease of Handling

For ease of lifting, extinguishers weighing less than forty pounds should be placed so that their tops are not more than five feet above the floor. Extinguishers weighing more than forty pounds should not be more than 3½ feet above the floor.

11.15.4 Individual Placement

To determine where portable extinguishers should be placed and what ratings are needed, it is necessary to look at what is in the shop and what is likely to burn. Areas with protection from automatic systems or hose stations will not need as many portable extinguishers as a completely unprotected area. If conditions permit, it better to put out a small fire with a portable fire extinguisher than to rely on a fixed system that if triggered could create a cleanup problem. Portable fire extinguishers can prevent a small fire from spreading and can rapidly extinguish a fire in its early stages.

Hazards should be itemized. Where is a fire most likely to start? What kind of fire is likely? How does the expense of the unit compare with the cost of cleanup? While it may be cheaper to install one dry chemical multipurpose unit (A, B, and C capabilities) instead of both a large water unit (A capability) and a small Halon 1211 extinguisher (B and C capabilities), the less expensive unit may create a much greater problem with cleanup than the two separate units.

When units are placed, adjacent hazards should be considered. If a flammable liquid fire could ignite surrounding combustible paper and wood, the fire extinguisher should have both A and B capabilities. If nearby electrical equipment may catch fire, then the agent inside the extinguisher should be nonconductive.

Ambient temperatures may be important. If the storage area is unheated or only minimally heated, water units may freeze. A multipurpose dry chemical unit that will operate at temperatures from 40°F to 120°F is a better choice in this circumstance.

Analyzing hazards often means that the shop will exceed NFPA minimum recommendations. The supervisor or employee should not hesitate to seek assistance from the local fire department or from a fire equipment distributor.

11.16 Training and Maintenance Requirements

Workers should be trained in the use of portable fire extinguishers. Such training includes:

- using a sweeping motion that extends at least six inches on each side of the near edge of the flames
- maintaining a proper distance from the fire to avoid splashing fuel or burning material
- holding portable units upright
- attacking flames at the fuel source
- keeping the agent flowing so that the fire cannot re-ignite

Workers should learn to react quickly. In addition to knowing how to operate extinguishers and how to apply agents effectively, they should be familiar enough with the equipment to know when a fire is beyond the capabilities of portable extinguishers.

Manufacturers of portable fire extinguishers work hard to make a product of high quality. Units and their components are tested prior to assembly. Most portable units are approved by Factory Mutual (FM) for the classes of fire for which they are suitable, or they are listed by Underwriters Laboratories (UL), and rated for class of fire and extinguishing potential. Many units bear the seals of both testing agencies whose quality control representatives periodically inspect manufacturers, exercising continual and careful control over the products they have approved or listed.

But any unit that is improperly maintained can pose a hazard. Carbon dioxide units may explode if subjected to severe neglect and corrosion or if exposed to extremely high temperatures when the relief valve fails to operate.

Mixing different types of multipurpose dry chemicals may produce a violent reaction. An extinguisher that has been even partially discharged must be recharged with the same kind of chemical.

11.16.1 Monthly Inspections

Extinguishers should be inspected monthly to be certain that:

- they are in their designated places
- access and visibility are unobstructed
- the operating instructions on the extinguisher nameplate are legible and facing outward
- any seals or tamper indicators that are broken or missing are replaced
- they have no obvious physical damage, corrosion, leakage, or clogged nozzles
- pressure gauge readings are in the operable range (water types without gauges should be hefted to determine fullness)

11.16.2 Annual Maintenance

Extinguishers should be maintained at least annually or according to nameplate instructions. A tag must be attached to show the maintenance or recharge date and the signature or initials of the person performing the service. Maintenance procedures should include a thorough examination of the following three basic elements of an extinguisher:

1. Mechanical parts
2. Extinguishing agent
3. Expelling means

Specific maintenance requirements for various types of extinguishers are included in NFPA 10, Portable Fire Extinguishers.

11.16.3 Hydrostatic Testing

Stored pressure extinguishers requiring a twelve-year hydrostatic test should be emptied and serviced every six years. The extinguisher sales representative usually perform this service at appropriate intervals.

11.17 Emergency Procedures

Except for incipient fires that can be extinguished easily in their first phase, the local fire department should be notified immediately in case of fire. A list of telephone numbers for local fire departments should be posted in a prominent place near the telephone.

Floor plans for designated areas should be posted, showing the locations of fire alarm activators, fire extinguishers, and exits.

While the fire department is being notified and while the fire is being combated, workers should shut off all power to machines and fans and then begin an orderly evacuation from the facility. A plan for orderly evacuation should include the following:

- an evacuation signal
- a well planned evacuation route for all areas of the shop
- designation of an assembly area for workers well away from the building
- an accounting procedure for all persons after assembly, keeping in mind those absent from the shop that day
- a search of the building to assure complete evacuation

Summary

Fires are classified into three categories, fires from ordinary combustible materials such as wood and paper (Class A fires), fires from flammable liquids, gases and greases (Class B fires), electrical fires (Class C fires), and fires from combustible metals (Class D fires).

Special attention must be paid to ensure that flammable and combustible materials are properly and safely stored. The National Fire Protection Association publishes standards for the safe storage of hazardous materials in storage containers, storage cabinets, storage rooms, and outdoor storage facilities.

The spread of fire can be controlled by fire walls, fire dampers, and adequate fire detection systems. There are four types of fire detection devices, each one able to detect a fire at one of its four stages. The four stages of fire are the ionization-incipient stage, photoelectric-smoldering stage, infrared-flame stage, and the thermal-heat stage.

Fires can be extinguished by automatic sprinkler systems and portable fire extinguishers. The wet-pipe and dry-pipe sprinkler system types are commonly used in small businesses. Portable fire extinguishers are the first line of defense against small fires. There are three basic types of portable fire extinguishers: water, gaseous, and dry chemical. Portable fire extinguishers must be:

- a reliable type
- the right type for each class of fire that may occur in the area
- in sufficient quantity to protect against the exposure in the area
- located where they are readily accessible for immediate use
- maintained in perfect operating condition, inspected frequently, checked against tampering, and recharged as required
- operable by the persons who are in the area, who can find them, and who are trained to use them effectively and promptly

The proper placement of portable fire extinguishers is essential to their effectiveness. The units must be accessible, visible, easy to handle, and appropriately placed for each specific location. Each unit must also be properly inspected and maintained.

Bibliography

Accident Prevention Manual for Industrial Operations, National Safety Council, Chicago, Illinois 60611. Latest edition.

Bochnak, Peter M. (1999) "How to Establish Industrial Loss Prevention and Fire Protection" Chapter 12 in the Handbook of Occupational Safety and Health, John Wiley & Sons, N.Y., N.Y.

Katzel, Jeanine A. (1979) "Selecting, Maintaining and Using Portable Fire Extinguishers," *Plant Engineering* (March 8, 1979), pp. 92—101.

National Fire Codes, 16 vols. (1979) National Fire Protection Association, Boston, Massachusetts 02110.

U.S. Department of Labor, Occupational Safety and Health Administration, *General Industry OSHA Safety and Health Standards (29 CFR 1910),* November 1978.

12

Health Hazards

Health hazards encountered in the work environment have the following characteristics:

- they cover a whole range of disorders involving many parts of the body including the lungs, liver, blood, kidney, skin, eyes, ears, brain, and nervous system
- they frequently escape detection
- they do not come with a label and therefore can be easily misdiagnosed
- they appear slowly, developing over months and years, and every person may not be affected
- new potential hazards are continually introduced through the use of new substances, new uses for old materials, new combinations of chemicals, and process changes

Health hazards are classified according to the following three categories:

- Category A — Biological Agents
- Category B — Physical Agents
- Category C — Chemical Agents

For years, safety and health professionals have been concerned with locating and eliminating or at least controlling safety and health hazards. But, because health hazards and their destructive impact are often not fully understood, many health hazards exist within the work environment without arousing the concern they deserve. Thus, workers are often exposed to excessive machinery noise, chemical agents which may cause dermatitis, and such airborne contaminants as fumes from welding operations, dusts from grinding, and vapors from solvents. Such exposure can have both short- and long-term effects on the health of both workers and supervisors.

The supervisor who is aware of the health hazards presented by the biological, physical, and chemical agents found in their work area has taken the first step in limiting these hazards. By applying the control measures detailed in this unit, he can limit the dangers presented by radiation, vibration, and noise. He can be sure that chemical agents in the shop are labeled adequately, used carefully, and disposed of properly. He can use such engineering controls as substitution, isolation, and ventilation to protect against the three modes of entry for hazardous substances. However, engineering controls are insufficient to protect against some physical and chemical hazards. Therefore, the supervisor needs to understand the uses of personal protective equipment in the safety and health program.

This chapter presents the health hazards that exist in the small business facility and how these hazards can be evaluated and controlled. The characteristics and classification of health hazards are discussed along with the dangers presented by such physical agents as radiation, vibration and noise. This Chapter will present the requirements for adequate labeling, careful use and proper disposal of chemical agents and the modes of entry for hazardous substances. The methods of evaluating and controlling health hazards are also presented in this Chapter.

12.1 Biological Agents

Biological agents including certain bacteria, fungi, parasites, and microorganisms are known to cause illness, extreme discomfort and, in some circumstances, even death. Biological agents are of particular concern to persons handling hides and skins, to butchers and others working with animal products, and to employees of sewage treatment facilities and waste areas.

12.2 Physical Agents

Physical agents are far more likely than biological agents to present health hazards in the average work place. In the chapter on illumination, we have seen how inadequate lighting and glare can contribute to eyestrain and fatigue and can cause accidents. Temperature extremes and humidity are other factors which affect safety and health. However, it is unlikely that shops deviate significantly from comfortable temperature ranges. We will now look at other physical agents which present health hazards in the shop environment: radiation, vibration, and noise.

12.2.1 Radiation

The electromagnetic (non-ionizing) radiation that may be encountered in the workplace can adversely affect the body. Infrared, visible, and ultraviolet radiation are manifestations of the same kind of electromagnetic radiation, differing from each other only in frequency, wavelength, or energy level. It is useful to discuss them as separate groups because of the physical effects which they produce.

Visible light is radiation that can be seen and includes the color spectrum from red, orange, yellow, and green to blue, indigo, and violet. The region beyond the red is infrared; the region beyond the violet is ultraviolet. Though radiation in these regions is invisible, it is a form of energy particularly dangerous to the human eye. The eye is an optical instrument equipped to receive radiation which is not limited to the visible portion of the electromagnetic spectrum. Because of its receiving ability and the delicate balance of its functional parts, it is easily injured.

12.2.1.1 Infrared Radiation (IR)

Infrared radiation does not penetrate below the superficial layer of the skin. Its only effect is to heat the skin and the tissues immediately below it. Except for thermal burns and damage to the eye, it presents a negligible health hazard. Infrared radiation is encountered in various shop operations:

- drying and baking of paints, varnishes, enamels, adhesives, printers' ink and other protective coatings
- heating of metal parts, especially through use of the electric arc and other flame-cutting devices
- dehydrating of textiles, paper and other materials

The major danger of IR, as with other forms of electromagnetic radiation, is to the eye. Low doses of IR over the years may not be felt but may cause serious permanent damage to the cornea, iris, retina, and lens of the eye. It can produce "heat cataract," an opacity of the rear surface of the lens which is particularly frequent among glassblowers and persons who work near industrial ovens and furnaces.

Goggles protect the eyes and regular clothing protects the skin against the dangers of IR. Ovens and other sources of IR can be shielded with shiny materials such as polished aluminum, which will reflect the heat back to its source.

12.2.1.2 Intense Visible Radiation

Intense visible radiation is emitted from the sun, artificial light sources, arc welding processes, and highly incandescent bodies. The physiological responses to intense visible light including adaptation, pupillary reflex, partial or full lid closure, and shading of the eyes are protective mechanisms to prevent excessive brightness from being focused on the retina.

In arc welding, the welder's eye protection equipment prevents exposure to intense visible light. Because others in the area can sustain retinal damage because they accidentally or carelessly view the electric arc, welding areas should be shielded and/or isolated.

12.2.1.3 Ultraviolet Radiation (UV)

Ultraviolet radiation is the portion of sunlight which causes sunburn and the most common exposure to UV is from direct sunshine. Symptoms of overexposure include reddening of the skin, blistering, and pain. UV radiation intensifies the effects of some chemicals. Long-term exposure to UV, especially when combined with such photosensitizing agents as the coal tars and cresols used in roofing, increases a person's chances of developing skin cancer. Certain protective creams contain compounds which minimize the effect of UV rays.

Many welding processes, especially the use of the electric arc, produce UV which can damage the eyes. Many arc welders are aware of the sensation of sand in the eyes which is known as "arc-eye" or keratitis. This painful condition occurs six to eight hours after exposure and is the result of excessive exposure to UV. Long-term exposure to UV can lead to vision loss. Welders must wear eye protection equipment with the appropriate shade lens.

UV reacts with chlorinated hydrocarbons such as perchloroethylene, trichloroethylene, and other chemicals commonly used as degreasers to form phosgene, a highly toxic nerve gas. To prevent such a reaction, welding operations should be shielded or isolated.

Conventional light sources produce random and disordered light wave mixtures of various frequencies. In contrast, lasers emit beams of coherent light of a single color or wavelength and frequency. Laser is an acronym for light amplification by stimulated emission of radiation.

12.2.1.4 Lasers

Lasers are relatively new and their use in industry is becoming more frequent. They are useful for projecting a reference line for construction work and for highly precise distance measuring in surveying. An all-purpose laser machine may be used for welding, cutting, drilling, and for micro-machining fine parts. Because the laser has a large energy density in a narrow beam, it can inflict serious injury, especially to the eye. Not only is it important to protect those persons who might view the direct beam but also those who might see a reflection. Hazard controls include barriers, shields, and protective equipment. The longer wavelengths, including power frequencies and broadcast and short wave radio, can produce general heating of the body.

12.2.1.5 Radio Frequency Waves

Radio frequency waves can be used as heating sources in various operations. Such heating equipment is used in metalworking for hardening gear teeth, cutting tools and bearing surfaces, and for annealing, soldering, and brazing. In woodworking, radio frequency heating equipment is used for bonding plywood, laminating, and general gluing. Other uses include molding plastics, curing and vulcanizing rubber, and thermo-sealing. The waves themselves are unlikely to emit sufficient exposure intensities to cause a radiation health hazard. The hazards of radio frequency heating are electrical shock and burns, hazards which will be discussed in a subsequent unit on electricity.

12.2.1.6 Microwaves

Microwaves are far more dangerous than radio frequency waves. Where microwaves are used for radar or communications, their hazards must be realized and necessary precautions taken.

12.2.1.7 Radar

Radar operates on the principle of microwave radiation echoing in a wavelength range from several meters to several millimeters. It can damage many parts of the body including the eyes, testes, gall bladder, gastrointestinal tract and certain other vital organs. Persons who work in or around high-power radar antennas or radar test equipment must minimize their exposure.

12.2.2 Mechanical Vibration

Another physical health hazard encountered in the shop is the vibration produced by such pneumatic tools as air hammers, compressed-air chisels, jack-hammers, riveting guns, pounding-up machines, and stonecutting hammers. The bodily response to vibration, which is often accompanied by noise, is a feeling of unease, fatigue, irritability, and discomfort.

A condition known as "dead fingers" or "white fingers" is produced by vibration of even fairly light tools while the fingers are held in a strained position. When the fingers are chilled at the same time that they are cramped, the condition is aggravated. Preventive measures include gloves, use of handles of comfortable size for the fingers, and directing the exhaust air from air-driven tools away from the hands so that they will not become unduly chilled. Because the condition is aggravated by gripping the vibrating tool too tightly, worker should be taught the proper way to hold pneumatic tools.

12.2.3 Noise and the Nature of Sound

Everyone at some time is exposed to noises that have the potential to damage the hearing. Ordinary shop noises, for example, those produced by compressors and circular saws, can cause hearing damage if there is sufficient exposure time. Noises at high levels of intensity do not require lengthy exposure time to cause hearing damage.

To understand how hearing can be damaged by noise, we first must understand something about both the characteristics of sound and the process of hearing.

Sound travels through the air in the form of a series of moving pressure disturbances or waves. These pressure waves, which are caused by minute back-and-forth movements of the air molecules, are formed by the vibration or motion of the sound source. A rough analogy to the motion of sound waves in the air is the motion of water waves on the surface of a pool of water when a rock is thrown into it.

As the energy is transmitted, the pressure variations reach the eardrum, and the vibrations are translated by the hearing system into a sensation called sound. A sound is not a sound until the brain identifies it as such.

12.2.3.1 Characteristics of Sound

The difference between sound and noise is subjective. Noise might be defined simply as unwanted sound. Whatever it is called, noise or sound can be a definite health hazard, interfering with job performance and safety and causing psychological distress and loss of hearing. Sound may be understood in terms of its two basic characteristics, pitch and intensity.

Sound travels through the air in the form of pressure disturbances or waves. The frequency with which the waves strike our ears determines the pitch of the sound. The higher the frequency of the waves, the higher the pitch of the sound.

Within a sound wave, each pressure disturbance or back-and-forth movement of the air molecules is referred to as a cycle of the wave. The frequency of sound waves can therefore be measured in terms of the number of cycles per second (CPS) that are generated by a sound source. The unit commonly used to describe frequency is the hertz (Hz). One hertz is equivalent to one cycle per second.

A sound source vibrating rapidly, for example, 10,000 times per second will produce a sound that strikes our ears at a frequency of 10,000 cycles per second (10,000 Hz). This is a sound of relatively high pitch, very near the upper limit of human hearing. A sound source vibrating slowly, for example at 200 times per second will produce a sound of 200 cycles per second (200 Hz), which is a sound of relatively low pitch.

Intensity, the second characteristic of sound, is commonly understand as loudness. While the pitch of a sound is determined by the frequency of the waves, the intensity of a sound is determined by the size of the air pressure disturbance. A larger pressure disturbance results in a sound of higher intensity while a smaller pressure disturbance results in a sound of lower intensity.

12.2.3.2 Decibels

Air pressure disturbance of sound waves is measured in units called decibels (dB). The higher the number of decibels, the greater the pressure disturbance and the more intense the sound. The sound produced by a gasoline-powered lawnmower, at about 90 decibels, would be considered of high intensity; the sound of leaves rustling, at about 20 decibels, would be considered very low intensity.

12.2.3.3 Threshold of Hearing

Everyone has what is known as a threshold of hearing, the sound level below which no sounds are heard. For most young people with normal hearing sensitivity, this threshold of hearing occurs near zero decibels. The decibel scale was developed so that its zero point would coincide approximately with the threshold of hearing.

Noise at high levels of intensity can raise this threshold so that we are unable to hear sounds at lower decibel levels, sounds that normally we can hear. Intense noise can raise the threshold on a temporary or permanent basis.

A Temporary Threshold Shift (TTS) is a condition in which we temporarily lose the ability to hear sounds at lower decibel levels. The TTS occurs during our exposure to potentially damaging noise. The TTS noticed after the noise has subsided or after we have removed ourselves from the noise. It is at this point that we may become aware that certain lower decibel sounds that are normally easy to hear are now more difficult to hear or perhaps cannot be heard at all.

This threshold shift is the result of damage to the tiny hair cells within the cochlea. These are the cells that ultimately transmit sound to the brain in the form of electrical impulses. When these cells are damaged, the brain does not receive sound signals. The sounds simply are not heard.

Intense noise damages the hair cells by over-stimulating or overloading them, thus weakening their ability to transmit signals to the brain. Given an opportunity to recover (being removed from the source of the damaging noise), the hair cells generally will do so. Following a recover period, which usually is a few hours, the threshold of hearing will return again to its normal level.

But this return to a normal threshold level does not always occur. When it does not, we experience a Permanent Threshold Shift (PTS). A PTS is a condition in which we permanently lose the ability to hear sounds at lower decibel levels. One of the most harmful effects of such a hearing loss is that we lose some of our ability to understand speech. A PTS can result from a single damaging exposure to very high intensity noise but most often results from exposure to moderately intense noise over an extended period of time.

A permanent hearing loss can occur over time without our even being aware of it. If exposed to sufficiently high levels of noise over time, we find our ability to hear diminished little by little, not enough at any one time for the loss of hearing to be noticeable. Unfortunately, hearing loss is often noticed only after permanent damage has been done.

A PTS can be the result of a series of temporary threshold shifts, each of which weakens the hair cells in the cochlea. The cumulative effect of the temporary shifts can be that the hair cells are actually destroyed. At this point, recovery is not possible. The lower decibel sounds never again can be heard. After the first permanent hearing loss has been detected, further hearing losses can occur so long as there continues to be exposure to damaging noise at higher levels.

No one can predict when a Temporary Threshold Shift will become a Permanent Threshold Shift. Our ears can, however, warn us when the danger of permanent hearing damage from relatively short exposure to intense noise is imminent.

12.2.3.4 Warning Signals

Warning signals, such as a ringing in the ears (tinnitus), a threshold shift which lasts more than a few hours or a tickling sensation in the ears (which actually is a mild form of pain) tell us that we should remove ourselves from exposure to high intensity noises or suffer the consequences. These warning signals also tell us that, before returning to the proximity of the high intensity noises, we should take steps to protect our hearing.

Unfortunately, permanent hearing loss often can result from long-term exposure to noise levels which are below the range where we perceive warning signals. It takes longer for hearing to be damaged by noise at these lower levels, but the result is the same.

Regardless of whether we have been receiving warning signals of hearing damage, hearing tests should be part of a routine physical examination. A hearing test can detect the early signs of a hearing loss and can alert us to a problem before more serious damage.

12.2.4 How Noise Damages Hearing

How intense must noise be before it has the potential to damage our hearing, on either a temporary or a permanent basis? There is no simple definitive answer to this question. There are too many variables involved. The four most important variables are:

1. the level of the sound, as measured in decibels
2. the length of time to which we are exposed to the sound

3. the number and length of quiet (recovery) periods between periods of sound
4. individual sensitivity to or tolerance for sound

The danger that noise poses to our hearing is a function of the interaction of these four variables.

12.2.4.1 Sound Level

Let us examine the first variable, the level of the noise.

- For most persons, the threshold of hearing occurs near 0 decibels, usually between 0 and 10 decibels.
- Sounds below approximately 40 decibels are considered low intensity noises. Examples include the rustling of leaves, a whisper and the ticking of a watch.
- Sounds between 40 and 70 decibels are considered moderate in intensity and include such things as conversational speech, a typewriter and the singing of birds.
- Sounds between 70 and 90 decibels are considered loud and include such things as a television, a dishwasher and a table saw.
- Sounds between 90 and 100 decibels are considered intense and include such things as a gasoline-powered lawnmower, a rock band and an emergency siren.
- Sounds between 110 and 130 decibels may induce pain in the ears. Examples include nearby thunder, sonic booms and jet plane takeoffs.

It is at the higher decibel levels (80-90 and above) that the likelihood of noise-induced hearing damage begins to increase if an individual is exposed to noise at or above these levels for a sufficiently long period of time. Machinery in the industrial shop produces these higher decibel noises (see Table 12.1 for examples).

12.2.4.2 Time Exposure

The second variable affecting potential hearing damage is the time variable. The higher the intensity of the noise, the shorter the time required for hearing damage to occur.

TABLE 12.1
Machinery Noise Production

Machine	dB
Punch Press	96--108
Hydraulic Press	130
Circular Saw	105
Wood Planer	98--110
Oxygen Torch	121

The Occupational Safety and Health Administration (OSHA) has established standards for occupational noise exposure. These standards describe the lengths of time beyond which a worker should not be exposed to noise at various levels of intensity during a normal eight-hour working day. Table 12.2 provides the general guidelines for the permissible sound exposures:

TABLE 12.2
Permissible Sound Exposures

Hours per Day of Exposure	A-Weighted Sound Level, dB
8	90
6	92
4	95
3	97
2	100
1-1/2	102
1	105
1/2	110
1/4 or less	115

Note that in this table sound level is designated "A-weighted." A-weighting is a sound measurement technique which filters out the low frequency sounds which the human ear does not hear well, thus roughly simulating the sensitivity of the human ear to sound frequency.

The table indicates that workers should not be exposed to a sound level which exceeds 90 dB on the average, for an eight-hour day. If such exposure cannot be avoided, steps must be taken to protect the worker's hearing.

Table 12.2 also indicates that workers should not be exposed to a sound level which exceeds 115 dB, on the average, for even fifteen minutes of a work day. Clearly, workers should never be exposed to steady sound levels above 115 dB.

It should be noted that the OSHA standards apply only to working environments. Our hearing is affected by the totality of the noise that we are exposed to in our daily lives.

12.2.4.3 Recovery Periods

This brings us to a third variable which determines how noise can damage hearing, that is, the number and length of quiet (recovery) periods between periods of sound. If we do work in a noisy environment in which it is possible to experience threshold shifts, we must avoid extended contact with noisy environments outside of our work. Using power shop tools or lawnmowers or attending rock concerts will not allow our hearing to recover from the effects of day-long noise exposure.

12.2.4.4 Individual Differences

The difficulty in stating precisely what an adequate margin of safety should be for all persons in all types of jobs brings us to the fourth variable affecting potential hearing damage, individual differences. Because all of us differ from one another in various ways, including hearing, safe level of noise exposure for one person may not be safe for another. The fact that others may not report any hearing difficulty resulting from work in a noisy environment does not mean that a given individual will not suffer hearing damage in the same environment.

12.2.5 Noise Control

There are three basic ways to control noise: at its source, along its path, and at the point of hearing through the use of protective equipment. Control of noise at its source is probably the best approach. The reason, quite simply, is that, if a piece of equipment or a tool is operating or is made to operate at a safe noise level, there is no hearing danger posed and no need to use additional approaches to noise control.

If source control is not possible, the next best approach is to control the noise along its path. Such control limits the number of persons exposed to the noise. However, it does not always eliminate the noise problem for all persons affected, especially those working directly with or very near the noise source, and it often requires that new noise control steps be taken whenever equipment is moved or work sites are changed.

The use of ear protection equipment is not as desirable as either source control or path control. It affords protection only to those on or near the site who are wearing the equipment, and workers must be willing to wear hearing protectors whenever they are exposed to potentially dangerous noise. Further, certain conditions and activities can reduce the effectiveness of the hearing protectors themselves.

12.2.5.1 Source Control

Source control begins with a careful analysis of the noise-producing equipment in order to isolate the major sources of noise within the equipment and to determine how the noise is being transmitted from these sources. The major noise source may be an engine or a motor, but the noise itself may be transmitted as vibration other parts of the equipment which, in turn, radiate the noise heard outside the equipment. In source control, both the major noise source and the secondary noise radiators must be examined and quieted to the greatest extent possible.

The U.S. Environmental Protection Agency has provided the following ten ways to control noise at its source:

1. Reduce impact noise produced when parts of equipment strike one another. This may be accomplished by:
    *   reducing the size or weight of the impacting mass
    *   reducing the travel of the impacting mass
    *   using small impact force over a longer period, rather than large impact force over a shorter period
    *   cushioning the impact with shock-absorbing material
    *   avoiding the use of metallic material on both impact surfaces
    *   applying smooth acceleration to impact mass

2. Reduce speed of moving parts and rotating parts. This may be accomplished by:
    *   Operating motors, turbines, fans, and so forth at lowest blade-tip speeds that meet job requirements.
    *   Using the largest diameter, lowest speed fans that meet job requirements.
    *   Using centrifugal or squirrel cage fans which are not so noisy as propeller or vane axial fans.

3. Reduce pressure and flow velocities in air, gas or liquid circulation systems. Reducing velocities lessens turbulence, which, in turn, reduces noise radiation.

4. Balance rotating parts. When shafts, flywheels, and pulleys are not in balance, they cause structural vibration, which transmits noise.

5. Reduce friction in rotating, sliding or moving parts. When friction is reduced, the smoother operation of parts translates into lower noise levels. Friction is reduced by lubricating moving parts, properly aligning moving parts, properly polishing smooth surfaces on moving parts, properly balancing rotating parts, and replacing eccentric or out-of-round rotating parts or any worn parts.

6. Reduce flow resistance in air and liquid circulation systems. By using large-diameter, low-velocity pipes and ducts and by ensuring that the inside surfaces of the pipes and ducts are smooth and free of obstruction and sharp corners, the flow will be streamlined and lower noise levels will result.

7. Isolate vibration within equipment. The following steps are recommended to prevent a vibrating component from transmitting all of its noise-producing vibration to other parts and surfaces of equipment:
    *   install the vibrating components (motors, pumps, fans and so forth) on the most massive part of the equipment
    *   install the components on vibration-absorbing, resilient mounts
    *   use belt- or roller-drive systems rather than gear trains
    *   use flexible, rather than rigid, hoses and wiring

8. Reduce the size of the surface radiating the noise. As a rule, the larger the vibrating surface, the greater the noise that is radiated. When vibrating surfaces are reduced in size, for example by removing excess material, cutting out portions of the surface, or using wire mesh in place of sheet metal; the noise output is reduced.

9. Apply vibration-damping materials to vibrating parts and surfaces. The concept of vibration damping is, quite simply, that reducing the vibration reduces the noise. Materials that can be applied to vibrating surfaces include liquid mastics (such as automobile

undercoating), pads (such as rubber, felt, adhesive tape, fibrous blankets), and sheet metal laminates or composites. The liquid mastics may be sprayed, the pads may be glued, and the sheet metal laminates may be bonded directly to the vibrating surfaces.

10. Reduce the leakage of noise from within equipment. Sealing noise within a piece of equipment is another simple noise control concept. This may involve:
    - sealing or covering all unnecessary holes and cracks
    - using gaskets around all electrical and plumbing penetrations
    - installing lids or shields with gaskets over functional or required openings
    - using mufflers, silencers or acoustically lined ducts for intake, exhaust, cooling or ventilation openings
    - directing openings away from the equipment operator, and to the greatest extent possible, away from other workers
    - using sound-absorbent linings on inner surfaces of equipment
    - using vibration-damping materials on vibrating inner surfaces of equipment

Approaches to control of noise at its source are simple and logical. Approaches such as those listed above should constitute the first line of defense against noise produced by equipment and tools. In situations where source control measures will not work or such methods will not lower the noise level to a safe point, control of noise along its path to the ear must be considered.

12.2.5.2 Path Control

In path control, the noise is blocked or reduced before it reaches the ears. This can be done by:

- containing or enclosing the noise
- absorbing the noise along its path
- deflecting the noise away from our ears
- separating the noise from the hearer

The approach that is chosen depends on the type of equipment or tool and the environment in which we are working.

Reducing the leakage of noise from within equipment is a basic approach, which applies both to source control and path control. Enclosing a noisy piece of equipment in a box or a room or covering a noisy pipe with a heavy sound-absorbing material can effectively quiet noise. Obviously, this may not be practical for highly mobile equipment. But for noisy equipment that is stationary or infrequently moved, it should be considered.

Absorbing noise is another approach common to both source control and path control. Sound-absorbing materials or acoustic lining in a box or room can be used to enclose equipment noise. Noise transmitted from its source through ducts, pipes, or electrical channels can be reduced through the use of sound-absorbing materials. The inside surfaces of these noise passageways can be lined with glass fiberboard, and the ducts, pipes, or channels can be wrapped with a glass fiber blanket. Baffles constructed of glass fiberboard can be installed inside the noise passageways.

Screens or barriers can be used to deflect the noise that is generated by equipment and tools. Much noise can be literally "walled-in" by barriers which surround the noise source. This can be aided by lining the barriers, which may be wood or metal panels, with sound-absorbing material.

But barriers do not have to surround the noise source to be effective in reducing noise transmission. If it is sufficiently large, a free-standing wall between the noise source and a hearer can reflect much of the noise and create a noise "shadow" to protect the hearer.

Putting distance between the noise source and the persons exposed to the noise is a simple and effective approach to path control. The further away from the noise source we work, the lower the noise level we receive.

12.2.5.3 Protective Equipment

The need for personal hearing protection arises when source control and/or path control are not present, when source and/or path control do not lower noises to safe levels, or when a worker cannot avoid direct exposure to noisy equipment and tools. There are three basic types of personal protective hearing devices:

1. Disposable pliable materials, such as fine glass wool, mineral fibers and wax-impregnated cotton, may be inserted in the ear. It must be fresh each day. Though such material offers some level of hearing protection, it does not provide the same benefits as do other devices.
2. Ear plugs may be inserted in the ear. These must be individually fitted to the wearer.

3. Cup-type protectors, like ear muffs, may be worn with the band over the head or around the back of the neck or may be incorporated into safety helmets.

The amount of noise protection afforded by these devices varies from one device to another at different sound frequencies. Although it is difficult to generalize for all of the devices available commercially, the wearer of a hearing protection device may expect noise reduction ranging from 10 dB to over 40 dB at certain frequencies. Hearing protective devices will be discussed further in Chapter 13.

12.2.6 Management of Noise Control

While source control, path control, and control at point of hearing are generally accepted as the three basic approaches to noise control, there is a fourth factor which directly affects the need for these three approaches and the amount of noise to which worker and supervisors are exposed. This fourth factor may be termed the management of noise control. The management of noise control refers to the administrative decisions that are made to purchase certain types of equipment and tools, to use certain procedures, and to schedule work during certain parts of the work period.

It is obvious that the purchase of equipment can affect the noise level on a work site. If relatively noisy equipment is purchased, the noise exposure will be higher, and the need for source, path and point-of-hearing controls will be greater.

Opportunities for decisions which have positive effects on noise control present themselves whenever a piece of equipment or a tool becomes damaged, worn out, or obsolete and must be replaced. Decisions to replace equipment and tools with the quietest models available should, over time, result in a much quieter and safer workplace.

Decisions to choose certain work procedures over others can also affect noise levels in some obvious ways. For example, if material can be either welded or riveted, the choice of welding would result in less noise generation. If concrete can be mixed off the site as well as on, the decision to mix off the site would result, obviously, in less noise on the site.

When noise is generated is as important a consideration as how much noise is generated. While this approach may be difficult, it may be possible to alter schedules for some noisy operations in order to minimize the number of persons exposed to high noise levels. Decisions also can be made to schedule noisy procedures for several short periods of time during a day or over a number of days rather than in one long, continuous period.

12.3 Chemical Agents

Having completed our examination of biological and physical agents, we come to the third category of health hazards, chemical agents. Chemical hazards may be defined as chemicals that may, under specific circumstances, cause injury to persons or damage to property because of reactivity, instability, spontaneous decomposition, flammability, or volatility. Included in this category are substances, mixtures, or compounds that are explosive, corrosive, flammable, or toxic.

- Explosives are substances, mixtures or compounds, which can enter into a combustion reaction so rapidly and violently that they can cause an explosion.
- Corrosives can destroy living tissues. Their destructive effect on other substances, particularly on combustible materials, may result in a fire or explosion.
- Flammable liquids are those liquids with a flash point of 100°F (38°C) or less. As we saw in the previous chapter on fire protection, combustible liquids (those with flash points above 100°F) also may be hazardous.
- Toxic chemicals are those gases, liquids or solids, which through their chemical properties, can produce injurious or lethal effects upon contact with body cells.

12.4 Processes and Operations which Involve Toxic and Corrosive Agents

Work in the shop requires processes and operations where contact with hazardous materials is inevitable. Four examples are metalworking, electroplating, painting and staining and the cleaning of metal parts.

12.4.1 Metalworking

Metalworking inevitably produces fumes and dust. Fumes are very small particles formed by the vaporization of metal during torch-cutting, burning or welding operations, whereas metal dust is generated by grinding. Special precautions such as exhaust ventilation and the wearing of respirators and eye protection need to be taken when cutting, burning or grinding scrap containing alloys of the more toxic metals, such as lead, zinc, cadmium, or beryllium.

12.4.2 Electroplating

Electroplating processes, used extensively for decorative purposes and for producing tarnish-resistant finishes, involve hazards of skin contact and inhalation. Because the skin may be exposed to strong acids and alkalis, emergency eye wash and shower facilities are required. Because mists or gases from plating solutions represent a respiratory hazard if dispersed into the air of the workplace, ventilation is required.

Chromic acid is used in chrome plating. Skin contact causes dermatitis and burns known as "chrome holes." The major hazard to health, however, is the inhalation of chromic acid mist or vapor, which causes irritation of the upper respiratory passages and may result in perforation of the nasal septum. Industrial poisoning has occurred from inhaling the mist of solutions containing as little as 5 percent chromic acid. Local exhaust ventilation should be used with all chromic acid tanks.

Copper plating baths are both acid and alkaline types. The cyanide salts in the alkaline bath are the greatest hazard in copper plating. These salt particles may become airborne when the tanks are charged. Cyanide solutions are readily absorbed, and skin contact must be avoided. Local exhaust ventilation systems are required to draw off the vapors, respirators may be needed, and workers must limit skin contact through the use of gloves which the cyanide cannot penetrate.

Good personal hygiene practices must be stressed, including frequent washing of exposed skin areas, particularly before eating or smoking. If a cyanide salt solution is mixed with acid, deadly hydrogen cyanide gas can result. All traces of acid must be rinsed away from parts before they are immersed in the cyanide vat.

Zinc and cadmium plating operations also use cyanide baths. As with copper plating, care must be exercised to avoid contact with the cyanide solution and to prevent a cyanide/acid mix.

12.4.3 Painting and Staining

Painting and staining are two of many shop processes that use solvents. Whenever solvents are used, health hazards are present. A solvent is a material used to dissolve another material. Although the physiological effects of different solvents vary, generally they include:

- dermatitis
- irritation of the respiratory tract
- interference with the central nervous system

Skin contact with solvents may cause dermatitis, ranging in severity from a simple irritation to actual damage to the skin. Even the most inert solvents can dissolve the natural protective barriers of fats and oils, leaving the skin unprotected. When these natural lubricants are removed, the skin becomes subject to disabling and possibly disfiguring dermatitis and the way to serious infection is opened. Some of the newer paints contain hardeners and other additives that can cause skin rashes.

Sometimes a worker washes their hands in such solvents as mineral spirits and turpentine. These solvents take the fats out of the skin, increasing the chance for skin rashes. In some cases, they can be absorbed through the skin. A worker must be instructed in specific procedures for cleanliness. They should remove with waterless hand cleaners any paints or stains which get on their skin. Solvent-resistant gloves and long-sleeved shirts worn while painting will prevent the paints or stains from contacting the skin in those areas.

Another principal mode of exposure to solvents is the inhalation of vapors. Such exposure may result in throat irritation and bronchitis and eventual damage to the blood, liver, kidneys and respiratory and gastrointestinal systems. Engineering controls such as ventilation, good work practices, and personal protection devices limit such exposures.

All organic solvents affect the central nervous system. Depending upon the degree of exposure and the solvent involved, these effects may range from mild narcosis to death from respiratory arrest. Because solvents act as depressants and anesthetics, they can cause drowsiness and loss of coordination, increasing the risk of accidents. Thinners used in most paints will have a narcotic effect on worker, and long-term exposure may cause irreparable liver and lung damage. Respirators should be worn in the spray area or paint booth, and ventilation should be provided.

12.4.4 Cleaning Metal Parts

Strong corrosive solutions are used for cleaning metal parts in dip tanks and reusable filters. When such caustics are used, controlled procedures are necessary. Because skin contact will cause severe burns, rubber gloves and a face shield or goggles should be worn when handling caustics. Any caustics that contact the skin must be washed off immediately. A safety shower and eye wash fountain should be installed where caustics are handled. Table 12.3 describes the hazards implicit in some other processes and operations common to the shop.

12.5 Examples of Toxic and Corrosive Agents

Many other toxic and corrosive agents are used in the workplace. This section contains brief descriptions of the following: aluminum, arsenic, asbestos, benzene, carbon monoxide, chlorinated hydrocarbon solvents, epoxy resins, fluorides, lead, ozone, refrigerants, and sulfuric acid.

Aluminum. In welding and cutting operations, aluminum is a major component of metals and filler metals. The inhalation of aluminum dust or its compounds, including aluminum oxide fumes, is not known to have any adverse health effects.

Arsenic. Arsenic may be encountered in welding and cutting operations as a component of various alloys, where it gives increased heat resistance and hardness. Welding or cutting on metals that are painted with arsenic compounds also can be hazardous.

Arsenic is a poison that accumulates in the body. It is deposited in many bodily tissues, especially in the liver and kidneys. Fumes and dust produce inflammation of mucous membrane surfaces, irritation of the eyes and exposed skin and a husky voice and cough. Because the effects of arsenic may not appear for weeks, months or even years after exposure, workers should be protected from all contact. Because serious skin irritation results from contact with rubber in the presence of arsenic, respirators used for protection should not be made of rubber.

Asbestos. Asbestos may be found in many locations in the workplace. Hand- and power-operated tools may produce or release asbestos fibers. Asbestos pads may be used for welding and soldering. Personal protective equipment (aprons, gloves, etc.) may contain asbestos. Workers repairing brakes or machine linings may be exposed to asbestos.

Once asbestos fibers are inhaled, they may remain trapped in the lung

TABLE 12.3
Industrial Operations and their Associated Health Hazards

Industrial Operation	Associated Health Hazard
Abrasive Machining	An abrasive machining operation is characterized by the removal of material from a work-piece by the cutting action of abrasive particles contained in or on a machine tool. The work piece material is removed in the form of small particles and, whenever the operation is performed dry, these particles are projected into the air in the vicinity of the operation.
Ceramic Coating	Ceramic coating may present the hazard of airborne dispersion of toxic pigments plus hazards of heat stress from the furnaces and hot ware.
Dry Grinding	Dry grinding operations should be examined for airborne dust, noise and ergonomic hazards.
Forming and Forging	Hot bending, forming or cutting of metals or nonmetals may have the hazards of lubricant mist, decomposition products of the lubricant, skin contact with the lubricant, heat stress (including radiant heat), noise and dust.
Gas Furnace or Oven Heating Operations (Annealing, Baking Drying, Etc.)	Any gas or oil fired combustion process should be examined to determine the level of by-products of combustion that may be released into the workroom atmosphere. Noise measurements should also be made to determine the level of burner noise.
Grinding Operations	Grinding, crushing or comminuting of any material may present the hazard of contamination of workroom air due to the dust from the material being processed or from the grinding wheel.
High Temperatures from Hot Castings, Unlagged Steam Pipes, Process Equipment, Etc	Any process or operation involving high ambient temperatures (dry-bulb temperature), radiant heat load (globe temperature) or excessive humidity (wet-bulb temperature) should be examined to determine the magnitude of the physical stresses that may be present.
Molten Metals	Any process involving the melting and pouring of molten metals should be examined to determine the level of air contaminants of any toxic gas, metal fume or dust produced in the operation.
Open-Surface Tanks	Open-surface tanks are utilized by industry for numerous purposes. Among their applications can be included the common operations of degreasing, electroplating, metal stripping, fur and leather finishing, dyeing and pickling. An open-surface tank operation is defined as "any operation involving the immersion of materials in liquids, which are contained in pots, tanks, vats or similar containers." Excluded from consideration in this definition, however, are certain similar operations such as surface-coating operations and operations involving molten metals for which different engineering control requirements exist.
Paint Spraying	Spray painting operations should be examined for the possibility of hazards from inhalation and skin contact with toxic and irritating solvents and inhalation of toxic pigments. The solvent vapor evaporating from the sprayed surface may also be a source of hazard, because ventilation may be provided only for the paint spray booth.
Plating	Electroplating processes involve risk of skin contact with strong chemicals and in addition may present a respiratory hazard if mist or gases from the plating solutions are dispersed into the air of the shop.
Vapor Degreasing	The removal of oil and grease from metal products may present hazards. This operation should be examined to determine that excessive amounts of vapor are not being released into the shop atmosphere.

TABLE 12.3 (Continued)

Industrial Operation	Associated Health Hazard
Welding--Gas or Electric Arc	Welding operations generally involve melting of a metal in the presence of a flux or a shielding gas by means of a flame or an electric arc. The operation may product gases or fumes from the metal, the flux, metal surface coatings or surface contaminants. Certain toxic gases such as ozone or nitrogen dioxide may also be formed by the flame or arc. If there is an arc or spark discharge, the effects of radiation and the products of destruction of the electrodes should be investigated. These operations also commonly involve hazards of high potential electrical circuits of low internal resistance.
Wet Grinding.	Wet grinding of any material may produce possible hazards of mist, dust and noise.

where the body tries to isolate them by producing scar tissue. Long-term exposure to high concentrations of asbestos fibers causes asbestosis, a disease of the lungs, and is thought to cause cancer. The main symptom of asbestosis is shortness of breath. Engineering controls include isolation, enclosure, exhaust ventilation and dust collection. Respirators and special clothing also may be necessary.

Benzene. Benzene (sometimes called benzol, phenyl hydride, or coal naphtha) is classified as a flammable liquid as well as a carcinogen. Inhalation of high concentrations can affect the central nervous system. High concentrations of benzene also are irritating to the mucous membranes of the eyes, nose and respiratory tract. Exposure to benzene also can lead to the development of leukemia. Its vapors can form explosive mixtures and burn with a smoky flame. Because it is one of the two or three most dangerous organic solvents in commercial use, benzene should not be used in the workplace. It can be replaced with toluene in most lacquers, synthetic rubber solutions, and paint removers. Benzene should not be confused with benzinc, a petroleum distillate.

Carbon Monoxide. Carbon monoxide is a gas usually formed by the incomplete combustion of various fuels. It is commonly found in automotive shops from engine exhaust. Portable gas-filled heaters also produce carbon monoxide, as do internal combustion engines and improperly maintained or adjusted burners and flues. Carbon monoxide is odorless and colorless and cannot be detected by the senses. Common symptoms of overexposure include pounding of the heart, a dull headache, flashes before the eyes, dizziness, ringing in the ears and nausea.

Chlorinated Hydrocarbon Solvents. Various chlorinated hydrocarbons such as trichloroethylene, perchloroethylene, and carbon tetrachloride are used in degreasing and other cleaning operations. They may injure the liver, cause dermatitis and depress the central nervous system. Skin contact should be avoided. A respirator should be used so that the vapors will not be inhaled. In welding and cutting operations, the heat and ultraviolet radiation from the electric arc will decompose the vapors of these solvents and form highly toxic and irritating phosgene gas.

Epoxy Resins. Materials which should be regarded as hazardous include wet or uncured epoxy resins and the chemicals used to harden, thin, strengthen, or make the resin flexible. Dermatitis can result from handling epoxy resins and the chemicals used to manufacture them or from sanding or polishing epoxy surfaces. The use of impervious plastic gloves and similar protection over other skin areas can help prevent this condition. Some of the symptoms of dermatitis include redness, itching, swelling, and blisters. Oozing, crusting and scaling of the skin also can occur. Respiratory irritation, headache, nausea, intestinal upsets and other conditions may result from breathing vapors or dust from the various epoxy manufacturing processes. The eyes also may be affected by vapors or by direct contact.

Fluorides. Fluoride compounds are found in the coatings of several types of fluxes used in welding. Exposure to these fluxes may irritate the eyes, nose, and throat and may produce painful skin burns. Repeated exposure over a long period to high concentrations of fluorides in the air may cause pulmonary edema (fluid in the lungs) and bone damage. Prolonged exposure to fluoride dusts and fumes also may cause acute systemic poisoning.

Lead. Lead was one of the first industrial materials to be recognized as a serious health hazard. The welding and cutting of lead-bearing alloys or metals whose surfaces have been painted with lead-based paint can generate lead oxide fumes. Inhalation and ingestion of lead oxide fumes and other lead compounds will cause lead poisoning. Symptoms include a metallic taste in the mouth, loss of appetite, nausea, abdominal cramps and insomnia. In time, anemia and a general weakness, chiefly in the muscles of the wrists, develop. Prevention of lead poisoning is almost entirely a matter of good housekeeping and dust control. Ventilation is very important.

Ozone. Ozone may be found either directly or indirectly. It is used to age wood rapidly, to bleach oils, and to dry varnish rapidly. It may be generated by such electrical equipment as copying machines and electronic air filters. It may be the result of electric arc welding, where enough energy is released into the atmosphere to change nitrogen and oxygen into both nitrogen dioxide and ozone. Without ventilation, the ozone concentration in the air gets up to about .06 ppm for flux-covered electrodes, and up to about .5 ppm for barewire, argon-shielded welding of aluminum.

Like nitrogen dioxide, ozone is irritating to the eyes and mucous membranes. Breathing ozone in low concentrations (above .05 ppm) may cause dryness of the mouth, headaches, coughing and pressure or pain in the chest, followed by difficulty in breathing. Ozone impairs the sense of smell, disguises other odors, alters tastes and reduces the ability to think clearly. It also depresses the nervous system, slowing the heart and respiration and producing drowsiness and sleep. Excessive exposure may produce pulmonary edema.

General ventilation alone may be sufficient to prevent the accumulation of ozone in inhabited spaces. Enclosures can isolate ozone processes, and respirators may be needed in some situations.

Refrigerants. Problems which may occur during installation, modification or repair of refrigeration units are leaks and, very infrequently, fire or explosion. Refrigerants may be considered in the following classes:

1. Nonflammable substances where the toxicity is slight, such as some fluorinated hydrocarbons (Freon). Although considered fairly safe, these refrigerants may decompose into highly toxic gases (e.g., hydrochloric acid, chlorine, phosgene) upon exposure to hot surfaces (sweating, welding, and so forth) or open flames.

2. Toxic and corrosive refrigerants (e.g., methyl chloride and ammonia) may be flammable in concentrations exceeding 3.5 percent by volume. Ammonia is the most common refrigerant in this category and is very irritating to the eyes, skin and respiratory system. If there are large releases of ammonia, the area must be evacuated. Re-entry may be made by wearing appropriate respiratory protective devices and protective impervious clothing. As ammonia is readily soluble in water, it may be necessary to spray water in the room via a water mist-type nozzle to lower concentrations of ammonia.

3. Highly flammable or explosive substances (e.g., propane, ethylene) must be used with strict controls, safety equipment, and administrative controls. If a refrigerant escapes, action should be taken for removal of the contaminant from the premises. If ventilation is used, exhaust from the floor area must be provided for heavier-than-air gases and from the ceiling for lighter-than-air gases.

Sulfuric Acid. Sulfuric acid is a strong dehydrating and oxidizing agent often found in pickling operations. It is rapidly destructive to tissues, producing severe burns. Concentrated solutions rapidly destroy any body tissue with which they may come in contact. Contact with the eye will result in almost immediate severe damage. Blindness may result if not promptly treated.

The inhalation of concentrated vapor or mist from hot acid will cause damage to the upper respiratory tract and possibly to the lung tissue. Continued inhalation of mist may cause a chronic inflammation of the upper respiratory tract and chronic bronchitis. Dermatitis may result from repeated contact with dilute solutions.

Sulfuric acid is not flammable. In high concentrations it may cause ignition of combustible materials by contact. The acid reacts with most metals to produce highly flammable hydrogen. The acid in its concentrated form reacts violently with organic materials and water, with evolution of heat. To determine if a hazardous mixture of hydrogen (produced by the action of the acid on metal) and air exists, commercial gas indicators are available.

Containers of acid should be isolated from organic or combustible materials or oxidizers such as nitrates, carbides, chlorates and metallic powders. Because hydrogen may be generated inside a drum or metal storage tank containing sulfuric acid, all means of ignition should be kept from these containers. Spills should be removed quickly by flushing the contaminated area with a large amount of water or by covering the area with dry sand, ashes or gravel if no water is available. Any remaining traces of the acid should be neutralized with soda ash or lime. Sulfuric acid is highly corrosive to most metals. NFPA 49, Hazardous Chemicals Data, provides safety information on this acid.

12.6 Labeling Hazardous Materials

Information concerning all hazardous materials used should be available to supervisors and workers alike. Proper labeling is a fundamental part of a safe and effective operation. Labeling has two principal functions:

- adequate identification, that is, what the material is and where it came from
- precautionary information for safe handling of all chemicals having significant material hazard

Purchased materials always have either the chemical or common, name of the materials, the name of the manufacturer, and a lot number. If the chemical is flammable, a flash point may be shown on the label accompanied by a precaution regarding fire hazard. If the material is corrosive, toxic, reactive, or unstable, other precautionary information is usually shown.

Absence of precaution on the label does not necessarily mean that there is no hazard. Some manufacturers are not diligent in providing such information and some purchasing agents are not conscientious in channeling information to the supervisor.

Absence of labels can be very dangerous. When hazardous materials are found in containers with no identifying label or with a label illegible from contact with chemicals, such containers must be removed until they can be identified and labeled. If such identification cannot be made, the material should be disposed. Liability suits are often based on inadequate labeling of a material involved in an accident. Good labeling practice helps protect the business from such litigation.

The Federal Hazardous Substances Act, now administered by the Consumer Product Safety Commission, requires precautionary labeling on all flammable, corrosive, reactive, toxic, or radioactive substances intended for non-industrial use.

The U.S. Department of Transportation requires certain shipping labels on packages of hazardous materials carried interstate. The Occupational Safety and Health Act has a general duty clause requiring an employer to provide a safe place of employment. Unlabeled or inadequately labeled hazardous materials could be construed as a violation of this act.

The National Fire Protection Association, in its NFPA 704M, Standard System for the Identification of the Fire Hazards of Materials, has developed a popular hazardous materials identification system. The NFPA System relies on a diamond-shaped diagram divided into four parts (see Figure 12.1). The top segment of the figure indicates the flammability hazard and is colored red. The segment to the left of the fire hazard indicates the health hazard of the material and is colored blue. The segment to the right of the fire hazard indicates the reactivity hazard and is colored yellow. The bottom segment is reserved to identify special information of which the user should be aware. For example, a W with a line drawn through it (W̶) indicates unusual reactivity with water. Oxidizing chemicals are identified by an OXY in the bottom segment, and radiation hazards are identified with the radiation symbol.

The hazard rating for each category of hazard pertaining to a given material is indicated by a number in the appropriate segment. The number scale ranges from 0 to 4, with materials designated as 0 presenting little or no hazard and materials designated as 4 presenting extreme hazard (see Table 12.4). It should be noted that the health hazard designations refer only to the immediate acute effects of exposure to the chemical. The chronic long-term health effects are not taken into account.

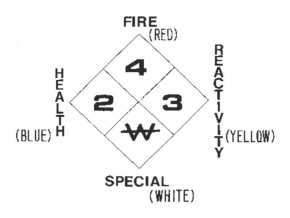

FIG 12.1 Hazardous Material Identification System.

What meaning do the health hazard designations have for shop operations? Because materials designated 4 and 3 present extreme danger, it is most unlikely that they will be present in the shop. A2 indicates materials which are hazardous to health; their use requires a full-faced mask which provides eye protection and a self-contained breathing apparatus. A1 indicates materials where a self-contained breathing apparatus (e.g., an approved canister type gas mask) may be desirable. A0 indicates that persons working with the material require no special clothing.

The NFPA hazard symbol is a method by which the employee can see at a glance the dangers presented by a particular substance. It helps to ensure proper storage and use of chemicals and solvents. Some manufacturers include the diagram on their labels. In order to make his own hazard symbol, the supervisor will need to refer to label statements and detailed Material Safety Data Sheets.

The Material Safety Data Sheet is a form by which manufacturers transmit hazard information to users. It contains relevant information about the physical and toxicological properties of the chemical. Labels generally list only the major components of chemical reagents or solvent mixtures. Therefore, the employee must depend on the MSDS sheet for information about chemical components which, though minor, may be toxic or dangerous.

Manufacturers and suppliers have developed MSDS. At the time that hazardous materials are purchased, the manufacturer should be asked to furnish toxicity and hazard information for each item. The MSDS forms should be distributed to supervisors, maintenance personnel and members of the Shop Safety and Health Committee.

TABLE 12.4
NFPA Hazard Coding System

Identification of Health Hazard Color Code: BLUE		Identification of Flammability Color Code: RED		Identification of Reactivity (Stability) Color Code: YELLOW	
Signal	Type of Possible Injury	Signal	Susceptibility of Materials to Burning	Signal	Susceptibility to Release of Energy
4	Materials which on very short exposure could cause death or major residual injury even though prompt medical treatment were given.	4	Materials which will rapidly or completely vaporize at atmospheric pressure and normal ambient temperature, or which are readily dispersed in air and which will burn readily.	4	Materials which in themselves are readily capable of detonation or of explosive decomposition or reaction at normal temperatures and pressures.
3	Materials which on short exposure could cause serious temporary or residual injury even though prompt medical were given.	3	Liquids and solids that can be ignited under almost all ambient temperature conditions.	3	Materials which in themselves are capable of detonation or explosive reaction but require a strong initiating source or which must be heated under confinement before initiation or which react explosively with water.
2	Materials which on intense or continued exposure could cause temporary incapacitation or possible residual injury unless prompt medical treatment is given.	2	Materials that must be moderately heated or exposed to relatively high ambient temperatures before ignition can occur.	2	Materials which in themselves are normally unstable and readily undergo violent chemical change but do not detonate. Also materials which may react violently with water or which may form potentially explosive mixtures with water.
1	Materials which on exposure would cause irritation but only minor residual injury even if no treatment is given.	1	Materials that must be preheated before ignition can occur.	1	Materials which in themselves are normally stable, but which can become unstable at elevated temperatures and pressures or which may react with water with some release of energy but not violently.
0	Materials which on exposure under fire	0	Materials that will not burn.	0	Materials which in themselves are normally stable,

In addition, the labels should contain the manufacturer's name and the lot number of the chemical and the age of the material. The first item is important because manufacturers sometimes need to recall batches of chemicals. When the manufacturer's name is on the label, the substance under question can be located quickly. Because some chemicals become unstable or ineffective with age, it is also necessary to identify the purchase date.

It may not be practical to put all the desired information on the label, especially if the container is small. The supervisor must use good judgment in selecting what information should appear. The most essential item is the chemical or product name, which never should be omitted. Figure 12.2 is a sample label which includes the information discussed in this section.

CHEMICAL NAME
_____

FORMULA   FLASHPOINT
_____

KINDS OF HAZARDS   CORROSIVE TO SKIN AND EYES ___   FLAMMABLE   ___
                      TOXIC BY INGESTION   ___   REACTIVE WITH WATER ___
                      TOXIC BY INHALATION   ___   UNSTABLE   ___
                      TOXIC BY SKIN CONTACT   ___   RADIOACTIVE   ___
_____

SEVERITY OF   HEALTH   0 1 2 3 4
HAZARDS   FIRE   0 1 2 3 4
                      REACTIVITY   0 1 2 3 4
                      SPECIFIC   _____
_____

EMERGENCY
TREATMENT

_____

SUPPLIER

_____

DATE   PREPARED _____   PURCHASED _____   RECEIVED _____
_____

USAGE

_____

OTHER
INFORMATION
_____

FIG. 12.2 Sample Label.

12.7 Disposal of Hazardous Materials

Toxic and hazardous wastes can be both difficult and expensive to dispose. The disposal of hazardous materials is strictly regulated by EPA at both federal and state levels. When a supervisor orders workers to dump hazardous materials into the sink and to flush them down or to pour toxic materials onto the ground, he may be violating applicable state or municipal laws. Municipal, state, and federal laws all regulate the discharge of toxic and hazardous materials into sewage systems, receiving waters, and the atmosphere.

Industrial shops have two basic disposal alternatives open to them. They can handle the disposal themselves, or they can pay a commercial disposal service. The first alternative is time-consuming, and the second is relatively expensive.

For assistance in learning what regulations apply and what disposal alternatives are best suited to their situation, supervisors should seek the help of the Environmental Protection Agency and other appropriate local and state government offices.

12.8 Toxicity

Toxicity is the capacity of a material to produce injury or harm. A hazard is the possibility that exposure to a material will cause injury when a specific quantity is used under certain conditions. The key elements to be considered when evaluating a health hazard are the:

- amount of material required to be in contact with a body cell in order to produce an injury
- total time of contact necessary
- probability that the material will be absorbed or come in contact with body cells
- rate of generation of airborne contaminants
- control measures in use

Not all toxic materials are hazardous. The majority of toxic chemicals are safe when packaged in their original shipping containers or contained in a closed system. As long as toxic materials are adequately controlled, they can be used safely. For example, many solvents, if not used properly, will cause irritation to eyes, mouth and throat. They are also intoxicating and can cause blistering of the skin and other forms of dermatitis. Prolonged exposure may cause more serious illness. But, if the solvents are used in a well-ventilated area and the worker is provided with protective equipment which prevents the substance from coming in contact with skin, then they can be used without being a hazard.

In this case, certain controls including ventilation and protective equipment have enabled the worker to use the solvent while minimizing the risk of illness. However, it must be understood that nontoxic materials should be substituted for toxic ones whenever possible.

The toxic action of a substance can be divided into acute and chronic effects. Acute effects involve short-term high concentrations which cause illness, irritation or death. They are the result of sudden and severe exposure, during which the substance is rapidly absorbed. Acute effects are usually related to an accident which disrupts ordinary processes and controls. For example, sudden exposure to a high concentration of zinc oxide fumes in the welding shop can cause acute poisoning.

Chronic effects involve continued exposure to toxic substances over long time periods. When the chemical is absorbed more rapidly than the body can eliminate it, accumulation in the body begins. Since the level of contaminant is relatively low the effects, though serious and irreversible, may go unnoticed for long periods of time. For example, breathing even low concentrations of carbon monoxide for long periods of time can cause damage to the heart muscles and blood vessels.

12.9 Sources of Information

Frequently, accidents involving hazardous materials occur because neither the supervisor nor the worker knows or is able to anticipate the effects of a particular chemical combination or the toxicity of a chemical compound. One of the primary goals of any safety and health program should be to minimize the frequency and severity of accidents which result from a lack of knowledge.

Information about chemical reactions and incompatible chemical compounds and elements should be used by supervisors whenever necessary to prevent bodily injuries and property damage resulting from unexpected chemical reactions. The reference publications should be consulted in case of doubt about a particular reaction and are excellent sources of information on hazardous chemical reactions.

Though safety hazards have been the subject of concern among employees for many years, many of the common chemicals found in the workplace have been largely ignored though they may present significant health hazards. For example, both benzene and asbestos have been designated as carcinogens.

The proper interpretation of toxicity information and data may require expertise that is beyond that ordinarily possessed by supervisors. Expert advice should be sought when needed either through the appropriate state officials or through the National Institute for Occupational Safety and Health, OSHA or EPA.

12.10 Modes of Entry for Toxic Materials

In order for a hazardous substance to exert its toxic effect, it must come into contact with a body cell. There are three modes of entry for chemical compounds in the form of liquids, gases, mists, dusts, fumes and vapors:

- ingestion (through the mouth)
- skin absorption
- inhalation (through the lungs)

Each of these modes of entry will be examined, with particular attention paid to dermatitis from skin contact and respiratory ailments arising from inhalation of toxic chemicals.

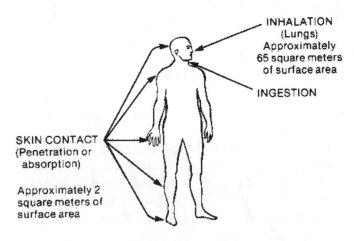

Fɪɢ. 12.3 Modes of Entry.

12.10.1 Ingestion as a Mode of Entry

Of the three modes of entry, ingestion is the least common. A harmful amount of toxic material can be swallowed accidentally, but this is uncommon. The likelihood of a worker ingesting toxic materials because of careless washing before eating, drinking, or smoking is much greater.

Inhaled toxic dusts can also be ingested in amounts that may cause trouble. Particles in inspired air which are insoluble in the mucous of the respiratory tract may be carried to the mouth where they are either spit out or swallowed. If they are readily soluble in body fluids, toxic materials can be absorbed in the digestive system and picked up by the blood

The fact that a substance has been swallowed does not necessarily mean that it will be absorbed. A certain selectivity tends to prevent absorption of unnatural substances or to limit the amount absorbed through the walls of the digestive tract. Materials not absorbed are eliminated directly through the intestinal tract. Food and liquid in the digestive tract dilute the toxic substance and may react with it to produce a harmless or insoluble substance. If the toxic substance is absorbed by the bloodstream, it will pass to the liver which may alter and detoxify it but possibly be damaged itself in the process.

The basic detoxification process involves the following steps:

- deposit in the liver
- conversion to a nontoxic compound
- transportation to kidney by way of the bloodstream
- excretion through kidney and urinary tract

12.10.2 Skin Absorption as a Mode of Entry

The second mode of entry for hazardous materials is through skin absorption. Some substances are absorbed by way of the openings for hair follicles while others dissolve in the fats and oils of the skin. Of all occupational diseases, skin ailments are the most frequent. Dermatosis is the name given to any disease of the skin and dermatitis refers to any inflammation of the skin.

The skin is composed of two layers, the epidermis and the dermis. The epidermis is of primary concern in the occupational dermatoses. The first layer of the epidermis is the lipid or fatty surface, sometimes called the acid mantle because it has an acid pH and the capacity to buffer certain weak alkaline substances. This lipid surface is composed of oil and sweat and is easily washed off.

Under the thin layer of lipids lies the stratus cornea or keratin layer. This layer is the chief barrier against water and aqueous solution and provides fair protection against attack by chemicals except for alkalis. Because it is composed of dead skin cells, this layer is constantly being replaced by new cells pushing to the surface from the deeper layer of the epidermis, the germinative layer.

The dermis, or "true skin," is tough and resilient. It contains connective tissue and, if injured, can form new tissue to repair itself. The dermis is the main natural protection against trauma.

12.10.2.1 Causes of Skin Irritation

There are five important causes of occupational dermatoses:

1. plants
2. biological agents
3. physical agents
4. mechanical agents
5. chemical agents

Plant poisons, which can cause dermatitis, are produced by several hundred plants, of which the best known are poison ivy, poison oak and poison sumac. Dermatitis from these three sources may result from bodily contact with clothing or other objects which previously have been exposed to the poison. Some woodworking shops have reported cases of dermatitis when worker were working with West Indian mahogany, silver fir and spruce, especially when sandpapering and polishing.

Biological agents, which cause dermatitis may be bacterial, viral, fungal, or parasitic. These agents are not often encountered in the workplace.

Physical agents, which lead to dermatosis include heat, cold, and radiation. Hot water softens the skin so that substances attack it more readily. X-rays and other ionizing radiation may cause dermatosis, severe burns and even cancer. Prolonged exposure to sunlight, the most common source of sun-damaging radiation, produces skin changes which may cause dangerous body alterations.

Mechanical causes of skin irritation include friction, pressure, and trauma, including abrasions. If the horny layers of the skin become softened by such mechanical causes or by high temperatures and excessive perspiration, dermatosis may result. Friction may result in an abrasion or, more commonly, a callus.

Chemical agents are the most frequent cause of dermatoses. Such chemicals can be divided into two groups: primary irritants and synthesizers.

Primary irritants react on contact, altering the chemistry of the skin. Such irritants cause a chemical reaction which may vary from complete destruction to burning to inflammation or irritation. About 80 percent of all occupational dermatoses are caused by primary irritants. The following are the five main categories of primary irritants.

1. *Organic solvents* irritate the skin because of theft solvent qualities. They remove the natural oil from the skin, leaving it dry, scaly, and subject to cracking and infection. Examples of such organic solvents are trichloroethylene, acetone, and petroleum distillates used for degreasing and hand cleaning operations.
2. Detergents also remove the natural oils from the skin or react with the oils of the skin to increase susceptibility to chemicals which do not ordinarily affect the skin. Such detergents include the alkalis and soap.
3. *Desiccators*, hygroscopic agents and anhydrides take water out of the skin and generate heat. Examples are sulfuric acid, potash, and sulfur dioxide.
4. *Protein precipitants* tend to coagulate the outer layers of the skin. Examples of protein precipitating agents are alcohol, tannic acid, formaldehyde, and phenol and such heavy metal salts as the salts of arsenic, chromium, mercury, and zinc.
5. *Oxidizers* unite with hydrogen and liberate nascent oxygen on the skin. Examples include chromic acid, hydrogen peroxide, chlorine, and ozone.

12.10.2.2 Sensitizing Agents

Some chemical substances produce no irritation on initial skin contact. After repeated or extended exposure, however, some individuals will develop an allergic reaction termed sensitization dermatitis. A once-immune person suddenly may develop an allergic reaction to a particular substance.

This allergic reaction may come in the form of small pimples or water blisters and look like contact dermatitis. However, the skin reaction may occur at a site quite different from that where contact occurred. The difficulty comes when determining just which agent is causing the sensitization. Once a person has become sensitized to a given material, the only sure way to prevent future allergic reactions is to avoid the product in the future. In some cases, medication and desensitization by a physician may be a solution.

Typical sensitizers include some epoxy resin amine hardeners, certain metals (e.g., chromium, nickel), and coal tar derivatives. Some substances act as both primary irritants and sensitizing agents. These include such organic solvents as turpentine and the chlorinated phenols, chromic acid, formaldehyde, and epoxy resin components.

12.10.2.3 Preventing Dermatosis

The best way to control dermatosis is to prevent skin contact with the offending agents. Where possible, chemicals of low toxicity and irritant potential should be substituted for more toxic chemicals. Contact can be minimized by good personal hygiene, personal protective equipment, and barrier creams. Frequent and thorough washing with appropriate cleansers removes irritating substances before they can cause trouble. Workers should be told where, how, and when to wash.

Wherever solvents are used, good personal hygiene is important. Spills and splashes should be cleaned up immediately with soap and water. Clothing which has been splattered by chemical solvents should be replaced with clean clothing.

In the case of local contact with acid, the most important treatment is immediately to wash the affected area with large amounts of water. For minor contact, such as a splash on the hand, running water from a tap may be sufficient, provided enough time is allowed for complete removal of the acid. In the case of large area contact, readily accessible, well-marked, rapid-action safety showers are required, and a minimum of fifteen minutes should be devoted to removing all traces of acid.

A special eye-washing fountain or bath also should be provided. Adequate time for removal of acid should be allowed. This may necessitate repeated fifteen-minute washings while medical attention is being sought.

The type of soap used is important. Even a good soap may cause irritation on certain types of skin. According to the National Safety Council, a soap should:

- be freely soluble in hard or soft, cold, or hot water
- remove fats, oils and other soil without harming the skin
- leave the natural fats and oils in the skin
- contain no harsh abrasives or irritant scrubbers

- be handy to use if in cake form or flow easily through soap dispensers if in granulated or powder form
- stay insect-free
- retain its properties in use

Workers sometimes contract dermatitis by washing their hands in the wrong kinds of solutions. They should not be permitted to use solvents, even if they may feel that these cleaning agents are the easiest to use and work the fastest. They should be required to use the soap-dispensing units provided.

To keep dermatitis-causing agents from contact with the skin, the worker may need impermeable protective clothing such as aprons, face shields, and gloves. But the worker must be aware that many solvents can penetrate rubber or neoprene gloves. Neoprene protects against most common oils and aliphatic hydrocarbons, but neither it nor rubber offers protection against the aromatic and halogenated hydrocarbons, the ketones, and many other solvents. For such uses, polyvinyl alcohol gloves provide protection, but these gloves must be kept away from water, acetone, and other solvents.

Regular periodic cleaning and drying of gloves is as important as using the proper type. Any equipment that cannot be decontaminated should be discarded.

Barrier creams are the least effective way of protecting the skin. They are not a substitute for gloves, except where there is only occasional and minor contact with a solvent. Barrier creams and lotions should be used to supplement good personal hygiene and personal protective equipment. Three main types of barrier creams and lotions are available:

1. Vanishing cream, which contains soap and emollients that coat the skin and cover the pores to make subsequent cleanup easier.
2. Water-repellant types (e.g., lanolin, petrolatum, organic silicones), which leave a water-insoluble film on the skin and thus repel water-dissolved irritants.
3. Solvent-repellent types (e.g., methyl cellulose, sodium silicate and tragacanth), which are insoluble in oils and solvents.

Barrier creams should be applied to clean skin. When the skin becomes soiled, both the cream and soil should be washed off and the barrier reapplied.

12.10.3 Inhalation as a Mode of Entry

The third mode of entry for chemical agents is inhalation. It is a particularly important mode of entry because of the rapidity with which a toxic material can be absorbed in the lungs, passed into the bloodstream and reach the brain. Had the same material been ingested instead of inhaled, it would have been considerably diluted with the contents of the stomach. Inhalation hazards arise from excessive concentration of mists, vapors, gases, or solids that are in the form of dusts or fumes.

12.10.3.1 Anatomy of the Respiratory System

The respiratory system consists of three main parts (see Figure 12.4):

1. Air passages or airways, including the nose, mouth, upper throat, larynx, trachea and bronchi.
2. The lungs, where oxygen is passed into the blood and carbon dioxide is given off.
3. The diaphragm and the muscles of the chest, which permit normal respiratory movement.

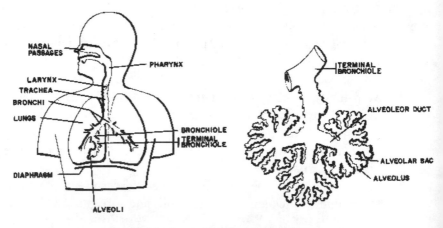

Fɪɢ. 12.4 Anatomy of the Respiratory System.

The respiratory system brings air containing oxygen into the lungs where oxygen transfers to the blood. The oxygen-enriched blood travels to all parts of the body to sustain life. At the same time, carbon dioxide transfers from the blood to the lungs and then out of the system. The system that delivers the air to the alveoli is extremely important to the healthy functioning of the lungs.

The nasopharyngeal passages serve as a heat exchanger and humidifier, warming and moisturizing inhaled air so that it will not constitute a shock to the delicate lung tissues. Some particles are removed by nasal hairs (cilia) at bends in the air path and by sedimentation.

These passages, as well as the trachea and bronchial tubes are covered with mucous membranes. The mucus secreted by these, membranes gives up heat and moisture and serves as a trap for particulate contaminants before they can reach the lungs. It also dilutes irritating substances.

The bronchi branching from the trachea divide into smaller and smaller bronchioles, forming an increasingly difficult route of entry for particulate matter. The respiratory bronchioles lead into several ducts, each of which ends in a cluster of air sacs, the alveoli.

Oxygen transfer takes place in the alveoli. These minute sacs are like clusters of grapes covered with a network of blood capillaries. The alveolar walls are membranes through which gases and even aerosols can pass easily. Through the thin walls of the alveoli, the blood takes on oxygen and gives up its carbon dioxide in the process of respiration.

12.10.3.2 Respiratory Hazards

Depending upon the solubility of the material, inhalation of chemical agents may irritate the upper respiratory tract, including the mucous membranes, or it may harm the terminal passages of the lungs and air sacs. Inhaled contaminants fall into three general categories:

1. Particulates which, when deposited in the lungs, may produce rapid local tissue damage, slower tissue reaction, disease or physical plugging (e.g., asbestos fiber).
2. Toxic vapors and gases that produce adverse reaction in the tissue of the lungs (e.g., hydrogen fluoride).
3. Toxic vapors and gases that do not affect the lung tissue locally but may either pass from the lungs into the bloodstream, where they are carried to other body organs (e.g., cadmium oxide fumes, solvents) or adversely affect the oxygen-carrying capacity of the blood cells themselves (e.g., carbon monoxide).

Dusts are solid particles generated by handling, crushing, grinding, rapid impact, detonation, and breaking apart by heating of organic and inorganic material. Dust normally contains a wide range of particle sizes, with the small particles greatly outnumbering the large ones. Consequently, when dust is noticeable in the air near an operation, probably more invisible dust particles than visible ones are present. A process which produces dust fine enough to remain suspended in the air long enough to be breathed should be regarded as hazardous unless it can be proved safe.

The respiratory tract, with its successive branches and passageways, is a highly efficient dust collector. All particles which enter the respiratory system and are smaller than four or five microns are deposited somewhere in the system. Larger particles, those greater than two microns in size, are deposited in the upper respiratory system (the nasal cavity, trachea, bronchial tubes and other air passages). Intermediate particles are deposited about equally in the upper respiratory system and in the alveolar or pulmonary air spaces. Smaller particles of one micron or less in size are generally deposited in the alveolar spaces.

Fumes are forms of particulate matter which differ from dusts only in the way they are generated and in their particle size. A fume consists of extremely small particles, less than a micron in diameter, and is generated by such processes as combustion, condensation, and sublimation.

Arc welding volatizes metal vapor that condenses in the air around the arc. These fumes, because they are extremely fine, are readily inhaled. Toxic fumes formed when welding galvanized metal may produce severe symptoms of toxicity. Either the fumes must be controlled by the local exhaust ventilation or the welder must be protected by respiratory equipment.

The proper evaluation of dust and fume exposures requires knowledge of the chemical composition, particle size, concentration in the air, and length of exposure time.

Potential health hazards from dust and fumes occur on three levels:

1. Regardless of its chemical composition, the inhalation of sufficient quantities of dust can cause a person to choke or cough.
2. Depending upon its chemical composition, the dust can cause allergic or sensitization reaction in the respiratory tract or on the skin.
3. Depending upon both its size and chemical composition, the dust can damage vital internal tissues.

Gases are formless fluids that expand to occupy the space in which they are confined. Examples are arc welding gases, sulfur dioxide, phosgene, ozone, and nitrogen dioxide. Vapors are the volatile form of substances that are normally in the solid or liquid state at room temperature and pressure.

Some highly reactive gases and vapors of low solubility can produce an immediate irritation and inflammation of the respiratory tract. They can cause chemical pneumonia and pulmonary edema. When lung tissue is burned, the injured tissue pours out fluid from the bloodstream. When this fluid accumulates in the lungs, oxygen exchange cannot take place and suffocation occurs. This condition is known as pulmonary edema.

12.11 Threshold Limit Values

Individual susceptibility to respiratory toxins is difficult to assess. Nevertheless certain safe limits can be established. A Threshold Limit Value (TLV) refers to airborne concentrations of substances and represents an exposure level under which most people can work, day after day, without adverse effect. Because of wide variations in individual susceptibility, an occasional exposure of an individual at or even below the threshold limit may not prevent discomfort, aggravation of a preexisting condition, or occupational illness. The term TLV refers specifically to limits published by the American Conference of Governmental Industrial Hygienists. These TLV limits are reviewed and updated each year. There are three categories of Threshold Limit Values:

1. Time-Weighted Average (TLV-TWA) is the time-weighted average concentration for a normal eight-hour day or forty-hour week. Nearly all persons may be exposed day after day to airborne concentrations at these limits without adverse effect.
2. Short-Term Exposure Limit (TLV-STEL) is the maximum concentration to which persons can be exposed for a period of up to fifteen minutes continuously without suffering:
   - irritation
   - chronic or irreversible tissue change
   - narcoses of sufficient degree to increase accident proneness, impair self-rescue, or materially reduce work efficiency. No more than four fifteen-minute exposure periods per day are permitted, with at least sixty minutes between exposure periods.
3. Ceiling (TLV-C) is the concentration that should not be exceeded even instantaneously.

The following points should be kept in mind when dealing with TLVs:

- Concentrations of chemicals rarely remain constant in the workplace.
- Most environments contain mixtures of chemicals rather than single compounds.
- Because individual susceptibilities vary, control measures must be provided for those persons whose sensitivity places them outside the average.

The first compilation of health and safety standards from the U.S. Department of Labor's OSHA appeared in 1970. Because it was derived from then-existing standards, it adopted many of the TLVs established in 1968 by the American Conference of Governmental Industrial Hygienists. Thus Threshold Limit Values, a copyrighted trademark of the American Conference of Governmental Industrial Hygienists, became, by federal standards, Permissible Exposure Limits (PELs). These PELs represent the legal maximum level of contaminants in the air of the workplace.

The methods of controlling respiratory hazards include those that have been outlined for controlling other health hazards: substitution, isolation, housekeeping, personal protective equipment and ventilation.

12.12 Appraising Health Hazards

Before recommendations for controls can be made, a thorough evaluation of health hazards is mandatory. Health hazard evaluation proceeds in two stages:

1. A preliminary survey is made to determine which operations and environmental physical agents may be hazardous.
2. A detailed study, including the use of air sampling and direct reading instruments, is made to determine the amount of air contaminants present; and the extent of exposure to physical agents.

These studies usually are made by an industrial hygienist or others specifically trained in this field. However, a skilled supervisor can be of valuable assistance to the industrial hygienist.

12.12.1 The Preliminary Survey

A preliminary survey is the first step in evaluating the shop environment. Following this procedure can save time and effort. By visual inspection, an experienced person can indicate those operations or conditions for which detailed studies are needed.

During this survey, processes and operations may be observed in which potentially harmful materials are handled, or equipment may be used in a manner that could result in excessive concentrations. The preliminary survey should consider:

- general sanitation
- raw materials, products and by-products
- physical agents
- control measures in use

One of the best guides on the subject of general sanitation is ANSI 24.1, Minimum Requirements for Sanitation in Places of Employment. This publication contains definitions of general requirements on waste disposal, housekeeping, ventilation and so on.

New substances and materials which require close evaluation to assess their potential health hazard are constantly being introduced. It is important to have a list of all substances and materials used in the workplace. The information will be obtained from the purchasing agent or the manufacturer. After the list is obtained, it is necessary to determine which of the materials are toxic and to what degree. Only when the toxicity and hazardous properties of the substances are known can the necessary safeguards be installed.

The preliminary survey should note sources of radiant heat, abnormal temperature and humidity, excessive noise, improper or inadequate illumination, and X-rays and gamma rays. The use of special instruments is necessary properly to evaluate these potential hazards.

The preliminary survey would not be complete unless the following types of control measures in use and their effectiveness were noted:

- local exhaust: effectiveness, hood design
- general ventilation
- respiratory protection devices
- protective clothing
- shielding from radiant or ultraviolet energy

Serious doubts about the effectiveness of the existing equipment would arise if dust was on the floor, holes were in duct work, fans were not operating and personal protective devices were being used improperly by the worker.

The health hazard of a given chemical or physical agent depends on several factors, including the:

- nature of the substance or agent
- intensity of the exposure
- duration of exposure
- individual susceptibility

The first three items will be determined during the preliminary study. The fourth item requires detailed studies. During the preliminary survey and detailed study, various sampling instruments are used.

12.12.2 Sampling

Where should the sampling begin? Should the sample be taken at the worker's breathing zone, out in the general air, or at the machine or process that is putting out the toxic gas or dust? The answer is that air at all three sites should be sampled. Should the sample be taken for two seconds, two hours, or a whole day? A combination of short- and long-term samples is the best way to get representative information.

There are two major types of samples:

1. The grab sample is taken over a short a period of time so that the atmospheric concentration is assumed to be constant throughout the sample. This time period is usually less than five minutes and will cover only part of an industrial cycle. Frequently, a series of grab samples will be taken in an attempt to define the total exposure.
2. The long-term sample, taken over a sufficiently long period of time so that the variations in exposure cycles are averaged. This kind of sample gives peak exposure concentrations only with very sophisticated recording equipment, and then only for certain materials.

Neither the grab nor long-term sample alone is sufficient; a combination of both must be used.

12.13 Methods of Controlling Health Hazards in the Workplace

Generally, health hazard control may be grouped into nine classifications:

1. Substitution of a less harmful material for one which is dangerous to health.
2. Change or alteration of a process to minimize worker contact.
3. Isolation or enclosure of a process or work operation to reduce the number of persons exposed.
4. Wet methods to reduce generation of dust in operations.
5. General or dilution ventilation with clean air to provide a safe atmosphere.
6. Local exhaust at the point of generation or local dispersion of contaminants.
7. Personal protective devices.
8. Good housekeeping, including cleanliness of the workplace, waste disposal, adequate washing, toilet and eating facilities, healthful drinking water and control of insects and rodents.
9. Training and education.

Replacement of a toxic material with a harmless or less toxic one is a practical method of eliminating a health hazard. In many cases a solvent with a lower order of toxicity or flammability may be substituted for a more hazardous one. For example, carbon tetrachloride can be replaced by such solvents as methyl chloroform, dichloromethane, or a similar substance. Wherever possible, detergent and water cleaning solutions should be considered for use in place of organic solvents.

A change in process often offers an ideal opportunity to improve working conditions. In some cases, a process can be modified to reduce the exposure to dust or fumes and thus markedly reduce the hazard. For example, brush-painting or dipping instead of spray painting will minimize the concentration of airborne contaminants from toxic pigments.

Some potentially dangerous operations can be isolated from nearby workers. Isolation can be accomplished by a physical barrier (such as sound-absorbing screens to reduce the noise from a piece of machinery), by time (such as providing semi-automatic equipment so that a person does not have to stay near the noisy machine constantly), or by distance (remote controls). Enclosing the process or equipment is a desirable method of control since the enclosure will prevent or minimize the escape of the contaminant or physical energy into the shop atmosphere.

Dust hazards frequently can be minimized by application of water or other suitable liquid at the source of dust when better methods, such as vacuum cleaning, cannot be applied.

General ventilation is an effective control for areas generating low concentrations of hazardous substances. It works by adding air to keep the concentration of a contaminant below hazardous levels. It uses either natural convection through open doors, windows, roof ventilators, and chimneys or artificial air currents produced by fans or blowers. Exhaust fans through roofs, walls, or windows constitute positive all-season dilution ventilation. Consideration must be given to providing make-up air, especially during winter months. Dilution ventilation is practical only when the degree of air contamination is not excessive and particularly when the contaminant is released at a substantial distance from the worker's breathing zone. Under other conditions the contaminated air will not be diluted sufficiently before inhalation.

General ventilation should not be used where there are major, localized sources of contamination, especially highly toxic dusts and fumes. Local exhaust is more effective and economical in such cases. When comparatively small amounts of the less toxic solvents are vaporized, general or dilution ventilation can be a satisfactory method of control. Several points must be kept in mind when using dilution ventilation:

1. Exhaust openings should be located as close as possible to the source producing the contaminant.
2. To keep the contaminants out of the breathing zone, the fresh air applied to the work space should first pass through the worker's breathing zone, then across the work space where the contaminant is produced and into the exhaust system as rapidly as possible.
3. Unless the exhausted air is discharged far away from the fresh air intake duct, the fresh air can become contaminated.

Examples of possible arrangements, both good and bad, for dilution ventilation of work space are depicted in Figure 12.5.

Local exhaust ventilation is the most effective means of control for airborne contaminants produced by welding or cutting. A local exhaust system works by trapping the air contaminant near its source so that a worker standing near the process is not exposed to harmful concentrations. This method is usually preferred to general ventilation, but should be used only when the contaminant cannot be controlled by substitution, changing the process, isolation or enclosure. Even though a process has been isolated, it still may require a local exhaust system. After the system is installed and set in operation, its performance should be checked to see that it meets engineering specifications, correct rates of air flow, duct velocities, and negative pressures. Its performance should be re-checked periodically as a maintenance measure.

GENERAL VENTILATION

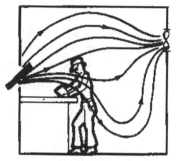

Poor General Ventilation

(Contaminant is driven into the worker's breathing zone and atmosphere)

Fair General Ventilation

(Incomplete flushing of the room; contamination of general atmosphere)

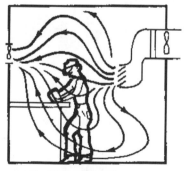

Good General Ventilation

(Air enters at breathing zone height and keep contamination away from worker)

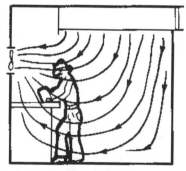

Best General Ventilation

(Low velocity diffusion through ceiling, immediate exhaust of contaminated air)

Fig. 12.5 Work Space Arrangements.

Personal protective equipment should not be used as a substitute for feasible engineering or administrative controls. However, workers who encounter physical and chemical hazards may need such personal protective equipment as hearing protectors, respirators, and special clothing, (aprons, gloves, etc.).

Good housekeeping in the shop goes a long way to reduce health hazards. Removing dust from ledges and on the floor prevents its dispersion into the shop atmosphere. A regular cleanup schedule using vacuum lines is the most effective method of removing dust from the work area. In the absence of a central vacuum cleaning system, portable vacuum cleaners should be used. An air hose for blowing away dust should not be used.

Good housekeeping is also essential where solvents are stored, handled and used. Leaking containers or spigots should be corrected immediately, either by transferring the solvent to sound containers or by repairing the spigots. All solvent-soaked rags or absorbents should be disposed of in airtight metal receptacles and removed daily from the shop.

Proper training and education are essential supplements to engineering controls. The worker must know the proper operating procedures that make engineering controls effective. If he performs an operation away from an exhaust hood, he not only will defeat the purpose of the control but also will contaminate the work area. Since new materials are constantly being marketed and new processes developed, re-education and follow-up instruction must be part of the training program.

Summary

The small business that is aware of the health hazards presented by the biological, physical, and chemical agents found in the workplace has taken the first step in limiting these hazards. By applying the control measures detailed in this chapter, the management can limit the dangers presented by radiation, vibration, and noise. They can also be sure that chemical agents in the workplace are properly labeled, used carefully, and disposed of properly.

Through the use of engineering controls such as substitution, isolation, and ventilation, workers can be protected against the three modes of entry for hazardous substances. However, engineering controls alone cannot protect against all physical and chemical hazards. The proper uses of personal protective equipment must be understood and incorporated into the overall safety and health program.

Bibliography

Accident Prevention Manual for Industrial Operations, 7th ed., National Safety Council, Chicago, Illinois 60611. 1974.

Diberardinis, Louis J. (ed.) (1999) "Handbook of Occupational Safety and Health" 2nd edition, John Wiley & Sons, N.Y., N.Y.

DiNardi, Salvatore R., (ed) (1997) "The Occupational Environmental – Its Evaluation and Control" AIHA Publication, Fairfax, VA.

Firenze, Robert J. (1978) The Process of Hazard Control, Kendall/Hunt Publishing Company, Dubuque, Iowa 52001.

13

Personal Protective Equipment

The last and sometimes the only practical way to reduce illnesses and injuries is to use personal protective equipment. Personal protective equipment (PPE) includes helmets, gloves, goggles, respirators, special footwear, and other items that protect workers against such hazards as flying particles, noise, dangerous chemicals, and electric shock. The shortcoming of personal protective equipment is major: PPE does nothing to reduce or eliminate the hazards. For this reason, such devices are normally considered as the last line of defense. PPE does not take the place of engineering controls such as substitution, isolation and ventilation. Protective equipment also creates an undue feeling of security. Its failure means immediate exposure to the hazard. Should a protective device become ineffective while being used, the wearer may be exposed to a very dangerous situation.

Standards governing personal protective equipment have been a part of the Federal Occupational Safety and Health Administration (OSHA) regulations since the original OSHA General Industry Standards were promulgated in 197 1. The OSHA PPE standards for head, face, eye, and foot protection in effect prior to July 05, 1994, were adopted from established Federal standards and national consensus standards (primarily the American National Standards Institute (ANSI) standards) that were current as of 1971.

The purpose of this chapter is to discuss the ways that personal protective equipment can reduce injuries and illnesses in the work place. The chapter will introduce the participant to factors involving the role of personal protective equipment in the total hazard control program, its selection, and the effectiveness of personal protective equipment, devices used to provide head, eye, face, hearing, respiratory, arm, hand, body, leg and foot protection, and the care and maintenance of protective devices.

13.1 Selecting Personal Protective Equipment

When a decision is made to employ protective equipment, those selecting the equipment should examine carefully six criteria.

1. The extent of the hazard's potential to cause harm must be evaluated. It is important to have a clear understanding of the nature of the hazard, where it originates, what harm it can cause and the likelihood of its occurrence. For example, if a respirator is needed to protect workers from dust, the size of the dust particles must be determined. Without such a determination, a respirator could be selected which may not perform its desired function.

2. The degree of desired protection is in direct proportion to the seriousness of the hazard. Two dangers are present in selecting PPE: the danger of "overkill" and the danger of selecting equipment, which offers inadequate protection. For example, greater face protection is required to guard against chemical splashes than is necessary for protection against the impact of flying objects. To offer the same device for both would mean either overprotection or under-protection, depending on which alternative is selected.

3. Protection must be considered along with the equipment's ability to interfere with the worker's task. Bulky gloves may decrease the dexterity required for certain tasks; full-face shields may be cumbersome during certain operations. Sometimes those selecting protective equipment must weigh the benefits of the most desirable alternative against the most practical one before making the final selection.

4. For most protective equipment (head, eye, face and foot protection, electrical protective devices), a label showing compliance with the appropriate American National Standards Institute (ANSI) standard will be affixed to the product. In the case of respiratory equipment, the National Institute for Occupational Safety and Health (NIOSH) and the Mine Safety and Health Administration (MSHA) have their label attached to the product. There are some types of protective equipment, for example, gloves and aprons for which no approvals are available.

5. Quality is an important factor to be considered at the time of purchasing protective equipment. Although many manufacturers are selling equipment which meets the ANSI standards and NIOSH and MSHA approvals, their products are not the same quality. Given the constraint of cost, equipment of the highest quality should be purchased. A product of high quality will be easier to maintain and can be expected to have a longer service life than less expensive equipment.

6. Last but not least is the cost of the equipment. Good protective equipment is not inexpensive. The highest quality equipment on the market costs proportionally more than its lower quality counterpart. When a decision must be made to determine how much will be spent on protective equipment, it is important to determine how long the equipment will be needed. If it will be used only for a short period of time while more suitable control measures are being installed, less expensive equipment may serve the purpose. However, if the protective equipment is intended to serve its purpose indefinitely, then strong consideration should be given to allocating ample funds to acquire the highest quality approved equipment.

13.2 Problems Interfering with PPE Effectiveness

When the foregoing criteria have been evaluated, the job of the purchaser is still far from over. Now consideration must be directed toward solving problems that interfere with the effectiveness of PPE.

When more than one variety of the same class of protective device is used, especially if each type has slightly different requirements for use, care and maintenance, it becomes difficult for supervisors to train workers in the use of the equipment and to provide adequate maintenance, monitoring and control. Ideally, one brand of a particular class of protective equipment should be used.

If protective equipment is to continue providing protection, it requires care and maintenance. A program of care and maintenance must be developed and must be in operation at the time when the equipment is distributed for use. Cleaning and minor repairs are usually done by the maintenance personnel. More complex repairs or repairs designated to be done by the manufacturer should not be made by in-house personnel. In such instances, the equipment should be returned to the manufacturer or the designated service representative.

Improperly fitted protective equipment discourages worker acceptance and in some cases causes a loss of protection. For instance, protective eye wear which does not fit properly may end up in the employee's pocket, and a respirator facepiece which does not seal properly on a worker's face will allow the toxic substance to enter the facepiece and be inhaled by the worker. When purchasing protective equipment, consideration should be given to the need for proper fit. In cases where a worker or supervisor cannot acquire a proper fit with conventional equipment, special fitting devices may be necessary.

Unless the persons who are required to wear the equipment are trained and educated in its necessity, proper use, care, and maintenance, protective equipment will do little to fulfill its intended purpose. Every supervisor needs to stress use, care, and maintenance as an integral part of job performance.

13.3 Head Protection

Types of head protection can be divided into three categories:

1. caps for long hair
2. bump caps
3. helmets or safety hats

Protective hair covering is necessary for workers with long hair who work at machines with exposed moving parts. Besides the obvious danger of hair becoming entangled in moving parts when the worker bends over, there is also the possibility that heavy charges of static electricity can lift hair into moving belts, rolls, gears, or rotating shafts.

Protective caps should cover the hair completely. Bandanas and turbans are not sufficient and pose their own hazards if they become loose. Although no standards have been accepted, protective fabric caps should be made of a durable, flameproof fabric rugged enough to withstand regular laundering. The cap should have a visor long enough and rigid enough to provide warning before the head itself comes into contact with a moving object, such as the spindle on a drill press.

Bump caps are thin-shelled, lightweight plastic headgear. They are not a substitute for helmets and do not provide adequate protection in the workplace. They only offer protection against bumping.

Helmets, also called safety or hard hats, are the best means for protecting workers and supervisors against impact blows and penetration from flying and falling objects and from limited electric shock. They can be designed to protect the scalp, face, and neck from overhead spills. They can also keep hair from becoming entangled in machinery and can shield the scalp from exposure to hear and irritating dusts.

The National Safety Council estimates that approximately 140,000 head injuries occur in a given year (see Table 13.1). Workers being struck by or against stationary objects or from falling materials cause most of the injuries.

Table 13.1
Part of Body Injured in Work Accidents

Body Part	# of Cases	% of Cases	% Compensation
Eyes	140,000	6	2
Head (except eyes)	140,000	6	7
Arms	210,000	9	9
Trunk	610,000	27	38
Hands	140,000	6	3
Fingers	370,000	16	9
Legs	300,000	13	13
Feet	140,000	6	3
Toes	70,000	3	1
General	180,000	8	15

Helmets are of two types, either full brimmed or brimless with peak. A brim provides the most complete protection for head, face, and back of neck. The cap type is used where a brim may interfere with a particular task.

13.3.1 Classes of Hard Hats

There are four classes of hard hats:

1. Class A, affording limited voltage resistance, for general service
2. Class B, high voltage resistance
3. Class C, no voltage protection (metallic helmets)
4. Class D, limited voltage protection, fire fighters' service only

All helmets must meet the requirements and specifications in the American National Standards Institute Z89.1, Safety Requirements for Industrial Head Protection. Helmets are identified on the inside of the shell with the manufacturer's name, ANSI designation, and the class.

Helmets (hard hats) are designed to transmit a maximum average force of not more than 850 pounds (385 kilograms) when tested in accordance with ANSI Z89.1. Hard hats that meet ANSI standards are designed with two major components, a shell and a suspension:

- The shell should be of one piece, seamless construction and designed to resist impact.
- The suspension gives the hard hat its impact-distributing abilities.

The suspension must be adjusted to fit the wearer, and a clearance of not less than 1¼" (3.18 cm.) must be maintained between the shell and the skull of the wearer.

A chinstrap will keep a hat from coming off and it affords full protection. A nape strap, provided with most helmets, helps to keep the headgear from falling off during normal use. Some headgear comes with removable type attachments for face shields and welding helmets.

13.3.2 Care and Maintenance of Hard Hats

To care for and maintain helmets effectively, the following recommendations should be followed:

1. Before each use, hard hats should be inspected for cracks and signs of impact. Once damaged, a protective helmet should be discarded. Alterations of any kind weaken the performance of the helmet.
2. Suspension systems should be inspected frequently to detect loose or torn cradle straps, broken sewing lines, loose rivets, and defective lugs. Once found, deteriorated systems should be replaced.
3. Chemicals, oils, and petroleum products must be removed from the shell as soon as possible. These agents will soften the shell materials and reduce its impact and dielectric protection. The manufacturer should be consulted as to what solvents can be used to remove such chemicals without themselves causing damage. If a hard hat needs painting, the manufacturer should be consulted so that a coating can be selected which will not harm the helmet.
4. Hard hats must be scrubbed and disinfected before being reissued to others. At least every thirty days the hat, sweatbands and cradles should be washed in warm soapy water or an approved detergent solution recommended by the manufacturer. Thorough rinsing should follow washing.
5. While in storage, hard hats should not be exposed to bright sunlight. Heat and light may reduce the degree of protection that they provide.

13.4 Eye Protection

The protection of eyes from damage by physical and chemical agents long has been an important part of a hazard control program. Some 140,000 disabling injuries occur annually, injuries that result in total or partial blindness.

In the workplace, the eyes are exposed to a variety of hazards including flying objects, splashes of corrosive liquids or molten metals, dust, and harmful radiation. OSHA requires that eye and face protection be designed to meet the performance requirements set forth in ANSI Z87.1, Practice for Occupational and Educational Eye and Face Protection.

13.4.1 Selecting Eye Protective Equipment

Before selection is made, consideration must be given to the:

- extent of the hazard to be guarded against
- ability of the eye-protective materials to afford this protection
- type of eye protection devices that fit the work objective

Table 13.2 illustrates recommended eye and face protectors for use in industry, businesses, and colleges, and indicates which type is recommended for particular operations.

13.4.2 Protection Against Flying Particles

Conditions requiring protection from flying particles caused by chipping, drilling, and grinding require the selection of one of three types of impact-protection equipment.

1. spectacles with impact-resistant lenses
2. flexible or cushion-fitting goggles
3. chipping goggles

Spectacles without side shields are not recommended because of the limited protection they provide. Normally, side as well as frontal protection is required. Full-cup side shields restrict flying particles from entering the wearer's eyes from the side. Semi- or fold-flat side shields may be used where only lateral protection is required.

Both flexible and cushion goggles have a single lens. These goggles give both frontal and side protection from flying particles.

Chipping goggles give maximum protection from flying particles. They come in two styles, either for use without other eyewear or for fitting over corrective lenses. Both types have contour-shaped rigid plastic eyecups.

TABLE 13.2
Selection Chart for Eye and Face Protectors

Operation	Hazards	Protectors (see key below)
Acetylene-Burning Acetylene-Cutting Acetylene-Welding	Sparks, harmful rays, molten metal, flying particles	7,8,9
Chemical Handling	Splash, acid burns, fumes	2 (for severe exposure add 10)
Chipping	Flying particles	1, 3, 4, 5, 6
Electric (arc) welding	Sparks, intense rays, molten metal	11 (in combination with 4, 5, 6, in tinted lenses, advisable)
Furnace Operations	Glare, heat, molten metal	7, 8, 9 (for severe exposure add 10)
Grinding-Light	Flying particles	1, 3, 5, 6 (for severe exposure add 10)
Grinding-Heavy	Flying particles	1, 3 (for severe exposure add 10)
Laboratory	Chemical splash, glass breakage	2 (10 when in combination with 5, 6)
Machining	Flying particles	1, 3, 5, 6 (for severe exposure add 10)
Molten Metals	Heat, glare, sparks, splash	7, 8 (10 in combination with 5, 6, in tinted lenses)
Spot Welding	Flying particles, sparks	1, 3, 4, 5, 6 (tinted lenses advisable; for severe exposure add 10)

KEY:
1. Goggles, flexible fitting, regular ventilation
2. Goggles, flexible fitting, hooded ventilation
3. Goggles, cushioned fitting, rigid body
4. Spectacles, without side-shields
5. Spectacles, eyecup type side-shields
6. Spectacles, semi-flat-fold side-shields
7. Welding Goggles, eyecup type, tinted lenses
8. Welding Goggles, Coverspec type, tinted lenses
9. Welding Goggles, Coverspec type, tinted plate lenses
10. Face Shield, plastic or mesh window
11. Welding Helmet

CAUTION:
Face shields alone do not provide adequate protection.
Plastic lenses are advised for protection against molten metal splash.
Contact lenses, of themselves, do not provide eye protection in the industrial sense and shall not be work in a hazardous environment without appropriate covering safety eyewear.

13.4.3 Protection Against Fumes and Liquids

Workers exposed to chemical fumes and liquids while handling volatile and corrosive chemicals and dipping in plating and pickling tanks, require chemical splash goggles or a face shield for adequate protection.

Chemical goggles have soft vinyl or rubber frames and lenses made of heat-treated glass or acid-resistance plastic. For exposures involving chemical splashes, the goggles are equipped with baffled ventilators on the side.

13.4.4 Protection Against Hot Splashing Metals

Where workers may come into contact with hot splashing metals (in casting, tinning, and pouring lead joints), cup goggles with impact-resistant lenses, metal screen face shields, or heavy plastic shields with impact-resistant spectacles beneath are required.

13.4.5 Protection Against Light Rays

Injurious rays are encountered in torch brazing, gas welding, cutting, and arc welding. Eye protection for welding operations should be chosen carefully to cope with the particular hazard as well as chemical and physical agents.

To exclude injurious flashes, hot metal and sparks, and to save the welding lenses from pitting, the worker who is flame welding, brazing or cutting should wear goggles with impact-resistant filter lenses. The goggles may have clear or colored glass depending upon the amount of exposure from adjacent welding operations.

According to federal statute, goggles or other suitable eye protection must be used during all gas welding or oxygen cutting operations. Spectacles with suitable filter lenses are permitted during gas welding operations on light work, for torch brazing or for inspection.

For electric arc welding involving exposure to ultraviolet rays, it is important that the filter lenses protect the worker from the ultraviolet rays as well as from the glare. Table 13.3 illustrates the proper shade number for eye protection in various welding operations.

13.4.6 Requirements for Eye Protectors

Protectors must meet the following minimum requirements man-dated in OSHA Safety and Health Standards (29 CFR 1910.133):

- They must provide adequate protection against the particular hazards for which they are designed.
- They must be reasonably comfortable when worn under the designated conditions.
- They must fit snugly and must not interfere unduly with the movements of the wearer. Supervisors should be trained in the fitting of goggles to assure proper fit.
- They must be durable.
- They must be capable of being disinfected. However, it is important that sterilizing solutions be removed so that skin irritation will not result.
- They must be easily cleaned.

Under conditions where goggles are apt to fog up, care must be given to the selection of eyewear designed to allow for ventilation. In extreme cases, anti-fog compounds should be used inside the lenses.

TABLE 13.3
Filter Lens Shade Number for Eye Protection

Welding Operation	Shade Number
Shielded metal-arc welding 1/16-, 3/32-, 5/32-inch diameter electrodes	10
Gas-shielded arc welding (nonferrous) 1/16-, 3/32-, 1/8-, 5/32-inch diameter electrodes	11
Gas-shielded arc welding (ferrous) 1/16-, 3/32-, 1/8-, 5/32-inch diameter electrodes	12
Shielded metal-arc welding 3/16-, 7/32-, 1/4-inch electrodes	12
5/16-, 3/8-inch diameter electrodes	14
Atomic hydrogen welding	10-14
Carbon-arc welding	14
Soldering	2
Torch brazing	3 or 4
Light cutting, up to 1 inch	3 or 4
Medium cutting, 1 inch to 6 inches	4 or 5
Heavy cutting, over 6 inches	5 or 6
Gas welding (light), up to 1/8-inch	4 or 5
Gas welding (medium), 1/8 inch to 1/2 inch	5 or 6
Gas welding (heavy), over 1/2-inch	6 or 8

13.4.7 Corrective Lenses

According to OSHA regulations, workers wearing corrective lenses must also wear one of the following:

- Spectacles with protective lenses which provide optical correction.

- Goggles that can be worn over corrective spectacles without disturbing the adjustment of the spectacles.
- Goggles that incorporate corrective lenses mounted behind the protective lenses.

Prescription eyeglasses or contact lenses worn in eye hazard areas require approved eye protectors.

The National Society for the Prevention of Blindness and the American Optometric Association advise the use of contact lenses in industry only in conjunction with proper safety shields, goggles or other protection. Two special precautions apply to wearers of contact lenses:

1. If caustic, chemical, or other solutions are accidentally splashed in the eye, the cornea may be burned. In the event of such an accident, first flush out the eye, remove the lens, and then flush the eye again.
2. If a foreign body (dust, for example) gets under a contact lens, it may scratch the cornea. In such an event, the lens should be removed immediately. It may also be necessary to consult a physician to ensure the foreign body is no longer present.

13.5 Face Protection

Several types of PPE shield the face (and sometimes the head and neck) from flying particles, chemical, or hot metal splashes and heat radiation. These types of PPE include:

1. face shields of clear plastic to protect the eyes and face of a worker who is sawing or buffing metal, sanding grinding, or handling chemicals
2. face shields with metal screens to deflect heat from a person working near furnaces or similar hear sources
3. babbitting helmets consisting of a window made of extremely fine wire screen a tilting support, an adjustable headgear, and a crown protector used to protect the head and face against splashes of hot metal
4. welding helmets, shields, and goggles to protect the eyes and face against both the splashes of molten metal and the radiation produced by arc welding
5. acid-proof hoods with a window of glass or plastic to protect the

head, face, and neck of persons exposed to possible splashes from
corrosive chemicals

6. air-supplied hoods, for work around toxic fumes, dusts, gases, or
mists

As a general rule, face shields should be worn over suitable eye protection.

13.6 Hearing Protection Devices

The need for personal hearing protection arises when source control
and/or path control are not present, when source and/or path control do not
lower noises to safe levels, or when a worker cannot avoid direct exposure
to noisy equipment and tools. There are three basic types of personal hearing
protection devices:

1. Disposable pliable material, such as fine glass wool, mineral fibers
and wax-impregnated cotton, may be inserted in the ear. It must be
replaced with new material each day.
2. Earplugs, individually fitted to the wearer, may be inserted into the
ear.
3. Cup-type protectors such as like earmuffs may be worn with the
band over the head or around the back of the neck or may be
incorporated into safety helmets.

To a great extent, selection of a protective device is governed by individual
preference. Factors to consider are effectiveness, comfort, and cost.

While cotton alone is a poor choice, if paraffin wax is mixed into the
cotton, the material becomes much more efficient. Glass down, also called
"Swedish wool," is another pliable material. Rubber and plastic types are
popular because they are inexpensive, easy to clean and give good
performance. Wax tends to lose its effectiveness because jaw movement
(talking, chewing or yawning) changes the shape of the ear canal breaking
the acoustic seal between ear and insert.

Earplugs specially molded to the individual's ear canal offer excellent
protection, generally reducing the noise reaching the ear by 25 to 30 decibels.
However, a trained, qualified person must fit them, and they should become
the property of the individual for whom they are molded. Neither of these
types should cause skin irritation or injured eardrums if they are properly
designed, well fitted, and hygienically maintained.

Cup- or muff-type devices cover the external ear to provide an acoustic barrier. They can reduce noise an additional ten to fifteen decibels. Their effectiveness if influenced by size, shape, seal material, shell mass, and type of suspension as well as by individual head size and shape. Glasses or long hair can break the seal over the ear. Combinations of earplugs and earmuffs give three to five more decibels of protection. Cup-type protectors cost more than other devices. However, this is a one-time cost and over a period of time, the cost of pliable insert material will exceed the cost of the cup-type protector.

Under dusty or dirty working conditions, cup-type protectors may be more hygienic than devices inserted in the ear. Under hot working conditions, pliable insert material or earplugs may be more comfortable than cup-type devices. It is imperative that anyone exposed to potentially dangerous noise levels have some kind of hearing protection.

13.7 Respiratory Protection

During our discussion of health hazards in a previous chapter, we examined the three modes of entry for hazardous materials: ingestion, skin absorption, and inhalation. The human respiratory system presents the quickest and most direct avenue of entry because it is intimately and inextricably connected with the circulatory system and the need to oxygenate tissue cells to sustain life processes.

The Occupational Safety and Health Administration (OSHA) and states working under an approved OSHA State Plan have established permissible exposure limits (PELs) for many airborne toxic materials. OSHA Standards (29 CFR 1910.134) state:

> In the control of those occupational diseases caused by breathing air contaminated with harmful dusts, fogs, fumes, mists, gases, smokes, sprays, or vapors, the primary objective shall be to prevent atmospheric contamination. This shall be accomplished as far as feasible by accepted engineering control measures (for example, enclosure or confinement of the operation, general and local ventilation, and substitution of less toxic materials). When effective engineering controls are not feasible, or while they are being instituted, appropriate respirators shall be used.

Respiratory protection is also required on the following special occasions:

- in oxygen-deficient atmospheres where the oxygen content in the breathable air is or could become insufficient
- for routine but infrequent operations
- for non-routine operations in which persons are exposed briefly to high concentrations of a hazardous substance during maintenance, repair, or emergency conditions

Those conducting a health hazard assessment will concentrate on locating four major categories of respiratory hazards:

1. oxygen-deficient atmospheres
2. gas and vapor contaminants
3. particulate contaminants (including dust, fog, fume, mist, smoke, and spray)
4. combinations of gas, vapor and particulate contaminants

It is important to determine what type or types of contaminants exist in order to select the most effective respiratory equipment. Because each respirator is designed to provide protection for a certain contaminant or group of contaminants, an arbitrary choice on the part of management could place their personnel and workers in a dangerous situation.

Before a respirator is selected, the following questions should be considered:

1. What is the contaminant concentration where the respirator will be used?
2. What is the permissible exposure limit (PEL) to the contaminant?
3. Is the contaminant a gas, vapor, mist, dust, or fume?
4. Could the contaminant concentration be termed immediately hazardous to life or health?
5. If the contaminant is flammable, does the estimated concentration approach the lower explosive limit?
6. Does the contaminant have adequate warning properties?
7. Will the contaminant irritate the eyes at the estimated concentration?
8. If the contaminants is a gas vapor, is there an available sorbent that traps it efficiently?
9. Can the contaminant be absorbed through the skin as a vapor or liquid? If so, will it cause serious injury?

13.7.1 Selecting Respirators

The person given the responsibility for selecting respirators will be guided by ANSI Z88.2, American National Standard Practices for Respiratory Protection. Of particular importance is the MSHA/NIOSH approval on the respirator. Selection should be based on the:

1. nature of the hazardous process or operation
2. type of respiratory hazard
   a. physical and chemical properties
   b. warning properties
   c. physiological effects on the body
   d. concentration of the material
3. period of time for which respiratory protection must be provided
4. location of the hazard area in relation to the nearest source of uncontaminated respirable air
5. function and physical characteristics of respiratory protective devices

13.7.2 Classes of Respirators

The three major classes of respirators are:

1. air-purifying respirators
2. supplied-air respirators
3. self-contained breathing devices

Each class has various characteristics, advantages, limitations, and subcategories. Federal regulations applying to respiratory protective devices are contained in 30 CFR 11 and 29 CFR 1910.134. Another valuable source is ANSI Z88.2, American National Standard Practices for Respiratory Protection.

13.7.2.1 Air-Purifying Respirators

Sometimes called dust, mist or fume respirators, air-purifying respirators by their filtering action remove contaminants from the atmosphere before they can be inhaled. Various chemicals can remove specific gases and vapors, and mechanical filters remove particulate matter.

There are four basic types of air-purifying devices:

1. mechanical filter respirators
2. chemical cartridge respirators
3. combinations of chemical cartridge and mechanical filter respirators
4. gas masks

Air-purifying devices are small, relatively inexpensive, and easily maintained. Because of the various combinations of facepieces, mouthpieces, filters, cartridges, and canisters available, the devices can be tailored to the particular shop situations.

Air-purifying respirators cannot be used in oxygen-deficient atmospheres or where the air contaminant level exceeds the specified concentration limitation of the device. Seals on quarter-mask, half-mask, and mouthpiece respirators are not always reliable, nor do these respirators protect the eyes or skin.

Mechanical filter respirators offer protection against particulate matter, including dusts, mists, metal fumes and smokes, but they do not provide protection against gases, vapors, or oxygen deficiency. They consist essentially of a facepiece (quarter mask, half mask or full-face) with a mechanical filter attached. Many kinds of mechanical respirators have filters specifically designed for the various classes of particulate matter.

Items to be considered in selecting mechanical respirators include:

*   resistance to breathing offered by the filtering element
*   adaptation of the facepiece to faces of various sizes and shapes
*   efficiency in removing particulates of specific size ranges
*   the time required to clog the filter

The single-use respirator is a respirator that is disposed of after use. Either the air purifier is permanently attached to the face-piece or the entire facepiece is made of filter material. At present, these respirators are approved only for pneumoconiosis, or fibrosis-producing dusts such as coal dust, silica dust, and asbestos.

The quarter-mask covers the mouth and nose, while the half-mask fits over the nose and under the chin. The half-mask produces a better facepiece-to-face seal than does the quarter mask and is therefore preferred for use against more toxic materials. Dust and mist respirators are designed for protection against dusts and mists whose TLV is greater than .05 mg/M3 or 2 mppcf.

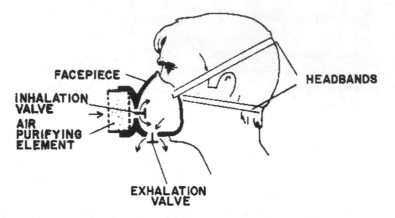

FIG. 13.1 Typical Quarter-mask Respirator.

Quarter- and half-mask fume masks, similar to the quarter- and half-mask dust and mist masks, utilize a filter element, which can remove metal fumes in addition to dusts and mists from the inhaled air. The filters are approved for metal fumes having a TLV above .05 mg/M3 or 2 mppcf.

Half-mask high efficiency masks are the same as the respirators mentioned above but use a high efficiency filter. Because of this filter, they can be used against dusts, mists, fumes, and combinations of those whose TLV is less than .05 mg/M3 or 2 mppcf.

Full facepiece respirators cover the face from the hairline to below the chin. In addition to providing more protection to the face, the full facepiece respirators give a better seal than do the half-or quarter-mask respirators. Depending upon the type of filter used, these respirators provide protection against dusts, mists, fumes, or any combination of these contaminants.

Air-purifying mechanical filter respirators offer no protection against atmospheres containing contaminant gases or vapors. They should not be used for abrasive blasting operations or in oxygen-deficient atmospheres. Another limitation is that the air flow resistance of a mechanical respirator filter element increases as the quantity of particles it retains increases, thus increasing breathing resistance.

Chemical cartridge respirators use various chemical filters to protect against certain gases and vapors. They differ from mechanical filter respirators in that they use cartridges containing chemicals to remove harmful gases and vapors. Consisting of a facepiece connected directly to one or two small canisters of chemicals, they offer protection against intermittent exposure to light concentrations (10 ppm to 1000 ppm by volume, depending upon the contaminant) of gases and vapors.

Many gases and vapors in extremely low concentrations may cause nausea and headache. Sometimes they may produce chronic disorders, which eventually may be fatal. Chemical cartridge respirators are particularly useful for guarding against:

- Organic vapors (acetone, alcohol, benzene, carbon tetrachloride, and gasoline)
- Acid gases (chlorine and sulfur dioxide)
- Other gaseous materials (ammonia gases and mercury vapor)

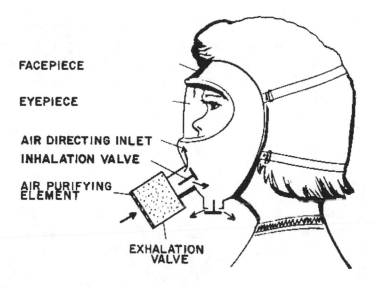

FACEPIECE

EYEPIECE

AIR DIRECTING INLET

INHALATION VALVE

AIR PURIFYING
ELEMENT

EXHALATION
VALVE

Fɪɢ. 13.2 Typical Full-facepiece Respirator

Chemical cartridge respirators have their limitations. They should not be used against gaseous material which:

- is extremely toxic in very small concentrations
- cannot be detected clearly by odor. Odor is necessary to alert the worker to the fact that the sorbent is saturated and that contaminated air is passing through the cartridge. Cartridges should be changed when the wearer smells the vapor.
- is in concentrations which are highly irritating to the eyes, unless supplementary eye protection is used
- is not stopped effectively, by the chemical fills utilized

The third category of air-purifying devices belongs to combination mechanical filter/chemical filter respirators, which utilize dust, mist, or fume filters with a chemical cartridge for dual or multiple exposure. Respirators with independently replaceable mechanical filters are sometimes used for this type of unit because the dust filter normally plugs before the chemical cartridge is exhausted. One type of unit has a back-mounted filter element. It is especially well suited for spray painting and welding operations where the air contaminant is concentrated in front of the worker.

Gas masks consist of a facepiece connected by a flexible tube to a canister. Chemicals in the canister purify contaminated air. Gas masks offer respiratory protection against specific gases, vapors and particulate matter. They are compact, economical, easy to operate, and easy to maintain.

Various gas masks have been tested and approved by the U.S. Bureau of Mines for respiratory protection against specific gases and vapors specified on the label. OSHA regulations require that each gas mask canister be labeled and color-coded to indicate the type of protection afforded (see Table 13.4).

TABLE 13.4
Colors or Combinations of Colors used for Gas Mask Canisters

Atmospheric Contaminants to be Protected Against	Colors Assigned*
Acid colors	white
Hydrocyanic acid gas	white with ½-inch green stripe completely around the canister near the bottom
Chlorine gas	White with ½-inch yellow stripe completely around the canister near the bottom
Organic vapors	black
Ammonia gas	green
Acid gasses and ammonia gas	Green with ½-inch white stripe completely around the canister near the bottom
Hydrocyanic acid gas chloropicrin	Yellow with ½-inch blue stripe completely around the canister near the bottom
Acid gasses, organic vapors, and ammonia gasses	brown
Radioactive materials, except tritium and noble gasses	purple (magenta)
Particulates (dusts, fumes, mists, fogs, or smokes) in combination with any of the above gases or vapors.	Canister color for contaminant, as designated above with ½-inch gray stripe completely around the canister near the top.
All of the above atmospheric contaminants	Red with ½-inch gray stripe completely around the canister near the top

*Gray shall not be assigned as the main color for a canister designed to remove acids or vapors.

NOTE: Orange shall be used as a complete body, or stripe coot to represent gases not included in this table. The user will need to refer to the canister label to determine the degree of protection the canister will afford.

Note that additional requirements for respiratory protection are now being included by OSHA in individual standards as, for example, in 1910.1001(d), Asbestos; 1910.1017(g), Vinyl Chloride; 1910.1028(g), Benzene Emergency Temporary Standard; and 1910.1029(g), Coal Tar Pitch Emissions.

Chin-type canisters, because of their smaller size, should be limited to concentrations not in excess of .5 percent by volume. They may also be used against dusts and mists having a time-weighted average threshold limit value of not less than .05 mg/M3 or two million particles per cubic foot.

The length of time that a gas mask provides protection depends upon the type of canister, the concentration of contaminants in the air, the breathing rate of the user, and the humidity.

Gas masks are subject to the following limitations:

1. Their use must be restricted to atmospheres which contain sufficient oxygen to support life.
2. Exposure concentrations must not exceed the specific limitations.
3. They should not be used against a gas or vapor with poor warning properties unless they have an indicator or timer that shows when the canister should be changed.

13.7.2.2 Supplied Air Respirators

The second class of respirator delivers uncontaminated air through a supply hose connected to the wearer's facepiece. There are essentially two major groups of supplied-air respirators: the airline respirator and the hose mask with or without a blower.

Airline respirators use a stationary source of compressed air delivered through a high-pressure hose. A trap and filter are installed in the compressed air line ahead of the masks to separate oil, water, grit or other matter from the air stream. Airline devices can be equipped with half- or full-face masks, helmets or hoods, or the devices can come as a complete suit. Airline respirators can be used for protection against particulates, gases or vapors.

A great advantage of airline respirators is that they can be used for long continuous periods. They provide a high degree of protection against contaminants. There are three basic classes of airline respirators:

1. constant flow - provides a regulated amount of air to the facepiece and is normally used where there is an ample air supply (such as that provided by an air compressor)
2. demand flow - delivers air only during inhalation and is used where air supply is restricted to high-pressure compressed air cylinders
3. positive pressure flow - provides a positive pressure during both inhalation and exhalation, avoiding both the possible inward leakage

caused by the negative pressure during inhalation which is part of the demand system and the relatively high air consumption of the constant flow system

Air line respirators cannot be used in atmospheres immediately dangerous to life or health because the user is completely dependent on the integrity of the air supply hose and the air source. If something happens to either the hose or air supply, the user cannot escape from the contaminated area without endangering his life. Extreme care must be taken to ensure that the source of air is free of hazardous substances such as carbon monoxide and oil mist.

Compressors or similar devices must be situated to avoid entry of contaminated air into the system. Alarms must be installed on compressors to indicate compressor failure and overheating, as well as excessive levels of carbon monoxide.

Airline hose never may exceed 300 feet in length. This requirement limits the wearer to a fixed distance from the air supply source. The hose must be protected from objects which could cut or puncture it during use. Connections on the airline must be airtight.

Hose masks supply outside air to the wearer through a length of hose. They are available either without a blower or with a power-driven or hand-operated blower. NIOSH/MSHA have approved hose masks with blowers for respiratory protection in any atmosphere, regardless of the degree of contamination or oxygen deficiency, if clean, breathable air can be reached within the distance of the permissible hose length (up to 300 feet), and if there is a person standing by with suitable rescue equipment.

The air hose must have a large inside diameter (approximately one inch) so that, in case of blower failure, the wearer can breathe through the hose while escaping from the contaminated area. The hose must be able to withstand crushing weight and be highly resistant to petroleum vapors. Hose masks without blowers are used when uncontaminated air can be reached within 75 feet. Such units carry only limited approval and cannot be used in atmospheres immediately dangerous to life and health.

13.7.2.3 Self-Contained Breathing Apparatus (SCBA)

The self-contained breathing apparatus (SCBA), the third class of respirator, provides complete respiratory protection against toxic gases and oxygen deficiency. The SCBA allows the user to carry a respirable breathing supply and eliminates the need for a stationary air source to provide breathable air. The wearer is independent of the surrounding atmosphere because he is breathing with a system admitting no outside air. This allows comparatively free movement over an unlimited area. In an environment containing a substance that is dangerously irritant or corrosive to the skin, a self-contained breathing apparatus must be supplemented by impervious clothing.

There are four basic types of SCBAs:

1. oxygen cylinder rebreathing
2. self-generating
3. demand
4. positive pressure

The first two are closed circuit and the second two types are open circuit systems. In the closed circuit devices, air is re-breathed after the exhaled carbon dioxide has been removed and the oxygen content re-stored. These devices are designed primarily for one- to four-hour use and consists of a relatively small cylinder of compressed oxygen, reducing and regulating valves, a breathing bag, facepiece or mouthpiece plus noseclip and a chemical container to remove carbon dioxide from the exhaled breath.

The self-generating type uses the principle of re-breathing, but it has no mechanical operating components. It has a chemical canister that evolves oxygen and removes the exhaled carbon dioxide. The wearer makes his own oxygen instead of drawing it from a compressed gas cylinder or liquid oxygen source.

Open circuit devices have an air supply which lasts from three to thirty minutes, depending on the respirator. When compared with re-breathing types, they are relatively inefficient because the exhaled air, which is still rich in oxygen, is released into the atmosphere instead of being used again.

All the different models of demand types apparatus designed for specific applications consist of a high-pressure air cylinder, a demand regulator connected either directly or by a high-pressure tube to the cylinder, a facepiece and tube assembly with exhalation valve or valves, and a method of mounting the apparatus on the body. After putting in the facepiece, the wearer opens the cylinder valve. The airflow is automatically regulated to the de-sired level to accommodate for varying breathing needs. The exhaled air passes through a valve in the facepiece to the surrounding atmosphere.

Positive pressure apparatus uses the same principle as positive pressure airline respirators. They are used in the same circumstances, namely where the toxicity is such that the potential facepiece leakage of demand apparatus presents an intolerable hazard.

13.7.3 Medical Aspects of Using Respiratory Equipment

The use of any type of respirator may impose some physiological stress on the user. Air-purifying respirators, for example, make inhaling more difficult because the filter or cartridge impedes the flow of air. Exhaling is more difficult because the expired air must force open a valve. The special exhalation valve on an open circuit pressure-demand respirator requires the wearer to exhale against significant resistance. The bulk and weight of a SCBA, up to 35 pounds, can also be a burden. Someone using an air line respirator or hose mask may have to drag around up to 300 feet of hose.

All of the above factors can increase significantly the user's workload. Those requiring respiratory equipment should have a medical examination to determine whether they have sufficient cardiovascular and pulmonary fitness to accommodate the additional physiological stress.

13.7.4 Maintaining Respirators

In large businesses where respiratory protective equipment is used routinely, respirators should be cleaned and disinfected daily. In small businesses where respirators are used occasionally, weekly, or monthly, cleaning and disinfecting are appropriate. Individual users who maintain their own respirators should be trained in the cleaning of respirators. The following suggestions are adapted form A Guide to Industrial Respiratory Protection (Cincinnati: NIOSH, 1076).

Two methods of cleaning can be adapted to the small business:

1. The respiratory protection equipment may be washed with detergent in warm water, using a brush, thoroughly rinsed in clean water and then air-dried in a clean place. Care should be taken to prevent damage from rough handling. This method is an accepted procedure for a small respirator program or a program where each user cleans his own respirator.

2. A standard domestic-type clothes washer may be used if a rack is installed to hold the facepieces in a fixed position. (If loose facepieces are placed in a washer, the agitator may damage them.)

This method is especially useful in large programs where respirator usage is extensive.

If possible, detergents containing a bactericide should be used. Organic solvents should not be used because they can deteriorate the rubber facepiece. If a detergent containing a bactericide is not available, a regular detergent may be used, followed by a disinfecting rinse. Reliable disinfectants may be made from available household solutions such as a(n):

- hypochlorite solution (50 ppm of chlorine) made by adding approximately two tablespoons of chlorine bleach per gallon of water. A two-minute immersion disinfects the respirators.
- aqueous solution of iodine (50 ppm) made by adding approximately one teaspoon of tincture of iodine per gallon of water. A two-minute immersion is sufficient and will not damage the rubber and plastic in the respirator facepiece.

Check with the manufacturer to find out the proper temperature for the solution.

If the respirators are washed by hand, a separate disinfecting rinse may be provided. If a washing machine is used, the disinfectant must be added to the rinse cycle, and the amount of water in the machine at that time will have to be measured to determine the correct amount of disinfectant to be added.

The cleaned and disinfected respirators should be rinsed thoroughly in clean water (140°F maximum) to remove all traces of detergent, cleaner, sanitizer, and disinfectant. This is very important to prevent dermatitis. The respirators may be allowed to dry by themselves on a clean surface. They also may be hung from a horizontal wire like drying clothes, but care must be taken not to damage the facepieces.

All the care that has gone into cleaning and maintaining a respirator can be negated by improper storage. Respiratory protective equipment must be stored so as to protect it from dust, sunlight, heat, extreme cold, excessive moisture, and damaging chemicals. Leaving a respirator unprotected, such as on a workbench or in a tool cabinet or toolbox among heavy wrenches, can damage the working parts or permanently distort of the facepiece, making the respirator ineffective.

After the respirators are cleaned and disinfected, they should be placed individually in heat-sealed or reusable plastic bags until reissue. They should be stored in a single layer with the facepiece and exhalation valve in a more or less normal position to prevent the rubber or plastic from taking a permanently distorted "set."

13.7.5 Inspection

An important part of the respirator maintenance program is the inspection of the devices. If carefully performed, inspections will identify damaged or malfunctioning respirators. All respiratory protective equipment must be inspected before and after each use and during cleaning. Equipment designated for emergency use must be inspected after each use, during cleaning, and at least monthly. Self-contained breathing apparatus must be inspected at least monthly. A record should be kept of inspection dates and findings for respirators maintained for emergency use.

13.7.6 Fitting the Respirator to the Wearer

It is essential that respiratory protective equipment be properly fitted to the user. All the care that went into the design and manufacture of a respirator to maximize protection will be lost if there is an improper match between facepiece and wearer. Fitting tests should be repeated at appropriate intervals, particularly when there is a change in the wearer's physical state, such as growth of facial hair or change in face contours.

Facial hair lying between the sealing surface of a respirator face-piece and the wearer's skin will prevent proper sealing. The sealing problem is especially critical when non-powered air-purifying respirators are used. The negative pressure developed in the facepiece of these respirators during inhalation can lead to leakage of contaminant into the facepiece when there is a poor seal. Some supplied-air respirators of the air line type, because of their mode of operation, can also lead to leakage at the sealing surface. Therefore, individuals who have stubble (even a few day's growth), a moustache, sideburns, or a beard that passes between the skin and the sealing surface should not wear a respirator.

Corrective lenses that have temple bars or straps should not be used when a full-facepiece respirator is worn since the bars or straps could pass through the facepiece-to-face seal. Manufacturers of respiratory equipment can provide kits for installing eyeglasses in their respiratory facepieces. A qualified individual to ensure proper fit must mount these glasses or lenses.

According to OSHA regulations (29 CFR 1910.134), the wearing of contact lenses in contaminated atmospheres with a respirator is not allowed. A properly fitted respirator (primarily a full-facepiece respirator) may stretch the skin around the eyes, thus increasing the possibility that the contact lens will fall out. Furthermore, contaminants that do penetrate the respirator may go underneath the contact lens and cause severe discomfort. The user's first reaction is to remove the facepiece to remedy the situation, a reaction which could be fatal in a lethal environment.

Scars, hollow temples, very prominent cheekbones, deep skin creases, and lack of teeth or dentures may cause respiratory face-piece sealing problems. Dentures or missing teeth may cause problems when sealing a mouthpiece in a person's mouth. Full dentures should be retained when wearing a respirator, but partial dentures may or may not have to be removed, depending upon the possibility of swallowing them. With full lower dentures, problems in fitting quarter-masks can be expected, as the lower part of the mask tends to unseat the denture.

13.7.7 Training and Education in the Respirator Program

Selecting the respirator appropriate to a given hazard is important, but equally important is its proper use. To ensure proper use and maintenance of respirators, a training program is needed for supervisors and workers alike. The content of the training program can vary widely, depending upon circumstances. However, OSHA 1910.134 requires that training programs for both users and supervisors include:

- an opportunity to handle the respirator
- proper fitting
- test of facepiece-to-face-seal
- a familiarizing period of wear in normal air

Furthermore, OSHA requires that the wearer receive fitting instructions, including demonstrations and practice in wearing, adjusting and determining the fit of the respirator. These requirements originated in ANSI Z88.2. Training of supervisors and workers also should include a(n):

- discussion of the engineering and administrative controls in use and the need for respirators
- explanation of the nature of the respiratory hazard and the consequences of improper use

- explanation of respirator selection criteria
- discussion of how to recognize and handle emergencies

13.8 Arm and Hand Protection

Statistics indicate that injuries to the arms, hands, and fingers account for more than a quarter of all disabling industrial mishaps. Most of the more than 720,000 hand injuries which occur each year are the result of accidents suffered when operating machinery, using tools, or handling materials.

Personal protective equipment can do little to prevent accidents when operating machinery. If protective gloves were worn to shield the hands from cuts and slivers when operating drills, lathes, and other machine tools, the gloves themselves would become a hazard. They could snag in the revolving parts and pull the hand into the machinery. Other means, including guarding devices and safe work procedures, must be found for protecting the hands of machine operators.

Gloves, mitts, and hand pads supplement good work practices to prevent hand injuries during the handling of materials and tools. There is a glove suitable for protection against many of the possible hazards including: abrasions, cuts, slivers, pinch points, oils, chemicals, radiation, electricity, cold, heat and flames. Table 13.5 offers a convenient summary.

Gloves which are especially suitable to the shop include:

1. Metal mesh gloves to protect against cuts and blows from sharp or rough objects and tools.
2. Rubber gloves to protect against electrical hazards (see ANSI J6.6, Rubber Insulating Gloves).
3. Rubber, neoprene, and vinyl gloves to protect against chemicals and corrosives. Leather gloves (including the chrome-tanned cowhide or horsehide leather gloves used by welders) to resist sparks, moderate heat, chips, and rough objects.
4. Knit cotton gloves to protect against dirt, slivers, chafing, or abrasion.
5. Disposable plastic gloves to protect against liquids, greases, oils, solvents, and dusts.
6. Aluminized gloves to protect against burns and discomfort when the hands are exposed to sustained conductive heat.

TABLE 13.5
Types of Hand Protection

Type of Hand Protection	Cuts, Abrasions	Light Materials Handling	Heavy Materials Handling	Pinch Points	Chemicals*	Electricity	Temperature Extremes	Liquids*	Flame	Radiation**	Sanitation
Knit Cotton	X	X					X				
Leather Gloves, Hand Pads	X	X	X				X		X		
Asbestos or Aluminized Gloves, Mittens, Pads							X		X		
Various Plastics (job-rated)	X	X			X			X			X
Rubber - Natural or Synthethic (job-rated)	X	X			X			X		X	X
Disposable Plastic (job-rated)		X			X			X			X
Plastic Coasted Glass Fiber Combination		X			X		X	X			
Electrical Lineman's Glove (must be worn with protective leather glove)						X					
Leather with Reinforced Metal Palm	X	X	X				X		X		
Mail or Woven Metal Gloves	X	X									
Cuff, Forearm Guards (fiber, metal, mesh, etc.)	X	X	X	X							X
Thumb Guards, Finger Cots, Protective Wrapping Tapes	X	X		X							
Protective Barriers	X										X

*Various plastic, rubber, and disposable rubber gloves must be job-related for working with specific alkalis, salts, acids, oils, greases, and solvents.
**For nuclear radiation hazards and X-rays, there are available leaded rubber, leaded plastic, and leaded leather apparel.

When choosing hand protectors for a certain job, the style of cuff is as important as the glove material. Gloves without a cuff offer no wrist protection. They can be used only for materials handling operations and general work where there is little danger of sparks, molten metal, or irritating chemicals (cither dry or liquid) getting into the glove.

A knit-wrist cuff offers only incidental wrist protection but keeps dust, dirt, and dry irritants from getting in the gloves. The gauntlet cuff protects the wrist and forearm and for special situations can extend all the way to the elbow. According to the capability of the glove material, it shields the lower arm against chemical, liquids, sparks, heat, flame, electricity, radiation, or sharp edges. Because they can be heavier and less flexible than gloves, hand leathers or hand pads may be more satisfactory than gloves for. protection against heat, abrasion, and splinters.

The purchasing of protective gloves must be done wisely. Because protective gloves are usually purchased in quantity, a purchasing agent can assume too easily that the shop can get along with just one or two types. Of course it is not economical to specify a different glove for every job, but there is a sensible and safe middle ground which involves:

- a shop operations hazard analysis
- a thorough knowledge of the types and uses of protective devices
- information about the budget, combined with knowledge of how gloves can be recycled

Glove suppliers and their agents can offer useful recommendations to control specific problems.

13.9 Body and Leg Protection

In the industrial shop, the workers' trunk and legs require protection from the hazards of molten metal, sparks, splashing liquids, heat, and cutting. Welders need aluminized aprons or aprons of fire-resistant fabric or leather. Because of the carcinogenic properties of asbestos, it is recommended that asbestos aprons not be used. The bib type aprons covers the chest, waist and knees or ankles, while the waist type lacks the chest-covering bib. Some welders prefer to combine cape sleeves (full sleeves attached to a bib or short jacket, which covers the chest) with leggings or split-leg aprons.

Kickback aprons and aprons made of metal mesh or leather rein. forced with metal studs offer another kind of body protection. The kickback apron shields the midsection against severe blows as in the kickback of stock during sawing operations. Metal mesh aprons prevent cuts from sharp tools or materials. Aprons made of leather studded with metal staples provide general protection against cuts, impacts, and abrasion.

Whenever there is the danger of splashing chemicals, all parts of the body that might be exposed must be protected. In addition to protecting the eyes, hands, and arms, the worker needs to protect the body with a coverall, overall, coat or apron. These garments are made of materials such as oiled fabric, natural or synthetic rubber, plastic, and plastic-coated fiber.

13.10 Foot Protection

Personal protective footwear can protect feet against the types of injuries which may result from falling objects, rolling objects, or accidental contact with edged tools or sharp sheet metal. Protective footwear falls into two main classes: safety shoes and foot guards.

The Office of Technical Services of the Division of Safety, U.S. Department of Labor, has designated five principal types of safety shoes:

1. Safety-toe shoes
2. Conductive shoes
3. Electrical hazard shoes
4. Explosives-operations (nonsparking) shoes
5. Foundry (molders) shoes

Because it is unlikely that the last two types would be used in the industrial shop, we will eliminate them from this discussion.

Safety-toe shoes provide toe protection by incorporating a steel toe box or its equivalent. The toe box adds little to the weight or cost of the shoe and a well made and properly fitted safety shoe is as comfortable as any other. Safety toes are incorporated in leather and rubber boots, oxfords, and leather shoes. Soles may be of leather, rubber, cord or wood. Plastic instep guards may be part of the shoe.

The ANSI label and class will be marked in the shoe. ANSI Z41.1, Men's Safety-Toe Footwear, divides footwear into three classes based on its ability to meet the minimum requirements for both compression and impact (see Table 13.6).

TABLE 13.6
Minimum Requirements of American Standard Z41

Classification	Compression (pounds)	Impact (pounds)	Clearance (inches)
75	2500	75	16/32
50	1750	50	16/32
30	1000	30	16/32

As with other types of protective devices, it is important that the extent of hazards be carefully assessed. Suppose that a supervisor makes an incorrect appraisal of the potential impact of a falling object, which is 70 ft./lbs. and purchases class 30 foot protection. What will happen? Under the stress of impact, the toe box may provide insufficient protection and the victim will be injured. For general shop use, a class 75 box toe capable of supporting a static load of 2,500 pounds with an impact strength of 75-foot pounds is recommended.

Conductive soles and heels drain off static charges so as to avoid the creation of static electricity in locations with a fire or explosion hazard. Electrical hazard shoes are made of leather. No metal is used in their construction except in the toe box, which is insulated from the shoe.

Basic types of foot guards include plastic, aluminum alloy, and galvanized steel coverings which temporarily attach to the shoes by means of a heel strap. Foot guards protect both the toe and instep against falling or rolling objects.

A combination foot and shin guard protects both the foot and shin against flying particle hazards. The shin guard is made of the same material as the foot guard. The shin guard is hinged to the foot guard and is held in place by straps around the leg.

Summary

Personal protective equipment is no substitute for engineering controls. As a supplement to safe work practices, however, PPE provides safeguards against hazards common in the workplace: flying particles, molten metal, chemical exposure, splashing liquids, excessive noise, and sharp objects. If shop operations and processes are to take place in an environment where hazards have been evaluated and controlled, equipment must be selected wisely, maintained properly, and used carefully.

Bibliography

Cohen, Howard (1999) "Respiratory Protection Devices", Chapter 17 in the Handbook of Occupational Safety and Health, John Wiley & Sons, N.Y., N.Y.

Mansdorf, S. Zack (1997) "Personal Protective Clothing" Chapter 35 in the Occupational Environment-Its Evaluation and Control, AIHA, Fairfax, VA.

Nill, Richard J. (1999) "How to Select and Use Personal Protective Equipment" Chapter 16 in the Handbook of Occupational Safety and Health, John Wiley & Sons, N.Y., N.Y.

U.S. Department of Health, Education, and Welfare, National Institute for Occupational Safety and Health, Respiratory Protection: An Employer's

Manual, October 1978.

U.S. Department of Labor, Occupational Safety and Health Administration, General Industry OSHA Safety and Health Standards (29 CFR 1910), November 1978.

14

Machine Guarding

The National Safety Council estimates that nearly 20% of all permanent partial disabilities result from injuries associated with machinery. Machine guarding is of the utmost importance in protecting workers from the hazards associated with machinery. In fact, the degree to which machines are guarded is a reflection of the company's interest in providing a safe workplace. The purpose of this chapter is to provide a general overview of OSHA requirements relative to guarding machinery and equipment and how guarding contributes to the total hazard control program.

Workers cannot always be relied upon to act safely enough around machinery in motion to avoid injury. Even the well coordinated and highly trained worker may commit an error which could lead to injury and death. While an effective guard will allow workers to be more comfortable around machinery, a poorly designed or inadequate guard is very dangerous because workers falsely believe that it will protect them.

The National Safety Council defines guarding as "any means of effectively preventing personnel from coming in contact with the moving parts of machinery or equipment which could cause physical harm to the personnel." Machine guarding prevents injury from the following sources:

- direct contact with the moving parts of a machine
- work in process making contact with personnel (kickbacks on a circular ripsaw, metal chips from a machine tool operation, or splashing of chemicals or hot metal)
- mechanical failure
- electrical failure
- human failure from any cause (curiosity, distraction, fatigue, worry, anger, illness, zeal, laziness, or deliberate risk-taking)

Machine guarding is a mandatory part of the safety and health program. With the information derived from this chapter, small businesses can evaluate whether or not the machines in their shops are adequately guarded, and if they are not properly guarded, they can take the necessary steps to comply with machine guarding regulations.

14.1 Mechanical Motions

All machinery movement employs one of three kinds of motion: rotary motion, reciprocating (back-and-forth) motion, or a combination of these. Each of these motions can produce crushing and shearing actions.

Rotary motion is found in simple rotating mechanisms, rotary cutting and shearing mechanisms, rotating mechanisms with in-running nip points, and screw or worm mechanisms. Rotary action is hazardous regardless of the speed, size, or surface finish of the shaft. Even smooth, slowly rotating shafts can grip clothing or hair or force an arm or hand into a dangerous position.

The hazard of reciprocating motion occurs when the moving part approaches or crosses a fixed part of the machine. A back-and-forth motion could also be called transverse motion depending on the position of the worker in relation to the machine.

14.2 Mechanisms Requiring Guards

Mechanisms using any of these three motions will need guarding if they are exposed. These mechanisms can be divided into the following groups:

- rotating mechanisms
- cutting or shearing mechanisms
- in-running nip points
- screw or worm mechanisms
- forming or bending mechanisms

A piece of equipment may employ more than one type of hazardous motion. For example, a belt and pulley drive is a hazardous rotating mechanism and also has hazardous in-running nip points. A rotating part is dangerous unless it is guarded. Common hazardous rotating machine parts include vertical or horizontal transmission shafts, pulleys, belts, rod or bar stock projecting from lathes, set screws, flywheels and their cross members, drills, couplings, and clutches. The danger increases when such items as bolts, projecting keys, or screw threads are exposed when rotating.

The hazards of cutting and shearing mechanisms lie at the points where a rotary cutting action is used or where the moving parts of a reciprocating mechanism approach or cross the fixed parts of the piece or machine. Examples of machines using cutting and shearing mechanisms are band and circular saws, milling machines, lathes, grinding machines, abrasive wheels, shapers and drills, and boring machines.

An in-running nip point is formed when two parts that are in contact with or close to one another rotate in opposite directions or a part rotates over, under or near a stationary object.

A nip point draws in objects or parts of the body and crushes, mangles, or flattens them. Once an object is drawn in, it is difficult if not impossible to withdraw it. Examples of nip points are the points of contact between a belt and pulley, chain and sprocket, gear and rack, and the squeeze spaces between shafts or rolls which are rotating close together and in opposite directions.

The hazards of screw or worm mechanisms are the shearing action set up between the moving screw and the fixed parts of the machine and the mangling or battering action created if a part is caught in the mechanism. Screw or worm mechanisms are used for conveying, mixing or grinding materials.

The hazard of all forming and bending mechanisms lies at the point where the punch or upper die approaches the lower die. In other words, the danger lies at the point of operation where stock is inserted, maintained, and withdrawn. Examples of such mechanisms are power, foot and hand presses, press brakes, metal shears, and forgoing machines.

14.3 Where Guarding is Necessary

Guarding should take place at two points: the point of operation; and the point where power is delivered to the machine.

14.3.1 Point of Operation

The point of operation is that area on a machine where material is positioned for processing or change by the machine. It is the place where work is actually performed upon the material. OSHA defines it as "that point at which cutting, shaping, boring or forming is accomplished upon the stock." OSHA regulations (in 29 CFR 1910.212) mandate:

> The point of operation of machines whose operation exposes an employee to injury shall be guarded. The guarding devices shall be in conformity with any appropriate standards therefore, or, in the absence of applicable specific standards, shall be so designed and constructed as to prevent the operator from having any part of his body in the danger zone during the operating cycle.

OSHA also has established requirements for the design, construction, application, and adjustment of point of operation guards (29 CFR 1910.212). In general, point of operations guards must:

1. prevent entry into the point of operation by hands or fingers reaching through, over, under, or around the guard
2. conform to the maximum permissible openings (see Table 14.1)
3. create no pinch point between the guard and moving machine parts
4. minimize the possibility of misuse or removal of essential parts by utilizing fasteners which the operator cannot remove readily
5. facilitate inspection
6. offer maximum visibility of the point of operation

Point-of-operation guards must be constructed with a feed opening that is limited in size so that stock or material can be admitted into the danger zone but hands and fingers cannot. Guard openings are necessary for:

- material-entrance, removal, scrap removal
- vision-cutting, line, inspection, selection
- lubrication-greasing, oiling, cleaning
- adjustments-repairs, alterations, changes

Maximum safe openings are placed so that they do not permit the operator's fingers to reach the point of operation (see Table 14.1). Suppose it is necessary to provide a ¾ inch opening between the bottom of the guard and the top of the feed table or between openings in a guard for the feeding of material. In this case, a 3/4-inch opening is permissible to within 5-1/2 inches of the point of operation.

TABLE 14.1
Maximum Permissible Openings

Distance of Opening from Point of Operation Hazard		Maximum Width of Opening	
(in inches)	(in centimeters)	(in inches)	(in millimeters)
½ to 1½	1.3 to 3.8	1/4	6
1½ to 2½	3.8 to 6.4	3/8	10
2½ to 3½	6.4 to 8.9	1/2	13
3½ to 5½	8.9 to 14.0	5/8	16
5½ to 6½	14.0 to 16.5	3/4	19
6½ to 7½	16.5 to 19.0	7/8	22
7½ to 12½	19.0 to 31.8	1 1/4	32
12½ to 15½	31.8 to 39.4	1 1/2	38
16½ to 17½	39.4 to 44.5	1 7/8	48
17½ to 31½	44.5 to 80.0	2 1/8	54

The average woman's finger (glove size 6½) will go through a 3/8-inch opening. Therefore, for openings in any guard, use ¼ inch as the allowable opening within 1½ inches of the hazards. When installing guards with openings over ¼ inch, tests should be carefully made by workers to make certain that the guard is effective before the machine is operated.

14.3.2 Point of Power Delivery

Distinct from point-of-operation guarding but complementary to it is guarding at the point where power is delivered to the machinery. This includes the entire power transmission apparatus. Because power transmissions are far more standardized and because it is not necessary to consider feeding material into the machine as with the point of operation situation, they are easier to guard effectively. The American National Standards Institute (ANSI) has established, in its Standard B15.1, a Safety Code for Mechanical Power Transmission Apparatus.

14.4 Types of Guarding

There are four basic types of guards which protect the worker from the motion of a machine:

1. fixed enclosure guards
2. interlocking guards
3. automatic guards
4. remote control, placement, feeding and ejecting guards

14.4.1 Fixed Enclosure Guards

The fixed enclosure guard is preferable to all other types and should be used in every practicable case. This type of guard prevents access to dangerous parts by completely enclosing the hazardous operation. It also confines flying objects. Fixed guards may adjust to different sets of tools and dies and to various kinds of work. Once adjusted, they should not be moved or detached.

Fixed guards may be installed at the point where material is being processed and at other places where there may be a hazard when inserting or manipulating stock. They also may be used to prevent contact with rotating or reciprocating motions of machine members. Fixed enclosure guards are found on such machines as power presses, drills, and milling machines.

14.4.2 Interlocking Guards

When a fixed enclosure guard cannot be used, an interlocking guard should be fitted onto the machine as the first alternative. The interlocking guard prevents operation of the control that sets the machine in motion until the guard is moved into position so that the operator is unable to reach the point of danger. When the guard is open and dangerous parts are accessible, the starting mechanism is locked. A locking pin or some other safety device prevents the main shaft from turning or other basic mechanisms from operating. When the machine is in motion, the guard cannot be opened. It stays closed until the machine has come to rest or has reached a fixed point in its travel.

An interlocked press barrier guard is required on mechanical power presses. It must be interlocked with a press clutch control so that the clutch cannot be activated unless the guard (or the hinged or movable sections of the guard) is in the proper position. An interlocking enclosure guard must do four things:

1. guard the dangerous point before the machine can be operated
2. stay closed until the dangerous part is at rest
3. prevent operation of a machine if the interlocking device fails
4. require the activation of a reset device to restart

When gate guards or hinged enclosure guards are used with inter-locks, they should be designed to completely enclose the point of operation before the operating clutch can be engaged.

14.4.3 Automatic Guards

When neither a fixed barrier nor an interlocking guard is practicable, an automatic protection device may be used. Such a device is less desirable than the two previously discussed. An automatic guard acts independently of the operator, repeating its cycle as long as the machine is in motion. The device is usually operated by the machine itself through a system of linkage, levers, or by electronic means. Such a device must prevent the operator from coming in contact with the dangerous part of a moving machine or must be able to stop the machine quickly in case of danger.

Sweep devices are no longer approved by OSHA for safeguarding at the point of operation. Pull-away or hand-restraint types remove the operator's hands or fingers from the danger zone as the ram, plunger or slide closes on the piece upon which the work is being done. These types of devices are being phased out and replaced with automatic electrical devices.

All electrical and electronic devices perform the same end function when energized. They interrupt the electric current just as if the "stop" button had been pushed. For example, a jointer must have an automatic guard to cover the section of the head on the working side of the fence or gauge. This guard automatically adjusts it self to cover the unused portion of the head.

All three of these devices: fixed enclosure, interlocking, and automatic can be used to guard points of operation. Only the fixed guard can guard power transmission components.

14.4.4 Other Types of Guards

Although they are not guards in the technical sense, certain methods can be used to accomplish the same effect; that is, to protect the operator from the hazardous point of operation. These methods may complement one of the other types of guards.

One example is a trip or control system which requires two hands to activate the machine. Such devices require simultaneous action of both the operator's hands on an electrical switch button, an air control valve or a mechanical lever. Because the operator must use both hands to give concurrent pressure, during the most hazardous part of the machine's operation, he cannot move his hands from the controls to the danger zone until the cycle has been completed. Removal of a hand from the control causes the machine to stop.

Stock may be fed automatically or semi-automatically by rolls, plungers, chutes, slide and dial feeds, and revolving and progressive dies in conjunction with ram enclosures. This method will not admit any part of the body to the danger zone.

Special jigs or feeding and holding devices such as pickers, grab bars, push sticks, long-handled tongs, and hand die holders may be used to manipulate stock at the point of operation and yet keep hands safe. Hand tools for placing or removing material are to be used in conjunction with fixed enclosure, interlocking, or automatic guards.

Mechanical or air-operated ejecting mechanisms may be used to remove parts, eliminating the need to place hands in the danger zone. The theory behind these methods is that, if it is impossible to completely enclose or isolate the hazard, the next most effective device or combination of devices should be used to keep exposure to a minimum.

14.4.5 Selecting Guards

Important considerations in selecting a method of machine guarding include the:

- type of operation
- size or shape of stock
- method of handling
- physical layout
- type of material
- job requirements or limitations

Some machines require specific methods of guarding. These machines and methods will be discussed later in this chapter. All machines are regulated by the general requirement that points of operation and power transmission must be guarded. Furthermore, Federal regulations (29 CFR 1910.212) mandate that:

One or more methods of machine guarding shall be provided to protect the operator and other employees in the machine area from hazards such as those created by point of operation, ingoing nip points, rotating parts, flying chips and sparks.

Chain drives, shafting, coupling, keys, collars, and clutches located seven feet or less above the ground, floor or working platform must be guarded to prevent accidental contact. V-belts and chain drives must be enclosed completely.

14.5 Guard Requirements

Good engineering principles should be applied to the design and construction of each machine in order to eliminate hazards and to permit safe and efficient operation. When it is not possible to design hazards out of machinery, then suitable safeguards should be provided. An acceptable guard should have the following characteristics:

1. It should give maximum protection to the operator.
2. It should protect others working close to or passing by the machine from coming in contact with moving parts.
3. It should be considered a permanent part of the machine or equipment. OSHA regulations (29 CFR 1910.212) require that guards be affixed to the machine where possible. The guard should resist tampering or easy removal. It should be designed for the specific job and the specific machine and its purpose and use should be evident, even to an uninformed worker.
4. It should be convenient. It should not interfere with the efficient operation of the machine, nor should it require continual adjustment or removal to accomplish certain work tasks. It should cause the operator no discomfort but should enable him to work with less tension.
5. It should prevent access to the danger zone during operation.
6. It should allow access to the machine for servicing. Provisions should be made for inspecting, adjusting, and making repairs on machine parts enclosed by guards without exposing the worker to moving parts. Lubrication of the machine should not require removal of the guard. Whenever possible, oil reservoirs should be located outside the guard, with the oil line leading to the point of lubrication.
7. It should be durable. It should be constructed of materials at least as durable as those used in any other part of the machine and should serve over a long period with minimum maintenance. It should resist normal wear and shock. The guard should be secured so that a blow to or vibration of the machine will not cause it to work

loose, break, or fall into or off the area being guarded. It should resist fire and corrosion.

8. It should be easy to repair.
9. It must "not offer an accident hazard in itself" (OSHA 29 CFR 1910.212). All edges should be rolled and bolted to eliminate sharp or rough edges and corners. There should be no shear or pinch points, splinters, exposed bolts, or other possible sources of injury.
10. It must conform to the requirements of OSHA and/or the state inspection department having jurisdiction. Purchasing agents should be advised against "pre-OSHA" bargains, which eventually will cost more to rectify than the savings realized at the time of purchase. Where American National Standards apply, the guard should conform to or exceed these requirements.

14.6 Guards for Special Situations

Many businesses, particularly older ones, may be using an older machine which, though still serviceable, is not adequately guarded and for which guards cannot be purchased because of the machine's age. Often the cost of newer, more modern, guard-equipped machines is prohibitive. It then becomes necessary for guards to be designed and built locally. It is essential that such guards meet the preceding requirements.

Some machines are of a standard type that has been converted or equipped to perform a special function. At the time of purchase, the machine may or may not have been equipped with guards. However, because of its specialized nature, the machine cannot carry a standard guard. Therefore, guards must be designed and built locally to provide adequate protection for the machine operators.

14.7 Commonly Guarded Machinery

Thus far, we have examined what mechanical motions need to be guarded, where guarding is necessary, and the types and characteristics of good guards. Now we will examine the guarding requirements of specific machinery found in industrial shops.

Throughout our discussion we will refer to the relevant OSHA standards. OSHA has granted many states the right to develop and enforce their own occupational safety and health standards under their own state plans. The basic criterion for approving such plans is that they be at least as effective as the federal program. Because all approved state plans equal or surpass OSHA regulations in stringency, these regulations are cited as minimal standards. Businesses must comply with the safety and health regulations of their own state plans. In states which have not adopted plans of their own, federal OSHA regulations provide guidance for the supervisor or employee wanting to create a safe shop environment. The following sections examine guarding requirements for the equipment found in woodworking shops. A useful source for the supervisor is ANSI 01.1-1971, American National Standard Safety Requirements for Woodworking Machinery.

Saws can be divided into many different categories and terminology may vary from one business to another. For this discussion, saws are divided into two broad classes: circular saws and bandsaws. The circular or table saw is one of the most versatile machines. It uses many special types of blades for its operations, the most common include the:

1. ripsaw blade with teeth filed straight across, used to cut with the grain of the wood
2. crosscut or cut-off saw blade, used to cut across the grain. The teeth are beveled and set, or bent, alternately right and left (that is, one tooth is bent to the right, the next to the left and so on).
3. combination saw blade, used to crosscut, rip and miter. The smaller crosscut teeth are filed and set as they are on the crosscut saw; the larger rake or rip teeth are filed straight across.

14.8 Hand-Fed Saws

The circular hand-fed saw can be either a ripsaw, a crosscut saw, or a combination saw. With this kind of power table saw, the operator first adjusts the blade height, then holds the stock and pushes it into the blade. A guide is used to maintain a straight cut. At the end of the cutting stroke, the operator must push the stock past the blade, or change positions at the saw so that he can pull the stock. General hazards are inherent in the power transmission, point of operation, kickbacks, and flying particles.

The saw may be driven by an individual motor or from a line shaft, comprising belts, gears, and pulleys. Though the power transmission is usually housed beneath the blade and enclosed, there are open units which expose the bottom portion of the blade. Accidental contact with the saw blade by the operator or others is possible, particularly because the operator is working very close to the point of operation.

Kickbacks occur when the saw blade seizes the stock being cut and hurls it back toward the operator. Serious injuries may result. Kickbacks can be caused by unsafe operating methods (height not properly adjusted), improperly maintained equipment (blade not properly sharpened) or by lack of proper physical safeguards. Ripping operations are more hazardous than cross-cutting because of the greater likelihood of kickbacks. Sawdust, splinters and chips are thrown off as a result of the cutting action of the blade and can come in contact with the operator or others.

OSHA regulations (29 CFR 1910.213) mandate the machine guarding requirements. They state that each circular hand-fed ripsaw and each circular table saw:

> ...shall be guarded by a hood which shall completely enclose that portion of the saw above the table and that portion of the saw above the material being cut. The hood and mounting shall be arranged so that the hood will automatically adjust itself to the thickness of, and remain in contact with, the material being cut... The hood shall be of adequate strength to resist blows and strains incidental to reasonable operation, adjusting and handling and shall be so designed as to protect the operator from flying splinters and broken saw teeth....
>
> The hood shall be so mounted as to insure that its operation will be positive, reliable and in true alignment with the saw; and the mounting shall be adequate in strength to resist any reasonable side thrust or other force tending to throw it out of line.

In order to use the hood guard effectively when cutting narrow strips on circular ripsaws, a filler piece should be used. This piece should be made of wood about two inches wide and should be about ¾ inch thick, or slightly thinner than the thickness of the material being cut. It should be provided with cleats or brackets at the ends so that it will either fit down over the front and back ends of the table or be quickly attached to the fence or gauge.

The same statute mentioned earlier deals with the hazards of flying particles by requiring a spreader. The statute requires that each circular hand-fed ripsaw and each circular crosscut table saw be furnished with a spreader to prevent material from squeezing the saw or being thrown back on the operator. The spreader shall be made of hard-tempered steel, or its equivalent, and shall be thinner than the saw kerf. It shall be of sufficient width to provide adequate stiffness or rigidity to resist any reasonable side thrust or blow tending to bend or throw it out of position. The spreader shall be so attached that it will remain in true alignment with the saw, even when either the saw or the table is tilted, and should be placed so that there is not more than ½ inch space between the spreader and the back of the saw when the largest saw is mounted in the machine.

Kickbacks on ripsaws are usually caused by one of the following:

1. failure to provide the required spreader
2. an improperly conditioned saw which allows material to pinch on the saw and rise from the table
3. improperly aligned gauge or rip fence
4. improperly conditioned or twisted grain lumber
5. improperly designed or mounted anti-kickback devices

OSHA regulations require guards on ripsaws to protect against both kickbacks and flying material. They require each hand-fed circular ripsaw shall be provided with non-kickback fingers so located as to oppose the thrust or tendency of the saw to pick up the material or to kick it back toward the operator. The saw shall be designed to provide adequate holding power for all the thickness of materials being cut.

Operators should be required to use proper protective equipment, such as goggles and/or face shield each time the saw is operated. Where there is a danger of kickback from any operation, anti-kickback aprons must be provided and worn. Long sleeves or other loose clothing should not be worn, nor should rings, bracelets or other jewelry. Long hair should be confined by hair nets or caps.

If saw parts are driven by belts that are exposed, belts and pulleys should be completely enclosed by sheet metal or heavy mesh guards, even though the saw may be partially fenced off or partially removed from other machines.

Operating controls should be recessed or shrouded. In addition, an emergency stop button is recommended. A main power electrical disconnect switch is required.

A push stick must be provided so that the operator will not use his hand to feed stock past the blade. This is especially important when cutting short or narrow stock.

For safety reasons, a crosscut saw should not be used for ripping nor a ripsaw for crosscutting. Using the wrong saw makes the work more difficult and requires additional force to feed the stock. Work that can be done on self-feed machines should not be done on hand-fed machines.

Long stock should not be crosscut. If the stock extends beyond one or both ends of the table, it may interfere with other operations or strike a worker. A swing or pull saw should be used on such stock.

14.9 Self-Feed Saws

The self-feed (or power-feed) saw is equipped with rollers or a conveyor system to hold the lumber and force-feed it into the saw blade. The hazards associated with this type of saw are identical to those of hand-fed saws with one additional point of operation hazard. Not only can the operator accidentally come in contact with the saw blade, but also there is the danger of being pinched between the stock and the in-running rolls.

OSHA requires a hood or guard to protect against this hazard:

Feed rolls and saws shall be protected by a hood or guard to prevent the hands of the operator from coming in contact with the in-running rolls at any point. The guard shall be constructed of heavy material, preferably metal, and the bottom of the guard shall come down to within 3/8 inch of the plane formed by the bottom or working surfaces of the feed rolls. This distance (3/8 inch) may be increased to 3/4 inch, provided the lead edge of the hood is extended to be not less than 5-1/2 inches in front of the nip point between the front roll and the work.

Feed rolls should be adjusted to the thickness of the stock being ripped. Insufficient pressure on the stock can contribute to kickbacks. The anti-kickback fingers required by OSHA should be checked regularly to make sure that they are sharp and that none of the fingers are bent. OSHA requires each self-feed circular ripsaw shall be provided with sectional kickback fingers for the full width of the feed rolls and that each ripsaw be located in front of the saw and so arranged as to be in continual contact with the wood being fed. The personal protective equipment requirements are the same as for hand-fed saws.

Because long stock is often ripped on self-feed ripsaws, the clearance at each working end of the saw table should be at least three feet longer than the longest material handled. Two emergency stop buttons should be provided, one for the operator and one for the offbearer.

14.10 Swing and Sliding Cutoff Saws

Swing and sliding cutoff saws are used for crosscutting operations. The swing saw, which is more common, is suspended from the roof or overhead. The operator pulls it forward like a pendulum. Hazards may be inherent in the power transmission and in the point of operation.

This type of saw is generally driven by an individual motor with the saw attached directly to the motor shaft. It also may be driven with an individual motor by belt and pulleys. Accidents occur at the point of operation when:

- the operator, while pulling the saw forward, accidentally comes in contact with the saw blade
- the operator reaches for sawed stock while the machine coasts or idles, before the saw returns to normal "rest" position
- an improperly adjusted saw swings beyond its safe limits into the body of the operator
- the operator may be struck by the saw as it bounces forward from the idle position or as it swings or drifts forward when the spring or counterweight fails

OSHA regulations for swing cutoff saws apply also to sliding cutoff saws mounted above the table. According to 29 CFR 1910.213:

Each swing cutoff saw shall be provided with a hood that will completely enclose the upper half of the saw, the arbor end and the point of operation at all positions of the saw. The hood shall be constructed in such a manner and of such material that it will protect the operator from flying splinters and broken saw teeth. Its hood shall be so designed that it will automatically cover the lower portion of the blade, so that when the saw is re-turned to the back of the table, the hood will rise on top of the fence, and when the saw is moved forward, the hood will drop on top of and remain in contact with the table or material being cut.

At the end of the cut, the lip of the hood should be in contact with the table surface.

Special regulations apply to inverted swing cutoff saws. Inverted swing cutoff saws shall be provided with a hood that will cover the part of the saw that protrudes above the top of the table or above the material being cut. It shall automatically adjust itself to the thickness of and remain in contact with the material being cut.

OSHA also requires an automatic device to return the saw to the back of the table:

> Each swing cutoff saw shall be provided with an effective device to return the saw automatically to the back of the table when released at any point of its travel. Such a device shall not depend for its proper functioning upon any rope, cord or spring. If there is a counterweight, the bolts supporting the bar and counterweight shall be provided with cotter pins; and the counterweight shall be prevented from dropping by either a bolt passing through both the bar and counterweight, or a bolt put through the extreme end of the bar, or, where the counterweight does not encircle the bar, a safety chain attached to it.

OSHA further requires:

> Limit chains or other equally effective devices shall be provided to prevent the saw from swinging beyond the front or back edges of the table or beyond a forward position where the gullets of the lowest saw teeth will rise above the table top.

A latch should be provided to catch and retain the saw at the rear of the table and to prevent it from rebounding. In some cases a non-recoiling spring or bumper is adequate.

Each swing and sliding cutoff saw table can be provided with a wood bumper or a pipe guard to prevent bodily contact with the saw blade when it is extended the full length of the support arm.

While operating the saw, it is important that the operator stand to the side of the saw on which the handle is located. The operator should use the hand nearest the handle. This keeps the operator's body out of line with the saw and makes it unnecessary to bring the hands near the saw while it is cutting.

Proper personal protective equipment, such as goggles or a face shield, shall be worn by the operator.

If an overhead drive is used, the entire drive line should be enclosed.

Operating controls should be recessed or shrouded, and an emergency stop switch is highly recommended. A main power electrical disconnect is necessary.

14.11 Radial Saws

Radial saws, developed from the old type of swing saw, are both versatile and dangerous. They cut downward and pull the wood away from the operator and against a fence. Many adjustments are required to permit its full use and these adjustments can create additional hazards. The principal hazards are those common to other power-driven saws: cutting injuries caused by the saw blade and injuries from flying materials and kickbacks.

OSHA requires that the upper half of the saw be guarded and that the lower half have a floating guard:

> The upper hood shall completely enclose the upper portion of the blade down to a point that will include the end of the saw arbor. The upper hood shall be constructed in such a manner and of such material that it will protect the operator from flying splinters, broken saw teeth, etc., and will deflect sawdust away from the operator. The sides of the lower exposed portion of the blade shall be guarded to the full diameter of the blade by a device that will automatically adjust itself to the thickness of the stock and remain in contact with stock being cut to give maximum protection possible for the operation being performed.

OSHA regulations further mandate that the unit be installed so that the front end of the unit is slightly higher than the rear. This will cause the cutting head to return gently to the starting position when released by the operator and prevent the cutting head from creeping toward the operator during crosscut operations. Furthermore, an adjustable stop shall be provided to prevent the forward travel of the blade beyond the position necessary to complete the cut in repetitive operations.

When the radial saw is used for crosscutting, the saw is pulled across the cutting area by a handle located on one side of the saw. For safe operations, the same stance should be assumed as was described in handling the swing cutoff saw.

When the radial saw is used for ripping, special precautions are necessary. A spreader must be provided, and OSHA requires:

> Each radial saw used for ripping shall be provided with nonkickback fingers or dogs located on both sides of the saw so as to oppose the thrust or tendency of the saw to pick up the material or to throw it back toward the operator. They shall be designed to provide adequate holding power for all the thicknesses of material being cut.

Furthermore, the direction of the saw blade rotation must be upward toward the operator:

> Ripping and ploughing shall be against the direction in which the saw turns. The direction of the saw rotation shall be conspicuously marked on the hood. In addition, a permanent label not less than 1½ inches by ¾ inch shall be affixed to the rear of the guard at approximately the level of the arbor, reading as follows: Danger: Do Not Rip or Plough from This End. Such a label should be colored standard danger red.

When using a radial saw, workers should wear safety glasses or face shields. As with other power saws, they should not wear gloves, rings, chains or loose clothing.

14.12 Portable Circular Saws

Guarding for portable circular saws is required by OSHA (29 CFR 1910.243):

> All portable, power-driven circular saws having a blade diameter greater than two inches shall be equipped with guards above and below the base plate or shoe. The upper guard shall cover the saw to the depth of the teeth, except for the minimum arc required to permit the base to be tilted for bevel cuts. The lower guard shall cover the saw to the depth of the teeth, except for the minimum arc required to allow proper retraction and contact with the work. When the tool is withdrawn from the work, the lower guard shall automatically and instantly return to covering position.

OSHA also requires that hand-held powered circular saws be equipped with a constant pressure switch or control that will shut off the power when the pressure is released.

14.13 Bandsaws

Up to this point we have been discussing circular saws. Now we come to the second category of saws, the bandsaw. The bandsaw is a machine which can be used for straight sawing as well as for cutting curved pieces.

It uses a thin, flexible, continuous steel strip with cutting teeth on one edge. The blade runs on two pulleys through a work table, where stock is fed into it. OSHA regulations require the following safety devices for band-saws and band resaws:

- an enclosure or guard for all portions of the saw blade except for the working portion between the bottom of the guide rolls and the table
- an enclosure for wheels
- effective brakes to stop the wheel in case of blade breakage (not required, but highly recommended)
- a tension control device to indicate the proper tension for the standard saws used on the machine
- a guard on the feed rolls of band resaws to prevent the hands of the operator from coming in contact with the in-running rolls

The American National Standards Safety Requirements for Woodworking Machinery (ANSI 01.1-1971.),.stress the importance of proper tension for the blade. Improper tension results in saw breakage. Additional operating practices which this standard requires or recommends include:

- adjustment of the back thrust shall to the normal position of the saw blade
- preventing the accumulation of dust on the rim of the bandwheels in order to secure satisfactory operation
- using the proper size saw blade. The saw blade should in all cases be as large as the nature of the work will permit. It is poor practice to use a small saw for large work or to force a wide saw to cut on a small radius.
- ensuring saws are not stopped too quickly or by thrusting a piece of wood against the cutting edge of the teeth when the power is off
- using brazed joints of the same thickness as the saw blade to avoid vibration
- periodically examining band saw blades to avoid use of cracked blades or blades which indicate probability of breakage

Other safe work practices include:

- setting the guard to just clear the stock to be cut

- if the stock being cut binds the blade, shutting off the machine and letting it stop before backing the work off the blade
- making release cuts before doing curves
- keeping the work area clean and clear of scraps
- using a brush to clean the saw table
- shutting off the machine and getting clear of it until it stops in the event of a blade breaking
- making all cuts when the power is on, never while the machine is coasting after power has been turned off

14.14 Jointers

A jointer is an electric machine used for facing or flattening wood by passing the stock over a cylindrical multiple-knife cutter head. It does the work of a hand plane, and when operated by an experienced person, it can do many other jobs.

There are separate OSHA regulations for hand-fed jointers with horizontal cutting head and for wood jointers with vertical head. For hand-fed jointers with a horizontal cutting head, the following guarding requirements apply.

Each handfed planer and jointer with horizontal head shall be equipped with a cylindrical cutting head, the knife projection of which shall not exceed one-eighth inch beyond the cylindrical body of the head.

The opening in the table shall be kept as small as possible. The clearance between the edge of the rear table and the cutter head shall be not more than one-eighth inch. The table throat opening shall be not more than 2½ inches when tables are set or aligned with each other for zero cut.

The opening between the table and the head should be just large enough to clear the knife. Deeper cuts should be avoided. Not only do they require a larger table opening, but such cuts also create kickback hazards. The regulation continues:

Each hand-fed jointer with a horizontal cutting head shall have an automatic guard which will cover all the section of the head on the working side of the fence or gauge. The guard shall effectively keep the operator's hand from coming in contact with the revolving knives. The guard shall automatically adjust itself to cover the unused portion of the head and shall remain in contact with the material at all times.

Each hand-fed jointer with horizontal cutting head shall have a guard which will cover the section of the head back of the gauge or fence.

Wood jointers with vertical heads must have either an exhaust hood or other guard so arranged as to enclose completely the revolving head, except for a slot of such width as may be necessary and convenient for the application of the material to be jointed.

Jointers are second only to circular saws as the most dangerous woodworking machines. Since jointers are used primarily in the jointing of small pieces of material, many serious accidents are frequently caused when jigs or similar holding devices are not used when working with small size blades. The minimum length of the piece jointed should be not less than four times the width of the bed opening.

Push sticks or blocks must be provided in the sizes and types suitable to the work being done. Use of push sticks prevents stock from tipping and also guards operators' fingers from coming in contact with knives.

Knives on jointers should be checked often for proper setting or adjustment, but knives or fence guards should never be adjusted unless the machine has come to a complete stop and the power has been turned off. Knives must be kept sharp.

Since a considerable amount of dust and chips is generated during the operation, the worker should be provided with proper and adequate personal protective equipment. A brush should be used to remove chips and dust from around knives or the area of the cutting head. Floors and the area around jointers should be kept free of debris which could cause stumbling or slipping.

An automatic feed device can be mounted above the stock, eliminating the need for the operator's hands to be near the cutter head. Operating controls should be recessed or shrouded. An emergency stop switch and a main power electrical disconnect are essential. An exhaust system is desirable. Neither gloves nor loose clothing should be worn by the operator or others working around jointers. Guards should be inspected frequently. Operators must never make a guard inoperative for any purpose.

14.15 Wood Planers

A wood planer is designed to dress and size rough sawed lumber on one or more sides. It planes boards to an even thickness. Stock passes under or between cylindrical, multiple-knifed cutter heads. These surfacing machines are less hazardous than jointers and shapers because they are power-fed and the operator's hands need not come close to the cutting head. The planer operator simply has to adjust for cut and then feed the stock into the infeed side of the machine. He then retrieves the surfaced board from the outrunning end.

Hazards may be inherent in point of operation, kickbacks, flying particles, vibration or clearances. Planers are often driven from a line shaft comprised of belts and pulleys running on the back side of the planer. Planer parts driven by belts and pulleys, even though they may be on the back side of the planer, should be completely enclosed by sheet metal or heavy mesh guards. Guards, should always be used regardless of the planer's location. Accidents arise from contact with blades while sharpening or adjusting and from pinching fingers and hands between materials and inadequately guarded in-running rolls.

OSHA requires that planing machines have all cutting heads, and saws if used, covered by a metal guard. This guard should be kept closed when the planer is running. Specific requirements apply to the material from which the guard is constructed:

If such guard is constructed of sheet metal, the material used shall be not less than 1/16 inch in thickness, and if cast iron is used, it shall be not less than 3/16 inch in thickness.

Where an exhaust system is used, the guards shall form parts or all of the exhaust hood.

OSHA regulations require feed roll guards. Feed rolls shall be guarded by a hood or a suitable guard to prevent the hands of the operator from coming in con-tact with the in-running rolls at any point. The guard shall be fastened to the frame carrying the rolls so as to remain in adjustment for any thickness of stock.

The worker must be sure that feed rolls, cutter heads and cylinders are stopped before reaching into the bed plate to remove wood fragments or to make adjustments. Sleeves must be tucked in or rolled up, and loose clothing should not be worn. Although gloves should not be worn while operating a planer, hand pads for handling rough wood may be necessary.

Kickbacks or throwing material toward the operator through action of the blade can be caused by:

- unsafe operating methods
- improper equipment maintenance or adjustment
- lack of physical safeguards

Kickbacks cannot be prevented by mechanical means. Therefore, the operator should always stand out of line of board travel. OSHA regulations require:

> Surfacers or planers used in thicknessing multiple pieces of material simultaneously shall be provided with sectional infeed rolls having sufficient yield in the construction of the sections to provide feeding contact pressure on the stock, over the permissible range of variation in stock thickness specified or for which the material is designed. In lieu of such yielding sectional rolls, suitable section kickback finger devices shall be provided at the infeed end.

Operators should never feed boards of different thickness at the same time because the thinner board will not be held by the feed rolls and can be kicked back from the cutter heads. Feed roll corrugations should be kept clean and free from dust, pitch, or any other impediments. They should be kept sharp by filing as needed so that they grip the material as tightly as possible.

Chips and splinters are thrown off as a result of the rapid cutting action of the blade. The operator should wear a face shield or goggles as protection against slivers and chips thrown back by the cutter heads. He should not bend over to watch the board being planed but should stand aside once the board starts through. A dust collection system is desirable to reduce the hazard of flying particles.

A planer is powerful, fast running and tends to vibrate excessively. Vibration also may be caused by dull or improperly sharpened blades. Vibration can be reduced by anchoring the planer on a solid foundation and, if necessary, by insulating it from the foundation with materials that absorb vibration.

The planer is a noisy machine, therefore, it should be isolated or enclosed in a soundproof area. If neither of these alternatives is possible, those working the immediate area will need ear protection.

Since materials passing through the planer are generally long and fast moving, workers can be caught between stock emerging form the planner and stationary objects or structural elements of the building. The space at the outrunning end should be fenced or marked off to keep workers out of the area. Insofar as possible, aisles should be located where workers will not need to pass in front of or to the rear of the planner.

14.16 Sanding Machines

There are three types of machines commonly used to sand surfaces: drum sanders, disk sanders, and belt sanders. Accidents at the point of operation may occur when operator's hands or fingers become caught between the work rest and the belt on a manually fed machine, between the feed rolls and the material on a self-feed machine or by accidental contact with the moving abrasive belt.

Because of the dangers presented to hands and fingers at the point of operation, OSHA requires machine guarding:

Feed rolls of self-feed sanding machines shall be protected with a semi-cylindrical guard to prevent the hands of the operator from coming in contact with the in-running rolls at any point. The guard shall be constructed of heavy material, preferably metal, and firmly secured to the frame carrying the rolls so as to remain in adjustment for any thickness of stock. The bottom of the guard should come down to within 3/8 inch of a plane formed by the bottom or contact face of the feed roll where it touches the stock.

All manually-fed sanders should be equipped with a work rest and be properly adjusted to provide minimum clearance between the belt and the rest and to secure support for the work being sanded. Pieces too small to allow the hands to be kept a safe distance from the work should be held in a jig or similar holding device.

The regulations for drum and disk sanders require an exhaust hood or, if no exhaust system is required, another guard arranged to enclose the revolving drum or disk, except for the portion above the table designed for the work feed. The distance between disk or drum and the table must be kept to a minimum.

Belt sanding machines require more guards because of the additional hazard created by nip points:

Belt sanding machines shall be provided with guards at each nip point where the sanding belt runs on to a pulley. These guards shall effectively prevent the hands or fingers of the operator from coming in contact with the nip points. The unused run of the sanding belt shall be guarded against accidental contact.

Considerable dust and bits of wood are thrown off in the sanding process. Eye protection is highly recommended.

Sanders should be provided with an exhaust system. The exhaust hood should cover all the sanding surface except the operating area. Exhaust intakes should be designed and placed so that the natural throw of the abrasive belt is directly into the exhaust hood. Intakes should be placed as near as possible to the point of contact of the wood with the abrasive belt.

Personnel operating sanders should wear goggles and dust-type respirators during sanding operation and while cleaning up afterwards. Loose or improperly tensioned belts cause undue wear and fraying and can tear or break, causing material on the jam or belt to fly toward the operator.

Abrasive belts used on sanders should be the same width as the, pulley-drum, and drums should be adjusted to keep the abrasive belt taut enough to turn at the same speed as the pulley-drum, yet not slip on the drum when material is brought into contact with the moving abrasive belt. Abrasive belts should be inspected before use, and those found to be cracked, frayed or excessively worn in spots should be replaced, even though the remainder of the belt appears to be in good condition.

The operating controls, while not a severe source of injury, should include an emergency stop button. A main power electrical disconnect is required.

14.17 Maintaining Woodworking Machinery

In the section on machinery and machine guarding, OSHA established nine inspection and maintenance requirements for woodworking machinery:

1. Dull, badly set, improperly filled, or improperly tensioned saws shall be immediately removed from service before they begin to cause the material to stick, jam, or kick back when it is fed to the saw at normal speed. Saws to which gum has adhered on the sides shall be immediately cleaned.

2. All knives and cutting heads of woodworking machines shall be

kept sharp, properly adjusted, and firmly secured. Where two or more knives are used in one head, they shall be properly balanced.

3. Bearings shall be kept free from lost motion and shall be well lubricated.

4. Arbors of all circular saws shall be free from play.

5. Sharpening or tensioning of saw blades or cutters shall be done only by persons of demonstrated skill in this kind of work.

6. Emphasis is placed upon the importance of maintaining cleanliness around woodworking machinery, particularly as regards the effective functioning of guards and the prevention of fire hazards in switch enclosures, bearings, and motors.

7. All cracked saws shall be removed from service.

8. The practice of inserting wedges between the saw disk and the collar to form what is commonly known as a "wobble saw" shall not be permitted.

9. Push sticks or push blocks shall be provided at the work place in the several sizes and types suitable for the work to be done.

14.18 Milling Machines

Milling is the process by which a piece of metal is machined by bringing it into contact with a rotating multiple-edged cutter. The horizontal milling machine has the spindle horizontal to the table, the work is fed into the cutter. The vertical milling machine has the spindle vertical to the table, stock can be fed to the tool or the tool can be fed to the stock. The plain machine is used with tables that cannot swivel, the universal machine is used with tables that can swivel up to 90 degrees of a horizontal position.

Milling machines may be driven by individual motors or from a line-shaft, comprised of a belt and pulleys. Power transmission components should be totally enclosed. On those machines using adjustable belt drives, an interlock arrangement is required.

14.19 Drill Presses

The drill press is a metal cutting machine which uses a multiple-cutting-edge rotating tool to remove metal and produce a hole in the stock. It can also be used for countersinking, reaming, boring, tapping, facing, and routing.

The most commonly used drilling machine is a single-speed, floor-mounted, belt-driven machine for non-production drilling. There are three main types of drill presses: upright (vertical spindle), multiple spindle, and radial, with a long arm which can be swung into any position around the column.

Drill presses may be driven by individual motors or from a line-shaft comprised of belts and pulleys. The gears, spindles, and counterweights also present hazards. Power transmission components should be entirely enclosed on those machines that have adjustable belt drives. The enclosure should have an interlocked access door to facilitate speed changes.

Counterweights should be enclosed with guards, preferably iron pipe or sheet metal and angle iron, from the floor to the top of the weight when in extreme upward position. Counterweight chains should be maintained in good condition.

An emergency stopping device should be provided within easy reach of the operator. On motor-driven drills, stop-and-start buttons are acceptable. A main power disconnect is necessary.

Hazards at the point of operation include:

- coming in contact with the spindle or tool
- work slipping or turning because it was not properly clamped
- attempting to clean chips while drill is turning
- being hit by flying pieces from metal chips or a broken drill

Where practical, a telescope guard should be installed over the drill and spindle to protect against accidental contact. All work should be firmly and securely clamped to the table before starting the drill press. The tool should not be touched while using a quick-change clutch. A brush or stick should be used to remove chips from a drill. Burrs should be filed or scraped from drill holes, but this should be done only when the drill is stopped.

When deep holes are being drilled beyond the flutes of the drill, the drill should be removed frequently and the chips cleaned out with a brush or stick. If chips are allowed to pile up in such an operation, the tool may jam, causing the drill to break and insecurely clamped work to spin.

The drill should not be operated at excessive speed or feed since it may break or shatter. The drill should not be used if it is dull. Chuck wrenches, keys, or drifts should not be left in chucks or on the table. All tools and loose material should be removed from the table before starting the machine.

When starting the drill, the operator should use a center punch mark. To avoid drilling into the table, the operator should position the work over an opening in the table or use a bottom piece under the work. To remove a drill bit from the chuck, the operator should lower the spindle so that the point of the drill is close to the table before loosening the socket. The machine must be shut off before work is set up or taken off. Operators should shut off power and be certain the machine has stopped before leaving.

Goggles should be worn when operating drill presses. Drill press operators should wear snugly fitted clothing, short sleeves, and no jewelry. Long hair should be confined in a proper hair net, hood or cap.

14.20 Grinding Machines

Grinding machines shape material by bringing it into contact with a rotating abrasive wheel or disk. Polishing, buffing, honing, and wire brushing are also classed as grinding operations. The most common grinding machines are:

- stand and bench grinders
- surface grinders
- cylindrical grinders

The major hazards with all grinding machines are related to the rotating abrasive wheel. ANSI B7.1-1970 forms the American National Standard Safety Code for the Use, Care, and Protection of Abrasive Wheels. Parts of this standard have been adopted in the General Industry OSHA Safety and Health Standards (29 CFR 1910.215, Abrasive Wheel Machinery). Hazards at the point of operation include:

- work getting caught between tool rest (or guard) and wheel
- hands coming into contact with the wheel
- clothing getting caught by wheel or spindle ends

Injuries involving the abrasive wheel and disks arise from:

1. Failure to use eye protection in addition to the eye shield mounted on the grinder.
2. Holding the work incorrectly.
3. Incorrectly adjusting the work rest or using the machine without a work rest.
4. Grinding on the side of the wheel.

5. Taking too heavy a cut.
6. Applying work too quickly to a cold wheel or disk.
7. Grinding too high above the center of a wheel.
8. Failing to use wheel washers (blotters).
9. Using bearing boxes with insufficient bearing surface.
10. Using a spindle with incorrect diameter.
11. Using a spindle with the threads cut so that the nut loosens as the spindle revolves.
12. Dressing the wheel incorrectly.
13. Using an abrasive saw blade instead of a grinder disk.

The hazards of flying fragments because of the disintegration or "explosion" of the abrasive wheel result from:

1. Improper mounting of the wheel.
2. Cracks or flaws in the wheel.
3. Incorrect wheel for the work.
4. Wheel being run too fast.
5. Flanges (lack of flanges, unequal size, etc.).
6. Vibration caused by the wheel's being out of balance, worn bearings, etc.
7. Side pressure on wheels not designed for that work.
8. Work being caught between tool rest or guard and the wheel.
9. Particles from the material being ground as well as wheel particles.

If the wheel starts vibrating or chattering, it must be stopped. Usually the vibration means that the wheel is not securely attached or is out of balance.

The power transmission including shafting, belts, and pulleys must be guarded in accordance with ANSI standards. Abrasive wheels must be guarded in accordance with ANSI standards, which have been adopted by OSHA. The guard should enclose the wheel as completely as the nature of the work permits. It should be adjustable so that as the diameter of the wheel constantly decreases, the protection will not be lessened. The maximum angular exposure varies with the type of grinder.

OSHA regulations state:

Bench and floor stands. The angular exposure of the grinding wheel periphery and sides for safety guards used on machines known as bench and floor stands should not exceed 90 degrees or one-fourth of the periphery....

Wherever the nature of the work requires contact with the wheel below the horizontal plane of the spindle, the exposure shall not exceed 125 degrees.

Cylindrical grinders. The maximum angular exposure of the grinding wheel periphery and sides for safety guards used on cylindrical grinding machines shall not exceed 180 degrees. This exposure shall begin at a point not more than 65 degrees above the horizontal plane of the wheel spindle.

Surface grinders and cutting-off machines. The maximum angular exposure for safety guards used on cutting-off machines and on surface grinding machines which employ the wheel periphery shall not exceed 150 degrees. This exposure shall begin at a point not less than 15 degrees below the horizontal plane of the wheel spindle.

Swing frame grinders. The maximum angular exposure of the grinding wheel periphery and sides for safety guards used on machines known as swing frame grinding ma-chines shall not exceed 180 degrees, and the top half of the wheel shall be enclosed at all times.

There are cases of grinding where only the top of the wheel is used. In these instances, OSHA regulations state:

Where the work is applied to the wheel above the horizontal centerline, the exposure of the grinding wheel periphery shall be as small as possible and shall not exceed 60 degrees.

Suitable racks, bins or drawers should be provided to store the various types of wheels used. Stored wheels should not be subject to extremes of temperature and humidity. Wheels can be damaged by high humidity and/ or freezing temperatures.

A method of inspecting abrasive wheels has been mandated by OSHA. The requirements describe the ring test to be used:

Immediately before mounting, all wheels shall be carefully inspected and sounded by the user (ring test) to make sure they have not been damaged in transit, storage, or otherwise. The spindle speed of the machine shall be checked before mounting of the wheel to be certain that it does not exceed the maximum operating speed marked on the wheel. Wheels should be tapped gently with a light nonmetallic implement, such as the handle of a screwdriver for light wheels, or a wooden mallet for heavier wheels. If they sound cracked (dead), they shall not be used.

Wheels must be dry and free from sawdust when applying the ring test; otherwise the sound will be deadened. It should also be noted that organic bonded wheels do not emit the same clear metallic ring as do vitrified and silicate wheels.

Tap wheels about 45 degrees each side of the vertical centerline and about 1 or 2 inches from the periphery. Then rotate the wheels 45 degrees and repeat the test. A sound and undamaged wheel will give a clear metallic tone. If cracked, there will be a dead sound and not a clear ring. This is known as the "ring" test.

Because most defective wheels break when first started, new wheels should be run at full operating speed for at least one minute before work is applied. During this test period, the operator should stand well away from the machine.

As the wheel wears down, the spindle speed is sometimes increased to maintain the surface speed. Therefore, when the worn wheel is replaced, the spindle speed must be adjusted. Otherwise the new wheel may break because it is operating at a surface speed that exceeds the manufacturer's recommendations.

Grinding machines should be provided with a means of limiting the diameter of the wheel which can be mounted. The safety guard is generally satisfactory for this purpose on single speed machines.

On variable speed machines, the speed shifting device should be connected with an adjustable guard or another diameter limiting device to prevent the mounting of a wheel which might run at higher than the recommended surface speed. When operating grinding wheels on equipment especially designed for high speed, it is the responsibility of the user to maintain this equipment in safe operating condition at all times. Rules for the safe operation of this equipment submitted by the builder should be observed.

If an existing machine is altered by the user to operate at special speeds, the user must assume all of the responsibility of the machine builder. The user should fully inform all operating personnel that only wheels identified for operation at special speed should be used and that at no time should the maximum speed marked on the wheel be exceeded. Protection to operating personnel, as well as adjacent areas, should be maintained at all times.

Grinding machines should be supplied with sufficient power to maintain the rated spindle speed under all conditions of normal operation. Stationary machines used for dry grinding should have provision made for connection to an exhaust system. Flanges are collars, disks or plates between which wheels are mounted. Grinding machines must be equipped with flanges in accordance with OSHA requirements:

All abrasive wheels shall be mounted between flanges which shall not be less than 1/3 the diameter of the wheel. Exceptions include:

- mounted wheels
- portable wheels with threaded inserts or projecting studs
- abrasive disks (inserted nut, inserted washer and stud type)
- plate mounted wheels
- cylinders, cup or segmental wheels that are mounted in chucks
- types 27 and 28 wheels
- certain internal wheels
- modified types 6 and 11 wheels (terrazzo)

Because the major stresses produced in an operating grinding wheel tend to combine and become greatest at the hole, stresses due to mounting and driving should act as far from the hole as practicable. This is best accomplished by using flanges at least as large as one-third the diameter of the wheel. OSHA recognizes three types of flanges: straight relieved flanges, straight unrelieved flanges, and adaptor flanges.

Straight relieved flanges shall be recessed at least 1/16 inch on the side next to the wheel at a distance which is specified in the dimensions for these flanges. Straight flanges of the adaptor or sleeve type shall be undercut so that there will be no bearing on the sides of the wheel within 1/8 inch of the arbor hole.

OSHA further requires that flanges be dimensionally accurate and in good balance. There shall be no rough surfaces or sharp edges. Although exceptions are made for Type 27 and 28 wheels and modified Types 6 and 11 wheels, OSHA requires that both flanges between which a wheel is mounted shall be of the same diameter and have equal bearing surface. All flanges must be maintained in good condition. When the bearing surfaces become worn, warped, sprung or damaged, they should be trued or refaced.

Blotters are used for several reasons. They tend to cushion the pressure of the flanges against high points or uneven surfaces and distribute the pressure evenly. They also prevent damage to the surfaces of the flanges from the abrasive surface of the wheel and provide better transmission of the driving power to the wheel.

OSHA requires that blotters (compressible washers) be used between flanges and abrasive wheel surfaces to insure uniform distribution of flange pressure. Exceptions are made for those kinds of wheels which do not require flanges.

On offhand grinding machines, OSHA requires work rests to be used to support the work. The work rests shall be of rigid construction and designed to be adjustable to compensate for wheel wear. The rest never should be adjusted while the wheel is in motion. The rest may slip, strike the wheel and break it, or workers may catch their finger between the wheel and the rest.

The supervisor who wants to promote safety will take some additional steps when grinding machines are used. He will make sure that the workers are trained in the hazards presented by grinding machines. He will be certain that, when a wheel has been mounted, the safety guard is properly positioned before the wheel is started. He will be certain that the grinding machine is run at operating speed with the safety guard in place for at least one minute before any work is applied. During this test period he will make sure that no one stands near the machine.

When a grinding wheel is broken in service, the supervisor must initiate an investigation immediately to find out why the wheel broke. In this way, not only will he be certain that the shop is in compliance with all state and federal regulations but also he will take steps to prevent breakage in the future.

Summary

This chapter examined the fundamental issues involved in machine guarding. The hazardous mechanisms which need to be safeguarded have been described. OSHA requirements for guarding at the point of operation and at the power source have been detailed along with suggestions for safe operation of machines commonly found in the shop. The basic types of guards have been described and the characteristics of good guards have been discussed.

As part of the safety and health program, machine guarding is not optional, it is mandatory. With the information derived from this chapter, the supervisor can evaluate whether the machines are guarded adequately and can take the necessary steps to comply with regulations devised for their well being and the welfare of the workers.

Bibliography

Perez, Graciela (1999) "Ergonomics: Achieving System Balance Through Ergonomic Analysis and Control" Chapter 23 in the Handbook of Occupational Safety and Health, John Wiley & Sons, N.Y., N.Y.

Supervisors Safety Manual, 5th ed., (1978) National Safety Council, Chicago, Illinois 60611.

U.S. Department of Health; Education, and Welfare, National Institute for Occupational Safety and Health, *Occupational Safety and Health in Vocational Education,* February 1979.

U.S. Department of Labor, Occupational Safety and Health Administration, *General Industry OSHA Safety and Health Standards (29 CFR 1910).*

15

Welding and Cutting Operations

Welding and cutting operations present several safety and health hazards in the shop. These hazards can be controlled by allowing the operations only in fire safe locations, providing adequate ventilation, using the proper protective equipment, and following the correct work procedures.

The hazards associated with welding and cutting arise from the associated toxic gases and fumes, radiation, electrical circuits, and flammable and combustible materials. This chapter describes safe operating procedures in welding and cutting which remove or at least minimize hazards while improving the quality of the work. Two commonly used terms in this chapter, gas-welding and oxygen-cutting, are defined as follows:

- Gas-welding process - A process which unites metals by heating them with the flame from the combustion of a fuel gas or gases. The process may include the use of pressure and a filler metal.
- Oxygen-cutting process - A process which severs or removes metal by the chemical reaction of the metal with oxygen at an elevated temperature maintained with heat from the combustion of fuel gases.

15.1 Welding and Cutting

The presence of oxygen is required to support any burning process. Oxygen must be combined with a "fuel" gas to produce the desired operating flame. Oxygen itself is not flammable or explosive; however, the presence of pure oxygen drastically increases the speed and force with which burning takes place. Its presence can turn a small spark into a roaring flame. Combustible materials burn much more rapidly in oxygen than in air. Oxygen also forms explosive mixtures in certain proportions with acetylene and other combustible gases.

Oxygen is ordinarily supplied in standard drawn steel cylinders. The

244-cubic foot cylinder is the most commonly used, but smaller and larger sizes are available. Full oxygen cylinders are pressurized from 2000 to 2600 pounds per square inch. Oxygen cylinder contents can be determined by reading the cylinder pressure gauge on the regulator when in use. Half of the full cylinder pressure rating indicates that half the volume (cubic feet) of oxygen remains. The maximum charging pressure is always stamped on the cylinder.

Oxygen must be labeled as "oxygen", never "air." Serious injury easily may result if oxygen is used as a substitute for compressed air. Oxygen should never be used in pneumatic tools, in oil preheating burners, to start internal combustion engines, to blow out pipe lines, to dust clothing or work, to create pressure, or for ventilation.

In cylinders of oxygen, there is as much as 2600 psig pressure. When the pressure is released from the cylinder through the regulator, the speed at which the oxygen travels exceeds the speed of sound, and heat and friction are generated. Oil and grease become highly explosive in the presence of oxygen under pressure. Every gauge made to be used with oxygen has this information printed on the side of it. OSHA requires (in 1910.252):

> Cylinders, cylinder valves, couplings, regulators, hose, and apparatus shall be kept free from oily or greasy substances. Oxygen cylinders or apparatus shall not be handled with oily hands or gloves. A jet of oxygen must never be permitted to strike an oily surface or greasy clothes or to enter a fuel oil or other storage tank.

Acetylene is a combination of Carbon and hydrogen (C_2H_2). It is produced when calcium carbide is submerged in water. The escaping gas from the acetylene generator is then trapped in a gas chamber and compressed into cylinders or fed into piping systems. Acetylene burned with oxygen can produce a higher flame temperature (approximately 6000°F) than any other gas used commercially. It ignites readily and in certain proportions forms a flammable mixture with air or oxygen. Its range of flammable limits is from 2.5 to 81 percent acetylene in air, a range greater than that of other commonly used gases.

Acetylene is an unstable gas when compressed in its gaseous state above 15 psig. OSHA states that under no condition "shall acetylene be generated, piped or utilized at a pressure in excess of 15 psig." This requirement does not apply to the storage of acetylene dissolved in a suitable solvent in cylinders approved by the U.S. Department of Transportation. Unlike oxygen, acetylene is too unstable to be stored in a hollow cylinder under high pressure. Therefore, acetylene cylinders are filled with a porous material, creating in effect a solid as opposed to a hollow cylinder. The porous filling is then saturated with liquid acetone. When acetylene is pumped into the cylinder, it becomes dissolved in the liquid acetone throughout the porous filling and is held in a stable condition. Since acetylene is highly soluble in acetone at cylinder filling pressure, large quantities of acetylene can be stored in comparatively small cylinders at relatively low pressure.

Acetylene for welding and cutting is usually supplied in cylinders having a capacity up to about 300 cubic feet of dissolved acetylene under pressure of 250 psi at 70°F. Other fuel gases such as propane, butane and their mixtures are used with oxygen in torches, primarily for oxygen cutting. These are supplied in cylinders in liquid form, generally under various trade names.

15.2 Handling Cylinders

Serious accidents can result from the misuse and mishandling of compressed gas cylinders. ANSI Z49.1-1967, Safety in Welding and Cutting, establishes standards for the marking, handling and storage of cylinders. This standard forms the basis for OSHA 29 CFR 1910.252, the federal regulations applying to welding, cutting, and brazing. OSHA states that:

Compressed gas cylinders shall be legibly marked, for the purpose of identifying the gas content, with either the chemical or the trade name of the gas. Such marking shall be by means of stenciling, stamping or labeling, and shall not be readily removable. Whenever practical, the marking shall be located on the shoulder of the cylinder.

It is illegal to tamper with the numbers and markings stamped into cylinders. Only cylinders which carry the approval of the U.S. Department of Transportation should be accepted. OSHA further requires that "all portable cylinders used for the storage and shipment of compressed gas shall be constructed and maintained in accordance with the regulations of the U.S. Department of Transportation." Cylinders meeting these specifications are manufactured under close supervision, subjected to severe testing and provided with proper safety devices.

Cylinders should be moved by tilting and rolling them on their bottom edges. Dragging and sliding cylinders across a surface should be avoided since this practice exposes the cylinders to unnecessary wear. It is preferable to move cylinders in a suitable cradle or cart. When cylinders are transported by vehicle, they should be secured in position. According to OSHA, cylinders "shall not be dropped" or struck, nor should they be permitted to strike each other violently because "rough handling, knocks or falls are liable to damage the cylinder, valve or safety devices and cause leakage." OSHA requirements state: that "all cylinders with a water weight capacity of over thirty pounds shall be equipped with means of connecting a valve protection cap or with a collar or recess to protect the valve."

Valve protection caps are designed to protect valves from damage. According to OSHA, valve protection caps shall not be used for lifting cylinders from one vertical position to another. Before raising oxygen cylinders from a horizontal to a vertical position, the cap should be properly in place and turned clockwise to be sure it is handtight.

A suitable cylinder truck, chain or steadying device should be used to keep cylinders from being knocked over while in use. If a cylinder truck is used, care must be taken to ensure its safe condition. Worn or bent wheels should be replaced; support chains must be present and in good condition; and, if the truck is equipped with a braking device, such a device must be kept in working condition. Cylinders must never be used as rollers or supports, whether full or empty. Their only purpose is to contain gas.

Full cylinders of oxygen and fuel gas should be used in rotation as received from the supplier. Unless otherwise marked, cylinders should be considered full and handled as such. Accidents can occur when containers under partial pressure were handled as thought they were empty. To avoid confusion, empty cylinders should be marked "empty" or "MT", segregated from full cylinders, and returned to the supplier as soon as possible. All valves must be closed, and valve protection caps must be in place.

According to OSHA, acetylene cylinders shall be stored valve end up. If acetylene cylinders are stored in a horizontal position, the acetone in which the acetylene is dissolved has a tendency to settle out to the end of the cylinder. An explosion may occur when the cylinder is opened and the oxygen and acetone are ignited. Storing cylinders in an upright position also minimizes external corrosion of the cylinder walls. Furthermore, OSHA requires that fuel gas cylinders shall be placed with valve end up whenever they are in use. Liquefied gases are to be stored and shipped with the valve end up.

When cylinders are stored inside a building, the storage area must be well protected, well ventilated, dry and at least twenty feet away from highly combustible materials such as oil. Indoor storage of fuel gas is limited to a total of 2,000 cubic feet or 300 pounds of LP gas.

Cylinders must be kept away from stoves, radiators, furnaces, or other hot places and they must not be located where they can become part of an electric circuit. Oxygen cylinders must be separated from fuel gas cylinders and combustible materials. OSHA requires a minimum distance of twenty feet or a noncombustible barrier at least five feet high having a fire-resistance rating of at least one-half hour.

Storage areas must be located where cylinders will not be knocked over or damaged by passing or falling objects and where no one can tamper with them. Cylinders should be secured by such means as chains or partitions. Where cylinders are stored outside, they should be protected from accumulations of ice, snow, and the direct rays of the sun.

Bars and similar devices shall not be used under valves or valve protection caps to pry cylinders loose when frozen to the ground or otherwise fixed. Warm (not boiling) water is recommended instead. A special T-wrench or key for opening or closing the cylinder valve on fuel gas cylinders must always be in position for use so the gas can be turned off quickly in case of emergency.

Filling cylinders is a delicate process requiring special equipment and training. Therefore, OSHA states that only the gas supplier is allowed to mix gases in a cylinder and that "no one, except the owner of a cylinder or person authorized by him, shall refill a cylinder."

15.3 Protective Equipment, Regulators, and Hoses

Fuel gas piping must have approved protective equipment installed to prevent the backflow of oxygen into the fuel gas supply system, the passage of a flashback into the fuel gas supply system, and excessive back pressure of oxygen in the system. Back-pressure protection requires an approved pressure-relief device set at a pressure not greater than the pressure rating of the backflow and flashback protection devices.

Regulators or reducing valves must be used on both oxygen and fuel gas cylinders to maintain a uniform gas supply to the torches at a correct pressure. Uncontrolled pressure is dangerous in itself. A properly adjusted regulator also acts as a safety device, tending to stop any flashback from entering the cylinder where it might cause serious damage.

Each regulator, whether oxygen or fuel gas, should be equipped with both a high pressure (contents) gauge and a low pressure (working) gauge. According to OSHA, "pressure-reducing regulators shall be used only for the gas and pressures for which they are intended." Pressure gauges should be tested periodically for accuracy. If the gauges have been strained so that the hands do not register properly, the regulator must be replaced or repaired before it is used again.

When regulators are connected but not in use, the pressure-adjusting device should be released. Cylinder valves should never be opened until the regulator is drained of gas and the pressure-adjusting device fully released.

The regulator is a delicate piece of equipment and must be handled carefully at all times. Hammers or wrenches must not be used to open or close cylinder valves. According to OSHA, the supplier shall be notified if valves cannot be opened by hand.

If a regulator "creeps," the cylinder should be closed and the regulator removed for repairs. "Creeping" is indicated on the low pressure (delivery) gauge by a gradual increase in pressure after the torch valves are closed.

A reverse-flow check valve on the regulator and torch handle reduces the possibility of mixing gases in the hoses and regulators. Once the torch is lit, mixed gases burn rapidly and can explode in the hoses, regulators, or cylinders. Such an explosion can result in injury to the welder and serious damage to the equipment. Reverse-flow check valves are screwed onto the regulator's outlet connection. They should be tightened securely with the proper wrench.

Oxygen and acetylene hoses should be color coded to prevent confusion. According to ANSI, the generally recognized colors are:

- Red for acetylene and other fuel gas hose
- Green for oxygen hose
- Black for inert gas and air hose

Hose connections must be checked for proper threading. Standard hose connections are threaded right-hand for oxygen and left-hand for acetylene or other fuel gas. This helps prevent an accidental switch of oxygen and fuel gas hoses. Figure 15.1 illustrates the proper connections and threading.

Oxygen and fuel gas hoses are not to be used interchangeably. A single hose having more than one gas passage should not be used. OSHA requires that when parallel sections of oxygen and fuel gas hose are taped together for convenience and to prevent tangling, not more than four inches out of twelve inches shall be covered by tape.

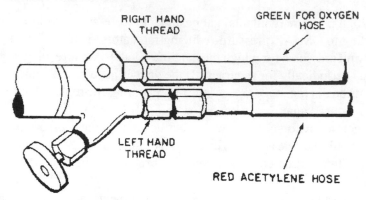

RIGHT HAND
THREAD

GREEN FOR OXYGEN
HOSE

LEFT HAND
THREAD

RED ACETYLENE HOSE

Fɪɢ. 15.1 Hose connections and proper threading.

Hose couplings must be of the type that cannot be unlocked or disconnected by means of a straight pull without rotary motion. The hose should be tested for leaks by immersing it, under normal working pressure, in water or by using soapy water (nonfat soap) or approved leak-test solution. Leaks and worn places in the hose must be repaired immediately by cutting the hose and remaking the joint with standard fittings. Leaks in the hose at the nipple connection should be repaired by cutting off the hose a few inches from the end and remaking the connection.

All leaks must be repaired at once. Escaping fuel gas may become ignited and start a serious fire. The OSHA standards require hose showing leaks, burns, worn places or other defects rendering the hose unfit for service to be replaced or repaired. Splices are the only acceptable way of making repairs. Repairs should not be made by taping.

After repairing, hose connections should be tested. According to OSHA 1910.252, hose connections shall be clamped or otherwise securely fastened in a manner that will withstand, without leak-age, twice the pressure to which they are normally subjected in service, but in no case less than a pressure of 300 psi. Oil-free air or an oil-free inert gas shall be used for the test.

Hoses must be protected from flying sparks, hot slag or other hot objects including grease and oil. Hoses should be stored in a cool place. Unnecessarily long lengths of hose are to be avoided since they are difficult to purge properly and they tend to become kinked or tangled.

15.4 Torches

Torches are constructed of metal castings, forgings, and tubing. Torches are usually made of brass or bronze although stainless steel may also be used. Torches should be designed to withstand the rough handling they sometimes receive and should be listed by Underwriters Laboratories or be approved by an agency such as the Factory Mutual System.

Torches should not be used as hammers or to knock slag from work. Such misuse can deform the torch or tips. Slag hammers and wire brushes should be available.

Gases enter the torch by separate inlets and move through the valves to the mixing chamber. From there, the gases move to the outlet orifice located in the torch tip. Several interchangeable tips are provided with each torch. The tip orifices come in various sizes to accommodate the specific task.

Unlike the welding torch, the cutting torch uses a separate jet of oxygen in addition to the jet(s) of mixed oxygen and fuel gas. The jets of mixed gases preheat the metal and the jet of pure oxygen is used for cutting. The flow of oxygen to the cutting jet is controlled by a separate valve.

15.5 Steps in Setting Up a Gas Welding and Cutting Operation

If the following procedure for setting up a gas welding and cutting operation is followed, workers can minimize the hazards to which they are exposed.

1. Inspect the Valve Threads. Inspect the cylinder valve threads. Remove dirt or dust with a clean cloth. If oil or grease is detected, do not use the cylinder.
2. Crack Cylinder Valve. "Crack" each cylinder valve to dislodge any dirt, dust or rust that may be present. To "crack" the valve, momentarily open it slightly and then close it immediately. On a fuel gas cylinder, first make sure that no source of ignition is near. Otherwise the gas may ignite at the valve. When "cracking" the valve, stand behind or to one side, not directly in front of the valve.
3. Inspect the Regulators. Remove dirt or dust with a clean cloth. If oil or grease is detected, the regulator must be cleaned by an authorized service representative. If threads are damaged, they must be repaired.
4. Attach Regulators to Valves. Connect the oxygen regulator to the

oxygen cylinder valve. Tighten securely (clockwise) with a regulator wrench. Attach the fuel gas regulator to the fuel gas cylinder. Tighten it securely. Do not use a pipe wrench or pliers. Be sure that the connections between the regulators and the cylinder valves are gas-tight.

5. Release Tension on Adjusting Screws. Release tension on the pressure-adjusting screws on the regulator by turning them counterclockwise until they are loose. This step keeps the regulator and gauges from being damaged when the cylinder valve is opened.

6. Open Cylinder Valves. Being careful not to stand in front or in back of the regulator, open the cylinder valve slightly. Never open a cylinder valve suddenly; the rush of gas might strain the cylinder pressure gauge mechanism. Let the hand on the high pressure gauge move up slowly until maximum pressure is registered. For an oxygen cylinder or any fuel gas cylinder other than acetylene, gradually open the valve to its full limit to completely seal the valve packing. For an acetylene cylinder it is best to open the valve no more than ¾ of a turn of the spindle; the valve must never be opened more than 1½ turns.

7. Examine Hoses. Before attaching hoses to the welding torch handle or regulator, examine them carefully. If cuts, burns, worn areas or damaged fittings are found, repair or replace the hose. If oil or grease is detected, do not use. If the hose is new, blow it out with oxygen to remove preservative talc. For a fuel gas hose, cup one hose end against the outlet connection of the oxygen pressure-reducing regulator. Open the regulator to about five psi pressure to blow out the hose. Then blow through the hose from the mouth to purge it of concentrated oxygen.

8. Connect Oxygen Hose to Oxygen Regulator. Connect the oxygen hose to the outlet of the oxygen regulator. Adjust the oxygen regulator to allow three to five psi to escape through the hose. Allow oxygen to flow five to ten seconds to clear the hose of dust, dirt or preservative. Then shut off oxygen flow.

9. Connect Fuel Hose to Fuel Regulator. Attach and clear the fuel hose in the same manner.

10. Inspect Torch. Inspect torch handle head, valves, and hose connections. Remove dirt or dust with clean cloth. Do not use the torch handle if oil or grease is detected or if parts are damaged.

11. Attach Hoses to Torch. Attach the oxygen welding hose to the oxygen inlet valve on the torch. Attach the fuel hose to the fuel

inlet on the torch. Fuel reverse-flow check valves should be used on the torch handle. Tighten securely with a wrench.

12. Check for Leaks. Check for leaks with an approved leak-detector solution. Bubbles will appear if the connection is leaking. Test the following points for leakage: cylinder valve stem, regulator inlet connection at the cylinder valve, all hose connections, and the torch valve.

   If fuel gas leaks around the valve stem when the valve is opened, close the valve and tighten the gland nut. This compresses the packing around the spindle. If this does not stop the leak, close the valve and move the cylinder outdoors. Attach a tag to the cylinder stating that it should not be used. Notify the cylinder supplier immediately.

   If fuel gas leaks from the cylinder valve and cannot be shut off with the valve stem or if rough handling should cause any of the fusible safety plugs to leak, the cylinder must be moved to an open place well away from any possible source of ignition and be plainly tagged as having an unserviceable valve or fusible plug. The cylinder should then have its valve opened slightly to let the acetylene escape slowly.

   While the fuel gas is escaping from the cylinder, a sign must be placed close by to warn everyone against coming near the cylinder with a lighted cigarette or other source of ignition. To make sure that no fire occurs, a responsible person should stay in the area until the cylinder is de-pressured. The supplier should be notified promptly and his instructions followed as to the return of the cylinder. Testing for hose leaks should be done in the manner described earlier in this unit.

13. Attach Proper Head, Tip, or Nozzle to Torch. Always inspect cone end, coupling nut, welding tip, and torch head before connecting. If damage, oil, or grease is detected, do not use. Connect the welding tip to the welding torch handle, and tighten the coupling nut. Some welding and cutting heads require only hand pressure when tightening, others require the use of a wrench. Follow the manufacturer's recommendation and always use the proper sized tip.

14. Adjust the Working Pressure and Purge Lines. With all torch valves closed, slowly open the oxygen cylinder valve. Open the torch oxygen valve. Turn in the pressure-adjust-mg screw on the oxygen regulator to the desired pressure. Close the torch oxygen valve.

Open the fuel gas cylinder valve (a maximum of 1-1/2 turns). With the torch fuel gas valve closed, turn in the pressure-adjusting screw to produce the desired pressure.

Purge each line separately. Open the oxygen torch valve and re-lcase oxygen to the atmosphere for a few seconds. Then close the valve. Do the same for the fuel gas, first making sure that there are no ignition sources nearby and that the area is well ventilated. Use the proper gas pressures for the size of the head, tip or nozzle selected.

In lighting the torch and adjusting the flame, always follow the manufacturer's directions for the particular model torch being used.

15.6 Lighting the Torch

The following precautions are necessary when lighting the torch:

1. Always wear safety goggles when working with a lighted torch.
2. Never use matches for lighting torches, hand burns may result. Instead of matches, use friction lighters, stationary pilot flames, or a similar ignition sources.
3. Never use acetylene at pressures above 15 psig. To do so is unsafe and violates OSHA standards, insurance regulations, and municipal and/or state laws.
4. Always check the area to make certain that ignition sources such as flames, sparks, hot slag, or metals are not present.
5. Never use equipment that has come in contact with oil or grease or that needs repairs.

In general, the following four steps are required when lighting a torch:

1. Open the torch oxygen valve to produce a small to moderate flow of gas (about ¼ turn).
2. Open the torch fuel gas valve to produce a flow somewhat greater than the oxygen flow (about ½ turn).
3. Using a friction lighter or stationary pilot flame, immediately light the mixture at the tip. Point the flame away from persons and cylinders.
4. Adjust the flame by opening the torch oxygen valve until a bright neutral flame is reached.

15.7 Backfires and Flashbacks

A backfire is a loud popping or snapping noise associated with the momentary extinguishment of the flame at the torch tip. It may be caused by touching the tip against the work, by overheating the tip, by operating the tip at pressures other than those recommended, by a loose tip or head, or by dirt on the seat. If the work is hot enough, the torch can be relighted at once. If it cannot be relighted instantly, a torch lighter must be used. Before relighting the torch, the cause of the backfire should be determined.

A flashback occurs when the flame burns back inside the torch or, if an explosive mixture is present in one of the lines, hoses, or regulators. Usually a flashback is accompanied by a shrill hissing or squealing. A smoky or sharp pointed flame may be present. A flashback indicates that something is very wrong, either with the torch itself or with the way it is being used. Flashbacks can be caused by failure to purge, improper pressures, distorted or loose tips or mixer seats, kinked hose, clogged tip or torch orifices, or overheating the tip or torch.

To stop the flashback, it is necessary to close the torch oxygen valve. Then the fuel gas valve should be closed and the torch allowed to cool off before relighting. Oxygen should be blown through the tip for a few seconds to clear out any soot which may have accumulated. In a cutting torch, oxygen should be blown through both the preheating and the cutting orifices. If a flashback burns the hose, the burned section must be discarded and any new hose purged before connecting it to the torch and regulator.

15.8 Shutting Down a Welding Operation

When the welding operation is completed, certain steps must be taken to safely shut down the apparatus. This procedure prevents leaks which could cause a serious fire and reduces the likelihood of a regulator fire when the oxygen cylinder valve is opened again. The following five steps should be followed:

1. Shut of Torch Valves. Shut off the fuel gas and oxygen valves in the order recommended by the torch manufacturer. If the oxygen valve is shut first the fuel gas flame enlarges appreciably and the welder can be burned. Unburned fuel gas also escapes into the work area, sometimes in the form of carbon "feathers." However, if the fuel gas valve is shut first, a pop or band may occur. This

noise can distract nearby workers, increasing the likelihood of accidents. It also throws carbon soot back into the torch, eventually partly clogging gas passages.

2.   Close both cylinder valves.

3.   Open Torch Valves, Turn Out Pressure-Adjusting Screw and Close Torch Valves. Both the oxygen and fuel valves need to be drained in order to release all pressure from the hose and regulator. Pressures should not be relieved simultaneously. Let the oxygen in the system drain out. Then close the oxygen valve. After the regulator gauge reading has reached zero, release the pressure-adjusting screw. Follow the same procedure for the fuel valve. When releasing the fuel pressure, care must be taken that a fire hazard is not created by the release of fuel gas.

4.   Uncouple the Regulator. When regulators are to be out of service for several weeks or longer, it is good practice to turn in the pressure-adjusting screw just enough to move the regulator valve off its seat.

5.   Remove Regulators Before Moving Cylinders. The regulators must be removed before moving the cylinders. Replace the cap over the cylinder valve when empty and mark the cylinder "MT."

When shutting down the apparatus for intervals of less than an hour, close only the torch valves. Leave the hose arid torch in an orderly fashion so that they will not be damaged. Never hang a torch or hose on a regulator or cylinder valve unless the cylinder and torch valves are closed and the hose is drained of gas. Never crimp hose to halt temporarily the flow of gases (for example, when changing a torch or tip). When shutting down for an extended period, all apparatus should be disconnected and stored.

15.9 Arc Welding and Cutting

Arc welding is a process of joining metals by means of the heat created by an electric arc. The pieces to be welded are placed in position and the intense heat of the electric arc applied to the joint melts the metal. Pressure may be applied and filler metal may be used. When the joint cools, it becomes one solid piece. Shielded welding uses gas and flux to blanket the welding. It is used for joining metals such as copper, aluminum, and stainless steel which oxidize readily at high temperatures.

Arc cutting has been replaced with arc-oxygen cutting, (especially useful for metals that do not oxidize readily), plasma arc cutting (for quality cuts), and carbon arc-air cutting (for smooth cuts). Arc welding and cutting require two welding leads, the electrode lead and the work lead, from the source of current. Usually one lead is connected to the electrode holder. The other cable, connected to the work, is the best means of providing the grounding circuit to the welding machine. The principal hazards presented by arc welding are:

- intense ultraviolet, visible and infrared radiation
- production of ozone and nitrogen oxides
- action of ultraviolet rays on chlorinated hydrocarbon vapors
- production of toxic fumes from melting toxic metals or metal alloy
- production of carbon monoxide
- splatter of molten metal
- handling high pressure gas in cylinders and manifolds
- electrical shock

The first five hazards in this list were discussed in Chapter 12, Health Hazards. In Chapter 13, Personal Protective Equipment, the hazards presented by molten metal were examined. Earlier in this chapter the requirements for handling compressed gas cylinders were discussed. Section 15.12 of this chapter includes a discussion of the hazards of electrical shock.

15.10 Radiation Hazards

Production of ultraviolet radiation is high in gas-shielded arc welding. For example, a shield of argon gas around the arc doubles the intensity of the ultraviolet radiation. With the greater current densities required (particularly with a consumable electrode), the intensity may be five to thirty times as great as with non-shielded welding.

Infrared heats the tissue with which it comes in contact. Unless the heat causes an ordinary thermal burn, there is no harm. Wherever possible, arc welding operations should be isolated so that other workers will not be exposed to direct or reflected rays. Walls, ceilings and other exposed inner surfaces should be painted with a finish of low reflectivity, such as zinc oxide and lampblack.

If the size of the work permits, arc welding stations for regular production work can be enclosed in booths. The inside of the booth should be painted with a finish of low reflectivity and provided with portable noncombustible or flameproof screens similarly painted or with curtains.

Booths and screens should be designed to permit circulation of air at the floor level. Welding should not be done near vapor-degreasing operations or spray booths. Such degreasing solvents as trichloro-ethylene can decompose under ultraviolet radiation and become dangerous.

According to OSHA (1910.252), helmets or hand shields shall be used during all arc welding or arc cutting operations, excluding submerged arc welding. Goggles also should be worn to provide protection from injurious rays from adjacent work, and from flying objects. Helpers or attendants shall be provided with proper eye protection. Goggles or other suitable eye protection are also required during all gas welding or oxygen cutting operations.

OSHA further requires that these helmets and hand shields "be made of a material which is an insulator for heat and electricity" and that they are not readily flammable. Specifically, helmets and hand shields must be arranged "to protect the face, neck and ears from direct radiant energy from the arc." If cracks occur in helmets or hand shields, they must be discarded immediately. Exposure to the arc rays can cause serious burns.

Protective shields are provided with a glass window, the standard size being 2 inches by 4 and 1/8 inches. This glass protects the eyes from molten metal splatter. The glass must be composed so as to absorb the infrared rays, the ultraviolet rays, and most visible rays emanating from the arc.

In selecting welding lenses and goggles, it is important to consider the manufacturer's reputation and his experience in the production of welding equipment, as well as the results of scientific tests of the lenses. Chapter 13 contains a discussion of eye and face protection, which illustrates various protectors and indicates which are recommended for specific operations, including welding. In cases of irritation or flash burn, a physician should be consulted immediately.

Clothing must not only keep off the splatter and molten particles but also must obstruct the rays of the arc. An arc-burn on the skin resembles a sunburn except that it is usually more intense. Dark-colored shirts are preferred to light ones because arc rays readily penetrate light-colored fabrics.

Gloves should be worn at all times to protect the hands and wrists. When extensive welding operations are to be performed in the vertical and overhead positions, leather sleevelets, aprons and, in some cases, leggings should be employed to prevent severe burns from splatter and molten metal.

Workers should be cautioned against picking up, either with bare or gloved hands, pieces of metal which have been welded or heated. The stub ends of discarded electrodes also can cause burns.

15.11 Production of Contaminants

Another major hazard created by welding operations is the production of contaminants, either as byproducts or as the result of the operation itself. OSHA requires (1910.252) that local exhaust or general ventilating systems shall be provided and arranged to keep the amount of toxic fumes, gases, or dusts below the maximum allowable concentrations.

For many welding and cutting operations, control by dilution ventilation is sufficient. That is, enough fresh air can be added to the contaminated air so that hazardous concentrations do not develop. However, the effectiveness of dilution ventilation depends on the following factors:

- the size of the space in which welding or cutting is done, especially the height of the ceilings
- the total number of welders working within the space
- the hazardous chemical or physical agents produced by the welding or cutting

OSHA requires mechanical ventilation when welding or cutting is conducted in areas meeting the following conditions:

- a work space of less than 10,000 cubic feet per welder
- a room less than sixteen feet high
- a confined space or a space containing partitions, balconies, or other structural barriers to obstruct cross ventilation

Otherwise, natural ventilation should be sufficient for most welding and cutting activities.

When dilution ventilation is used, the system must move at least 2,000 cubic feet of air per minute per welder, unless local exhaust hoods and booths are used to control fumes where they are produced. Chapter 12 includes a discussion of several methods of local exhaust ventilation were described including fixed enclosures, freely movable hoods, and down-draft benches.

A recent development in local exhaust ventilation is the extractor nozzle. In this system, a slotted exhaust chamber is installed as part of the welding equipment itself. The slotted exhaust chamber is positioned to allow the welder a clear view of the electrode. The contaminated air from the welding operation is drawn through the chamber to an exhaust system. If gasoline-driven welding machines are operated indoors, exhaust gases must be piped outside in order to avoid carbon monoxide poisoning.

15.12 Electrical Hazards

Arc welding processes depend on the intense localized heat from applied electrical energy, not from a chemical reaction. Electrical hazards are many, but careful operation can prevent most accidents.

15.12.1 Open Circuits

The polarity switch changes the flow of electric current from one terminal to another, that is, either from positive (reverse polarity) to negative (straight polarity) or the reverse. The range switch or tap switch helps the operator of a DC welding generator equipped with a system of tapping into the welding circuit to obtain the desired current setting. These switches should be operated only when the machine is idling and the welding circuit is open.

Arcing is apt to occur if the circuit is open at high current, resulting in burns to the person throwing the switch and the contact surfaces of the switch.

15.12.2 Grounded Circuits

Every power circuit should be grounded to prevent accidental shock. OSHA requires that the "grounding of the welding machine frame shall be checked. Special attention shall be given to safety ground connections of portable machines." Otherwise a stray current may give a severe shock if one hand is placed on the motor and the other on the switch box or other grounded equipment. Conduits containing electrical conductors and pipelines must not be used to complete a work-lead circuit.

15.12.3 Electrodes and Electrode Holders

Arc welding is done with either a metallic or a carbon electrode. For gas-shielded metal arc welding, the electrode is a solid or flux-cored wire. For shielded metal arc welding, the electrode is a covered wire.

The electrode holder consists of a heat-resistant handle and a clamping device for holding the electrode. It is so designed that the electrode can be gripped firmly at any angle and held in that position. A fully insulated electrode holder reduces the likelihood of accidentally striking an arc. Electrode holders will become hot during welding operations if holders designed for light work are used on heavy welding or if connections between the cable and the holder are loose.

The jaws of the electrode holder should be maintained tight and the gripping surfaces in good condition to provide close contact with the electrodes. Defective jaws will permit the electrode to wobble and render control of the welding operations difficult. The connection of the electrode lead to the holder should be brazed. If the older type mechanical connection is used, it should be maintained tight at all times.

The voltage between the electrode holder and the ground, during the "off" arc period, is the open circuit voltage. Unless the welder uses the equipment provided for his protection, he can become exposed to this voltage while changing electrodes, setting up work, or changing working position. This danger is particularly great during hot weather when he is perspiring.

The welder should keep his body insulated from both the work and the metal electrode and holder. He should never permit the bare metal part of an electrode, the electrode insulation, or any metal part of the electrode holder to touch his skin or damp clothing. He should never change electrodes with wet gloves or bare hands or when standing on wet floors or grounded surfaces.

When welding is to be interrupted for more than an hour, OSHA requires that the machine be disconnected from the power source, that all electrodes be removed from the holders, and that the holders be "carefully located so that accidental contact cannot occur." In addition, OSHA regulations state that electrode holders when not in use shall be so placed that they cannot make electrical contact with persons, conducting objects, fuel or compressed gas tanks.

When not in use, the electrode holder must never be left in contact with the table top or other metallic surface in direct contact with the welding ground. An insulated hook or holder should be provided for the electrode holder when not in use. When in contact with the ground circuit, the holder causes a dead short circuit on the welding generator. Should the machine be started up, this short circuit would cause an excessive load on the motor and could damage the insulation and fuses. Great care must be taken to prevent contact of the electrodes, electrode holders, or other live parts with compressed gas cylinders. Serious explosions or fire may occur from such contact.

15.12.4 Welding Cables

Welding cables must not be subjected to currents in excess of their rating capacity. Otherwise, overheating and rapid deterioration of the insulation can occur. OSHA requirements state that work and electrode lead cables should be frequently inspected for wear and damage. Cables with damaged insulation or exposed bare conductors shall be replaced. Joining lengths of work and electrode cables shall be done by the use of connecting means specifically intended for the purpose. The connecting means shall have insulation adequate for the service conditions.

If exposed sections of cable come in contact with metallic objects grounded in the welding circuit, an arc may result. If flammable materials such as oil or grease are in the vicinity, a fire may result.

All cable connections should be tight. OSHA requires that coiled welding cables "be spread out before use to avoid serious over-heating and damage to insulation." OSHA also forbids the use of cables "with splices within ten feet" of the electrode holder. A welder should not coil or loop welding electrode cable around parts of his body. OSHA regulations require that welding cable and other equipment be placed "clear of passageways, ladders and stairways." All wiring must be in compliance with the NFPA Code and local requirements.

The voltage across the welding arc varies from fifteen to forty volts, depending on the type and size of electrode used. To strike the arc, the welding circuit must supply somewhat higher voltage. This is called the open circuit or "no load" voltage. OSHA requires that open circuit voltages be as low as possible consistent with satisfactory welding or cutting. Table 15.1 summarizes this OSHA requirement. Proper switching equipment for shutting down the equipment must be provided.

TABLE 15.1
Open Circuit Voltages

Type	Operation	
	Manual	Automatic (mechanized)
Alternating Current	80 volts	100 volts
Direct Current	100 volts	100 volts

15.13 Purchasing, Installing, and Training

Equipment used for welding should be approved either by the National Electrical Manufacturing Association (NEMA) or by the Underwriters

Laboratories. The equipment must be installed by a qualified electrician, who is following the appropriate requirements of the National Electrical Code. Workers who are to do welding must be properly instructed in hazards of the operation and the safe practices which must be followed.

15.14 Prohibited Locations for Welding and Cutting

Because of the hazards created by welding and cutting operations, great care must be taken in locating work sites which do not create fire hazards or endanger nearby workers. Because of the potential for fire and explosion, OSHA lists four locations where cutting or welding are not permitted:

1. In the presence of explosive atmospheres (mixtures of flammable gases, vapors, liquids, or dusts with air), or explosive atmospheres that may develop inside uncleaned or improperly prepared tanks or equipment which have previously contained such materials, or that may develop in areas with an accumulation of combustible dusts.
2. In areas near the storage of large quantities of exposed, readily ignitable materials.
3. In sprinklered buildings while such protection is impaired.
4. In areas not authorized by management.

15.14.1 Prohibition of Welding and Cutting Operations Near Combustible and Flammable Vapors, Liquids, Gases, or Dusts

Welding and cutting operations should not be permitted in or near rooms containing flammable or combustible vapors, liquids, gases or dusts. They are also prohibited on or inside tanks or other containers which previously held such materials unless these are first properly purged. When fire and explosion hazards have been eliminated from areas formerly containing these flammable or combustible materials, thorough ventilation is necessary. Local exhaust equipment should be provided for removing the hazardous gases, vapors, and fumes that ventilation fails to dispel.

OSHA requirements state that no welding, cutting or other hot work is to be performed on used drums, barrels, tanks or other containers until they have been cleaned so thoroughly as to make absolutely certain that there are no flammable materials present or any substances such as greases, tars, acids, or other materials which, when subjected to heat, might produce flammable or toxic vapors. A cleansing agent should be used which is appropriate for the gas or liquid which was in the container. Then the container must be cleaned again, using either water or steam.

In cooperation with the American Welding Society, ANSI A6.0-1965 details Safe Practices for Welding and Cutting Containers That Have Held Combustibles.

15.14.2 Prohibition of Welding and Cutting Operations Near Ignitable Materials

Welding and cutting operations should not take place near large quantities of exposed, readily ignitable materials. Such materials should be moved at least 35 feet away from the work site. If neither the work nor the material can be moved, then combustibles must be protected with flameproof covers or sheet metal. According to OSHA, "edges of covers at the floor should be tight to prevent sparks from going under them."

Floors must be swept clean for a radius of 35 feet of such combustible materials as paper scraps, wood shavings, and sawdust. It is best to cover floors with metal or other noncombustible material. Covering the floor with damp sand or wetting it down is a less desirable solution. The moisture increases the electrical shock hazard for arc welders and necessitates special protection.

Where cutting or welding is done near walls, partitions, ceilings, or roofs of combustible construction, fire-resistant shields or guards shall be provided to prevent ignition. When welding is to be done on metal walls, partitions, ceilings or roofs, combustibles on the other side should be relocated. If combustibles cannot be relocated, OSHA requires that a fire watch on the opposite side from the work be provided.

When it is not possible to move flammable materials a safe distance from the cutting or welding work, suitable protection must be used to keep back sparks. Areas should be inspected for floor openings or cracks through which sparks could fall or in which they could lodge. Guards must be large enough and tight enough so that they do not permit sparks to roll underneath or slide through openings. Curtains should be weighted down against the floor or ground so that sparks cannot possibly get underneath. Recommended weights include angle irons, pipes, bricks, or sand. Only fire-resistant guards should be used for shielding sparks. Tarpaulins should not be used since they may catch fire.

Ducts and conveyor systems that could carry sparks to distant combustibles must be protected or shut down during welding or cutting. Because cutting produces more sparks and hot slag than does welding, safeguards against sparks are particularly important during cutting.

When welding or cutting is done near ignitable materials, a worker should be assigned to see that sparks do not lodge in floor cracks. OSHA regulations require fire watchers whenever welding or cutting is performed in locations where other than a minor fire might develop. In addition, fire watchers are required where:

- there is appreciable combustible material (either in materials or structural elements) closer than 35 feet to the welding or cutting operation
- appreciable combustibles are more than 35 feet away but are easily ignited by sparks
- wall or floor openings within a 35-foot radius expose combustible material in adjacent areas
- combustible materials are adjacent to the opposite side of metal partitions, walls, ceilings, or roofs and are likely to be ignited by conduction or radiation

Fire watchers must have some specific training, according to OSHA:

Fire watchers shall have fire extinguishing equipment readily available and be trained in its use. They shall be familiar with facilities for sounding an alarm in the event of a fire. They shall try to extinguish fires only when obviously within the capacity of the equipment available, or otherwise sound the alarm. A fire watch shall be maintained for at least a half hour after completion of welding or cutting operations to detect and extinguish possible smoldering fires.

It is not realistic to expect the welder or cutter to be the fire watcher. His attention should be on his work, and his goggles will obscure his peripheral vision.

It is good practice to provide each welding booth with a Class B and C fire extinguisher of a dry chemical, multipurpose, or carbon dioxide type. Pails of water and buckets of sand may be useful, particularly to catch the dripping slag from any cutting that is being done.

15.14.3 Prohibition of Welding and Cutting Operations Near Sprinklered Buildings

If a sprinkler system needs to be shut down for a time, this should be done when no welding or cutting is in progress. OSHA prohibits cutting or welding in sprinklered buildings while such protection is impaired.

15.14.4 Prohibition of Welding and Cutting Operations in Unauthorized Areas

According to OSHA standards, cutting or welding shall be permitted only in areas that are or have been made fire safe. Such areas should be:

- of noncombustible or fire-resistive construction
- essentially free of combustible and flammable contents
- suitably segregated from adjacent areas

15.15 Responsibilities of Supervisors

OSHA requirements assign specific responsibilities to supervisors. These responsibilities include:

1. the responsibility for the safe handling of the cutting or welding equipment and the safe use of the cutting or welding process
2. determining what combustibles and hazards are present in the work location
3. protecting combustibles from ignition by relocation of work, relocation or shielding of combustibles, and the careful scheduling of operations
4. giving the cutter or welder assurance that conditions are safe before work is begun
5. determining that fire protection and extinguishing equipment are properly located
6. ensuring fire watchers are on the site when conditions require their presence

Summary

There are may hazards presented by welding and cutting operations in the shop. These hazards as associated with the use of compressed gas cylinders and torches, fuel gas piping, arc welding and its associated electrical hazards, and from the contaminates produced by the operation.

These hazards can be controlled by allowing the operations only in fire safe locations, providing adequate ventilation, using the proper protective equipment, and following the correct work procedures.

Bibliography

American Welding Society (1967) *USA Standard Safety in Welding and Cutting,* 3rd ed., USAS (ANSI) Z49.1-1967, New York, New York 10017.

Union Carbide Corporation (1974) *Precautions and Safe Practices in Welding and Cutting with Oxygen-Fuel Gas Equipment,* New York, New York 10017.

U.S. Department of Health, Education, and Welfare, National Institute for Occupational Safety and Health, *Safety and Health in Arc Welding and Gas Welding and Cutting,* January 1978.

16

Electrical Safety

ELECTRICITY IS essential to everyday operations. When safety is viewed as an important component of the total facility program, the benefits of electricity can be enjoyed while its dangers can be recognized and controlled. It is critical that supervisors and employees be fully trained to recognize and correct any electrical hazards. In this way fires can be prevented, electrical shock avoided, and safe operations can be maintained. Lockout/ tagout is covered in Chapter 18.

In recent years, the hazards associated with electricity (burns, shock, falls, electrocution, and fire) have been greatly increased due to the increasing numbers of power tools and equipment used by workers. When these employees are unfamiliar with the safe practices necessary to ensure electrical safety, the dangers become even greater. Also, in any given year over a thousand deaths result from direct contact with electrical current. This figure does not include deaths in fires caused by electrical malfunctions, nor does it reflect injuries which are not fatal such as burns from electrical flashes.

Finally, electrical code violations are cited by OSHA approximately five times more frequently than the next most common class of violations. OSHA standards are those of the National Electrical Code, the universally accepted minimal requirements published mutually by the National Fire Protection Association and the American National Standards Institute. Since all electrical installations should comply with the requirements of the code, the number of violations is particularly noteworthy.

In order to better understand electrical hazards, it is important to define several key terms:

1. Current is the rate at which electricity flows through a circuit. It is measured in amperes. Because an ampere is a very large quantity in relation to its effect on the body, the milliampere (.001 amp) is used to measure electrical shock. Alternating current (AC) is a current (measured in amperes) which alternates in frequencies

measured in cycles per second (hertz, abbreviated Hz).

2. Voltage is the pressure which causes electricity to flow through a circuit. Every electrical circuit requires an electrical path from one terminal of the power source to the load (the device that uses the power) and a return path from the load to the other terminal of the power source or to the ground. The voltage is the difference between the two power source terminals. The ground is literally the earth and is always at 0 voltage. Voltage is equal to the product of current (amperage) and resistance (ohmage) and is measured in volts.

3. Power is the amount of electricity that flows through the circuit. It is equal to the product of the voltage and the current (amperage) and is measured in watts.

4. Resistance is anything that impedes or retards the flow of electricity. It is measured in ohms. Sometimes the term "resistance" is used to apply to direct current (DC), while the term "impedance" is applied to alternating current (AC).

16.1 Factors Affecting the Severity of Shock

The severity of electrical shock is determined by the amount of current flow. The amount of current flow at which an individual can still let go of an object held by the hand (before "freezing") varies from 10 to 16 milliamperes. Respiratory paralysis can occur with 30 milliamperes, and values greater than 75 milliamperes can cause ventricular fibrillation, a discoordinated heart action which usually results in death.

A lower voltage can be more dangerous than a higher one. Higher voltages cause such violent muscular contractions that the victim is thrown away from the circuit. Lower voltages may "freeze" the victim to the circuit. Since the victim is unable to let go, they are exposed to a longer current flow. The mild shocks caused by even lower voltages can cause accidents. Because of muscular contractions, a person may lose their balance and fall. One researcher observed that the current drawn by a 7.5-watt, 120-volt lamp, passed from hand to hand or foot, is enough to cause a fatal electrocution. This amount of current is readily obtained on contact with low-voltage sources of the ordinary lighting or power circuit.

Current flow depends on voltage and resistance. Resistance to current flow is found mainly on the skin surface. When the skin is moist, resistance is dramatically decreased. Once skin resistance is overcome, current flows unimpeded through body tissues and blood. Table 16.1 lists the levels of human resistance to electrical current. The protection offered by skin resistance decreases as voltage increases.

TABLE 16.1
Human Resistance to Electrical Current

Body Area	Resistance (ohms)
Dry Skin	100,000 to 600,000
Wet Skin	1,000
Internal-hand to foot	400 to 600
Ear to ear	(about) 100

The severity of shock is determined by amperage, which in turn depends on the following variables:

- voltage – the amount of electricity involved
- resistance – the number of ohms impeding the flow of voltage
- duration of current flow
- frequency of electrical waves (in an alternating current)
- part of the body exposed

16.2 Accident Circuit

This section addresses some of the more common hazards of electrical circuits. The term "circuit" suggests the continuous flow of electricity in the electrical system. The typical system consists of two conductors or wires which transmit the electrical energy. The electricity in one wire is under pressure and is trying to flow to the other wire. An electric current will flow through any conductive object which becomes connected between the two wires of the power system. The object of electrical safety is to prevent any part of the body from becoming part of the conductive path.

In the conventional power system, one of the conductors is physically connected to the earth. That portion of the power system is commonly referred to as the neutral conductor. Contact between the human body and the neutral side of the power system is usually established by contact with or connection to a grounded object. Workers can easily come into contact with the neutral side of the power system by touching such grounded objects as metal machinery, equipment, or pipes. Grounded objects in direct contact with the human body provide an attraction or "drain" for stray currents from other pieces of equipment.

The hot or energized side of the power system is the conductor which is not connected to the earth. In cases where the human body becomes part of the electrical circuit, the hot side is usually a piece of energized equipment or a wiring device. When an electrical current flows between the power circuit and any part of the equipment in contact with the body, then the human body which is in contact with another grounded object becomes the path for the current. The body itself is grounded.

16.3 Means of Protection

Protective measures used to prevent current from coming into contact with the body include isolation, insulation, grounding and overcurrent protection. If the power system and the body do not come in contact, there can be no accident circuit. Barriers and enclosures can prevent accidental contact with electrical equipment. Live parts should be enclosed whenever possible. Adequate machine guarding is an important part of an electrical safety program.

Most electrical equipment contains insulation between the power circuitry and other parts of the equipment. Currents cannot flow through the nonconductive materials used for insulation. When the insulating material deteriorates, very high and dangerous currents maybe released. The breakdown in the insulation of cords and plugs is a frequent cause of electrical accidents.

Double insulated tools are accepted in the National Electrical Code as an alternative to grounding. Such equipment has two complete and independent systems of insulation. Each system would have to break down in the same spot for shock to occur. The limitations of double insulation include:

1. Employees may develop bad habits. For example, if workers have been trained to look for the grounding pin on all attachment plugs, this "exception to the rule" may contradict rather than reinforce the practice which the supervisor has advocated.
2. Few types of devices are available in double-insulated form.
3. Double insulation only protects the user from faults within the device itself. It does not protect the user from faulty cords or plugs.

Equipment grounding directs possibly unsafe amounts of electricity to ground. Both equipment and electrical systems are grounded. Equipment grounding connects exposed non-current carrying metal parts of electrical equipment to ground. This grounding prevents a voltage above ground on these parts. Electrical systems are grounded to limit excessive voltage from lightning, line surges or unintentional contact with higher voltage lines.

Both circuits and enclosures are grounded to cause over-current devices to operate in case of insulation failure or ground faults. Over-current protection devices such as fuses and circuit breakers open the electrical circuit automatically in case of excessive current flow from ground, short circuit or overload.

16.4 Guarding of Live Parts

Supervisors and employees should be aware that parts connected to conductors of electrical circuits should be considered "live." Current will flow over all possible paths between live parts or from a live part to ground. Hazards and possible injury and death can result from:

- damaged insulation and/or connectors
- equipment installed without enclosure or protective devices
- enclosures or covers removed

Persons or objects are always in danger of coming into contact with live parts which are not enclosed. Persons becoming part of the path will experience electric shock. When metallic objects come into contact with live parts, the resulting short circuit can cause arcing (including blinding sparks) or molten metal splatter, both of which can cause burns and/or fires.

16.4.1 OSHA Standards

The Occupational Safety and Health Administration (OSHA) in its Safety and Health Standards (1910.309) specifies that the requirements in the National Electrical Code shall apply to the guarding of live parts.

Except as elsewhere required or permitted by this Code, live parts of electrical equipment operating at 50 volts or more shall be guarded against accidental contact by approved cabinets or other forms of approved enclosures, or any of the following means:

1. By location in a room, vault or similar enclosure that is accessible only to qualified persons.
2. By suitable, permanent, and substantial partitions or screens so arranged that only qualified persons will have access to the space within reach of the live parts. Any openings in such partitions or screens shall be so sized and located that persons are not likely to come into accidental contact with the live parts or to bring conducting objects into contact with them.
3. By location on a suitable balcony, gallery or plat-form so elevated and arranged as to exclude unqualified persons.
4. By elevation of 8 feet or more above the floor or other working surface.

In locations where electrical equipment could be exposed to physical damage, enclosures or guards shall be so arranged and of such strength as to prevent such damage.

Entrances to rooms and other guarded locations containing exposed live parts shall be marked with conspicuous warning signs forbidding unauthorized persons to enter.

16.4.2 Hazard Prevention

Several specific precautions can prevent contact with live parts and the resulting injuries. These precautions include:

1. Covers, screens, or partitions that are not easy to remove should be used.
2. If covers are removed from equipment such as panels, motor covers, or fuse boxes, they should be replaced as quickly as possible.
3. When live parts cannot be enclosed completely, guards or barriers should be provided.
4. When live parts are elevated the required eight feet, workers should be cautioned against using metal ladders or long metal rods.
5. Workers who see exposed live parts should be instructed to report the condition immediately so that it can be corrected.
6. Unused conduit openings in boxes should be closed to prevent accidental contact.

16.4.3 Working Clearances

In addition to requirements to protect employees from accidental contact with live parts, the NEC has requirements to protect the person, presumably a trained electrician, who is qualified to work on the equipment. The NEC requires that sufficient access and working space be provided and maintained about all electric equipment to permit ready and safe operation and maintenance of such equipment. Briefly summarized, the code requires:

- a work space at least thirty inches wide in front of the electrical equipment
- adequate illumination
- sufficient access area to the working space
- minimum headroom of 6¼ feet

The code forbids storage of materials in the work space. Because masonry surfaces can be conductive under various circumstances, the code requires that concrete, brick, or tile walls be considered as grounded. The NEC should be consulted for the minimum clear distances required for specific conditions involving exposed live parts on such equipment as switches, panelboards, circuit breakers, and motor starters.

16.5 Flexible Cords

Because it can be designed for the particular type of service and location, fixed wiring is preferable to flexible cords which can be misused and are more vulnerable to damage. Wiring methods which can be used in certain circumstances are armored cable, rigid metal conduit, flexible metal conduit, raceways, nonmetallic sheathed cable, and concealed knob-and-tube work. The type of wiring method used will depend on several factors, including:

- the building materials themselves
- the size and distribution of the electrical load
- exposure to dampness
- Exposure to corrosives (oil, grease, vapors, gases, fumes, liquids)
- exposure to temperature extremes
- the location of equipment

Permanent wiring by itself cannot satisfy all the needs of an industrial facility. Electrical cords and fittings provide the flexibility required for:

- maintenance
- portability
- isolation from vibration
- temporary power needs

The NEC recognizes electrical cords and fittings as a supplement to permanent installations. However, their selection, use, and maintenance must be supervised carefully, and workers must be warned that each use of a cord creates an additional hazard.

There are two general classifications of portable cords:

1. Electrical cord sets with fittings used as extension cords. These have an attachment plug at one end and a cord connection with from one to six outlets at the other end.
2. Power supply cords, either
   a. non-detachable—a flexible cord terminating at one end in an attachment plug cap and permanently attached at the other end to some utilization equipment (e.g., a hand-held power saw).
   b. detachable—a length of flexible cord with an attachment plug cap at one end and an appliance coupler at the other end (e.g., some portable drills).

16.5.1 Uses of Flexible Cord

The following situations allow the use of flexible cords:

- electrical fixtures suspended from the ceiling
- wiring of fixtures
- connection of portable lamps and appliances
- wiring of hoists and cranes
- connection of stationary equipment to facilitate frequent interchange
- facilitating the removal or disconnection of fixed or stationary appliances for maintenance or repair (e.g., water coolers, exhaust fans)
- prevention of the transmission of noise or vibration

The NEC prohibits the use of flexible cords in the following situations:

- as a substitute for the fixed wiring of a structure

- where run through holes in walls, ceilings, or floors
- where run through doorways, windows, or similar openings
- where attached to building surfaces
- where concealed behind building walls, ceilings, or floors

16.5.2 Hazards of Temporary Wiring

The temptation in using temporary wiring is to allow it to become a permanent solution. Cord should not be extended to some distant outlet simply to avoid providing a fixed outlet. When new electrical needs are anticipated, time should be allowed for the proper installation of fixed wiring. When temporary wiring is necessary, a schedule for its removal should be established.

In all cases, both the number and length of extension cords should be kept to a minimum. Permanent receptacle outlets should be installed at convenient locations in order to limit the length of cord required for the job.

Extension cords should be listed by the Underwriters Laboratories. Receptacles and attachment plugs, which provide a connection for the equipment grounding conductor of the cord, should also be listed for equipment grounding service by Underwriters Laboratories.

Cord that is used in ways prohibited by the code is likely to be damaged by abrasion from adjacent materials, edges, or clamps. Cord that is not visible for its entire length cannot be inspected for damage or deterioration. Over a period of time, damaged cord will partially expose conductors creating a potential danger of shock, burns, or fire.

The tripping and falling hazard created by cords can be minimized by keeping the cord off the floor. The cord should be suspended above the floor without abusing the cord materials. Special fittings for this purpose are available. Such an arrangement must be high enough to allow safe clearance below. When it is not practical to string cords overhead, a rubber treadle can be snapped over the cord. Molded in a light yellow color, the treadle protects the cord and minimizes the tripping hazard. A cord should not be pulled or dragged over nails, hooks, tools or other sharp objects that can cut the insulation.

The NEC permits flexible cord only in continuous lengths without splice or tap. If the cord is damaged, the defective portion can be cut out and the remaining cord joined with the use of an additional attachment plug and conductor.

16.5.3 Special Hazards of Portable Head Lamps

The NEC requires that portable lamps be equipped with a substantial guard attached to the lampholder or the handle (410-42). Plastic or rubber insulated guards should be used wherever the guard could come in contact with an electrical circuit and wherever deteriorating agents such as oil and grease may be present in the work area

Uninsulated guards should be electrically continuous with the equipment grounding conductor. Metallic guards must be grounded by means of an equipment grounding conductor run with circuit conductors within the power supply cord.

The NEC requires that portable lamps be equipped with a handle of molded composition or other material approved for the purpose (410-42). Handles should be made of a high-grade rubber compound or a similar material that gives maximum insulation and durability and resistance to oil and to softening at high temperatures. At the end where the cord enters the handle, there should be a dust seal. Where the flexible stem enters the base or stem of a portable lamp, a bushing or its equivalent must he provided. The bushing shall be of insulating material unless a jacketed-type of cord is used (410-44).

Metal shell, paper-lined lampholders are specifically prohibited by the NEC. Portable lamps used in locations similar to the auto shop are subject to special regulation. The NEC states that such lamps must be equipped with handle, lampholder, hood, and substantial guard attached to the lampholder or handle. The regulation also states that:

All exterior surfaces which might come in contact with battery terminals, wiring terminal or other objects shall be of nonconducting material or shall be effectively protected with insulation. Lampholders shall be of unswitched type and shall not provide means for plug-in of attachment plugs. Outer shell shall be of molded com-position or other material approved for the purpose.

16.5.4 Choosing the Proper Cord

The types of cords commonly used in an industrial shop are listed in Table 16.2. The letter J in a designation stands for Junior. Residential-type cords are not designed to withstand the rigorous usage encountered in the shop and should not be used.

TABLE 16.2
Types of Flexible Cord

Trade Name	Type Letter	Size AWG	Number of Conductors	Insulation on Each Conductor	Outer Covering	Use		
Junior Hard	SJ	18 to 14	2,3,or 4	Rubber	Rubber	Pendant or Portable	Damp Places	Hard Usage
	SJO				Oil-Resistant Compound			
Service Cord	SJT			Thermoplastic Or Rubber	Thermoplastic			
	SJJO				Oil Resistant Thermoplastic			
Hard Service	S	18 to 2	2 or more	Rubber	Rubber	Pendant or Portable	Damp Places	Hard Usage
	SO				Oil-Resistant Compound			
Cord	ST			Thermoplastic Or Rubber	Thermoplastic			
	STO				Oil Resistant Thermoplastic			

Note: Flexible cords and cables shall be marked by means of a printed tag attached to the coil reel or carton. The tag shall contain the information required in Section 310-11 (a) of the Code.

Types S, SO, ST, and STO cords are rated for extra hard usage. They have conductors that range in size from 18 to 2 AWG (American Wire Gauge) and are insulated with a material suitable for voltages up to 600 volts. Types SJ, SJO, SJT, and SJJO are made only in conductor sizes 14, 16 and 18 AWG. They are rated for 300-volt insulation and are smaller than the extra hard usage types. Both types depend upon the chemical composition of the jacket for the protection they provide against abrasion, water, oil, chemicals, and temperature extremes.

The type of cord and plug required is related to the type and use of the equipment. A cord and plug adequate for a piece of equipment which is never moved may not meet the greater wear demands of a cord and plug used for portable tools.

The voltage impressed between the conductors or between conductor and ground should not exceed the voltage rating of the cord itself. Before the cord is put into use, the power supply voltage should be determined and checked against the rating of the cord.

The amperage required by the equipment that is to be connected to the cord can be determined from its nameplate. The voltage, frequency, current, and phase characteristics of the circuit should match the nameplate characteristics of the fittings. The proper cord set can be selected by matching the nameplate information against the rating information on the cord set label or package. Units that draw large initial starting currents require cords of sufficient wire size to minimize the voltage drop which will be produced.

Unless the tool is double-insulated, extension cords must contain a separate equipment grounding conductor. Because the metal frames of portable electric equipment should be grounded, cord with a green-covered ground conductor should be used with a polarized plug and receptacle.

Sockets should be covered with porcelain, composition, or rubber. Extension cords with brass shell sockets should not be used. The socket may become energized through contact with loose wires inside the socket, through abrasion of the insulation where the cord enters the socket, or through moisture in the socket insulation.

16.5.5 Cord Maintenance and Storage

Before inspecting, repairing, and servicing portable equipment and fittings, cords should be removed from the electrical power source. Electric cords and fittings should be inspected regularly. Cords should be wiped clean with a dilute detergent and examined for small breaks, abrasions, and defects in the jacket. Fittings should be inspected for wear, looseness, arcing conditions, or other mechanical defects.

Cord reels are recommended in locations where a power source or portable light is needed frequently. Reels pull the cord out of the way when not in use while still keeping it readily available for service. Where reels are not used, cords should be coiled or hanked for storage. Care should be taken to avoid kinking or abrading the cord. Power supply cords should not be wound tightly around portable tools or hand lamps. This practice can damage the insulation and break conductor strands at the point where the cord is bent sharply.

16.6 Plugs and Attachments

Plug and connector housings should be made of a material that will protect internal parts and connections from mechanical damage (see Table 16.3). For cords size 14 AWG or larger, heavy duty attachment plugs are recommended.

Nylon is the material recommended for standard use in the shop. It is durable, an excellent insulator, and highly resistant to attack by chemicals. It is resilient enough to resist shattering when dropped or struck, yet hard enough to retain its shape even under great pressure. In the unlikely event that a nylon device is damaged, visual inspection easily reveals the damage. This is not true of rubber and neoprene devices, where the housing springs back into shape and conceals internal damage.

TABLE 16.3
Chemical Resistance of Materials Commonly Used in Writing Devices

Chemical	Nylon	Melamine	Phenolic	Urea	Polyvinyl Chloride	Polycarbonate	Rubber
Acids	C	B	B	B	A	A	B
Alcohol	A	A	A	A	A	A	B
Caustic bases	A	B	B	B	A	C	C
Gasoline	A	A	A	C	A	A	B
Grease	A	A	A	A	A	A	B
Kerosene	A	A	A	A	A	A	A
Oil	A	A	A	A	A	A	A
Solvents	A	A	A	A	C	C	C
Water	A	A	A	A	A	A	B

A-Completely resistant. Good to excellent for general use
B-Resistant. Fair to good, limited service.
C-Slow attack. Not recommended for use.

Cord terminations are far more vulnerable than those in fixed-wiring because such cord is exposed, flexible, and in some cases not secured. The finely stranded wiring inside the cords is necessary for continuous flexing, but strands can escape from under terminal screws, creating a hazard. According to NEC:

> Flexible cords shall be so connected to devices and to fittings so that tension will not be transmitted to joints or terminal screws. This shall be accomplished by a knot in the cord, winding with tape, by a special fitting designed for that purpose, or by other approved means which will prevent a pull on the cord from being directly transmitted to joints or terminal screws.

All plugs that are attached to cords must have the terminal screw connections covered by suitable insulation.

Workers should be taught to disconnect a connection by pulling on the plug, not on the cord. A non-metallic cord gripper incorporated in plugs and connectors can prevent strain on the ground and power connections when someone pulls the plug out by the cord. Locking-type attachment plugs, receptacles, and connectors pro-vide additional protection against accidental disengagement.

Before an attachment plug is inserted or withdrawn from a receptacle, the control switch on electrical equipment should be in the off position. This precaution prevents arcing at the plug.

Isolated pockets are provided inside the body of the plug for each wire and its terminal. This dead-front construction is preferable to merely securing mechanically the front covers for wire terminals. It eliminates the possibility of the common fiber disk cover coming off or allowing wire strands to protrude. It prevents the conductors from contacting a metal wall plate. It keeps out metal chips or other foreign substances which can enter when the disk is loose or improperly fitted.

16.7 Grounding of Cord- and Plug-Connected Equipment

Cord- and plug-connected equipment is subject to more abuse than equipment supplied by fixed wiring. If cord- and plug-connected equipment are not properly grounded, the voltage between the metallic parts and other surfaces in the vicinity may be sufficient to cause harmful or lethal shock. If a piece of conductive material becomes part of a ground-fault path, sparking and burning may result. Therefore, the grounding of such equipment, whether stationary or portable, is required by law for the safety of those exposed to it.

A current leakage might be defined as an electrical current (generally less than one ampere) that has deviated from its normal path to ground and is seeking an alternate route. Such leakages are sometimes called "low current faults" and occur in all electrical equipment, even when it is new. A good grounding system, that is, the green wire or a rigid metal conduit usually carries off this current leakage. Therefore, no shock is perceived when the current leakage of tools is below one ampere and the grounding current has a low resistance. Over a period of time, however, the dielectric properties of insulation wear down, and a greater amount of current can be measured on the metal frame or case of the equipment. Should the grounding conductor become inoperative, for example, through a broken prong or increased resistance, the user of the equipment is placed in grave danger.

This hazard due to loss of ground occurs far too frequently with portable and cord-connected tools and equipment. Such equipment may be used in damp or wet locations where good earth contact exists. Another hazard is created by the fact that portable and cord-connected tools usually require a full-hand grasp making it difficult for the worker to release his grasp in the event of insulation failure in the tool or wire.

16.7.1 NEC Standards

The NEC specifies that "exposed noncurrent-carrying metal parts of cord- and plug-connected equipment likely to become energized" must be grounded in certain circumstances. These situations include several likely to occur in small businesses. Double insulated tools are exempted from the grounding requirement. Specifically, the requirement applies to:

1. portable, hand-held, electrical-operated tools such as drills, sanders and saws
2. portable tools likely to be used in wet and conductive locations
3. equipment operated at more than 150 volts to ground (an exception is made for guarded motors)
4. hazardous locations including areas where flammable liquids and gases are used, areas where combustible dust is present, and areas where materials producing easily ignitable fibers are handled or produced

16.7.2 Typical Grounding System

The typical grounding system for cord- and plug-connected equipment consists of a:

- third wire contained in the power cord
- three-prong plug with a grounding pin
- three-wire grounding type receptacle
- third or grounding conductor, which connects the receptacle grounding contact to the power distribution system ground (neutral) at the main service switchboard

16.7.3 Three-pronged Connectors

The basic safety feature on plugs is the three-pronged connection. Two prongs conduct electricity and the third prong goes to the ground. This system gives a continuous ground to the frame of the tool. An insulation failure which energizes the frame, causes a short through the equipment grounding conductor. This short actuates overcurrent protection devices such as fuses and circuit breakers.

This system provides excellent protection as long as the third-wire prong is kept intact and the grounding system from the outlet is unbroken. Workers should be instructed never to use a plug with the third prong broken off.

Some attachment plugs are manufactured of high-impact transparent plastic material. Such plugs permit inspection of terminations without the necessity of disassembly.

16.8 Grounding

The purpose of the equipment grounding conductor system is to provide a low impedance (resistance) path to ground for currents resulting from faults or the inherent leakage of electrical apparatus. If the framework and cabinets of electrical equipment are purposely grounded, a path to ground for stray currents is provided before they reach and pass through a human body. The general rule for electrical safety is to ground the equipment, but not the human body.

The characteristics of an adequate grounding system are described in the National Electrical Code:

Effective Grounding Path. The path to ground from circuits, equipment, and conductor enclosures shall be permanent and continuous, have the capacity to conduct safely any fault current likely to be imposed on it, and have sufficiently low impedance to limit the voltage to ground and to facilitate the operation of the circuit protective devices in the circuit.

In addition to requiring the grounding of equipment connected by cord and plug, the NEC requires the grounding of exposed noncurrent-carrying metal parts of fixed equipment likely to become energized under the following circumstances:

- within eight feet vertically or five feet horizontally of ground or grounded metal objects
- in a wet, damp, nonisolated location
- in electrical contact with metal

- in a hazardous location
- where supplied by a wiring method that is metal-clad, metal-sheathed, or a metal raceway
- where the equipment operates with any terminal in excess of 150 volts to ground

The grounding of hoists and cranes is required by NEC. The standard states that:

All exposed metal parts of cranes,... hoists and accessories including pendant controls shall be metallically joined together into a continuous electrical conductor so that the entire crane or hoist will be grounded. Moving parts, other than removable accessories or attachments having metal-to-metal bearing surfaces,...shall be considered to be electrically connected to each other through the bearing surfaces for grounding purposes. The trolley frame and bridge frame shall be considered as electrically grounded through the bridge and trolley wheels and its respective tracks unless local conditions, such as paint or other insulating material, pre-vent reliable metal-to-metal contact. In this case a separate bonding conductor shall be provided.

16.8.2 Grounding Type Receptacles

In 1975, the NEC was revised to require the installation of only grounding type receptacles in new construction. The right-hand (short) slot is the hot lead, the terminal from which the current flows. The left-hand (longer) slot is the grounded neutral connector, the terminal for the return path. The third slot is for the grounding connector. In a 240V receptacle, the grounding connector slot is shaped like the right- and left-hand slots.

A matching male connector with a fixed grounding prong is required for use with the receptacle. Pigtail connectors are not allowed because they permit the following three hazardous conditions to exist:

1. The grounding wire may be left unconnected.
2. Even when the grounding wire is connected, it may be attached to an object that is itself ungrounded or is highly resistant to ground.
3. Electrocution can result if the pigtail grounding connector is accidentally inserted into the hot slot of the receptacle.

Non polarized standard plugs can be used in the receptacle.

Over a period of time, the ability of the grounding circuit in the receptacle to make contact with the plug grounding pin can be lost. The detection of this condition requires determining the contact force or tension between the receptacle contact and the plug grounding pin. Such tests should be scheduled on a regular basis and be conducted by qualified persons aware of the potential hazards. These inspections will enable the maintenance department to replace receptacles before they produce an ineffective equipment grounding contact or cause a fire in the power contacts.

16.9 Protection Against Ground Faults

A ground fault occurs when electricity, seeking to maintain continuity with the rest of the circuit, uses the ground as a conductor. If the body becomes part of the circuit, electrical shock will occur. If the ground contact is poor, the voltage will not be able to force a harmful amount of current (amperage) through the resistance. But if the resistance is lowered through contact with grounded metal or damp concrete floors or through perspiring hands, the voltage will be able to force a larger, more damaging current through the body.

Grounding protects against ground faults in equipment. Ground faults are usually caused by abrasion, aging, or other types of damage which breaks through the insulation of conductors and allows the metal of the conductor to touch the enclosures or adjacent metal parts. An insulation breakdown can occur in the windings or other internal functioning parts of the equipment or in the wiring brought to or from the equipment.

Whenever a current-carrying conductor makes contact in this manner, the exposed metallic surfaces become energized at the same voltage as the conductor involved. The nature of voltage is that it will force current to flow from one terminal of the source, out through the circuit conductor, through any available fault-current paths, and back to the other source terminal.

If the metallic surfaces are not bonded together and to ground by a low-impedance equipment grounding conductor such as a conduit, dangerous voltage can exist between electrical enclosures and from some enclosures to plumbing, building steel and other grounded surfaces. Because the fuse or circuit breaker will not open the circuit unless there is a large ground-fault current flow, the dangerous voltage may continue to exist.

With the grounding-type receptacles and the matching male plugs, the condition known as "reversed polarity" is unlikely to occur. Reversed polarity exists when the current exits through the hot wire instead of coming into the equipment through the switch and exiting through the neutral wire. This produces heat in the internal wiring up to the switch, resulting in a shock hazard. If a circuit tester indicates reversed polarity, the internal hot and neutral connections of the outlet should be interchanged immediately. Every cord set and power supply cord should be checked for ground continuity and correct polarity before being placed in service.

16.10 Overcurrent Devices

Overcurrent devices, such as fuses and circuit breakers, open the circuit automatically in case of excessive current flow from ground, short circuit, or overload.

16.10.1 Ground Fault Circuit Interrupters (GFCI)

The ground fault circuit interrupter (GFCI) is a fast-acting circuit breaker which senses small imbalances in the circuit caused by current leakage to ground and, in a fraction of a second, shuts off the electricity. The GFCI continually matches the amount of current going to an electrical device against the amount of current returning from the device along the electrical path. If the amount going differs from the amount returning by approximately two to five milliamperes, then the GFCI interrupts the electric power within as little as 1/40 of a second.

There are two types of ground fault circuit interrupters:

1. Differential - Current carrying wires go through a differential transformer. If a minimum of five milliamperes do not flow through but trickle to the ground instead, the circuit breaker is tripped and the flow of electricity stopped.
2. Isolation - This type combines the safety of an isolation system with the response of an electronic sensing circuit. Ground fault passes through the electronic sensing circuit which has enough resistance to limit current flow to as little as two milliamperes.

The GFCI can operate on both two-wire and three-wire (equipment grounding) systems. It protects both two-wire and three-wire equipment and continuously monitors the system.

According to the National Safety Council, there has not been a single recorded electrocution from ground fault in any installation employing a GFCI. The ground fault circuit interrupter also provides protection against fires, overheating, and the destruction of insulation on wiring. A GFCI does not protect a person from line-to-line hazards, such as those developing when a person is holding two hot wires or a hot and neutral wire. It is not a substitute for good electrical safety procedures. Portable models are available.

In the industrial shop GFCIs should be used in the following instances:

- where water and electricity are used in close proximity, that is, where motors or other electrical apparatus are located near sinks and basins
- where the user of electrical equipment cannot avoid being grounded
- on circuits providing power to outdoor receptacles

16.10.2 Fire Hazards Created by Overloading

Electrical conductors and machinery are designed to carry a rated load. Their safe current-carrying capacity is determined by their size, the material of which they are made, the type of insulated covering, and the manner of installation. If they are forced to carry loads greater than their capacity they will overheat.

For example, if a ¼ horsepower motor is given a load of one horsepower, it will try to carry that load but probably will burn itself up in the process. Because the voltage is constant, the motor will draw more current. The excess current will heat the electrical conductors to the point where they will break, creating a fire hazard. Such overheating frequently causes the insulation to burn, exposing live parts. Fires sometimes start because electrical conductors within a wall raise a combustible material to its ignition temperature, causing it to burn.

16.10.3 Overcurrent Devices

Overcurrent devices open the circuit automatically when triggered by excessive current flow from overload, accidental ground, and short circuits; and by a circuit interrupter causing a circuit breaker to open. Overcurrent devices are basically passive. They operate only when something goes wrong. They are not a substitute for ground fault circuit interrupters but may act as a supplement to them.

Fuses and circuit breakers operate on a time-versus-current principle. The larger the amount of current, the shorter the time required to break the circuit. So that large amounts of current can flow quickly thereby activating overcurrent devices and protecting the circuit from damage, it is important for the grounding conductor to be continuous, have low impedance, and have sufficient ampacity.

16.10.4 Fuses

A fuse is a part of a conductor in a circuit. When too much current flows through, the fuse heats up within a fraction of a second. The fusible metal melts and interrupts the circuit. There are three types of fuses:

1. Link fuse - a fusible metal forms a strip between the two terminals of a fuse block.
2. Plug fuse - used on circuits which do not exceed 30 amperes at not more than 150 volts to ground, the fusible metal is completely enclosed.
3. Cartridge fuses - the fusible metal strip is enclosed in a tube.

Each type should be used only in the circuit for which it was designed. Fuses of the wrong type or size can injure personnel and damage equipment.

Fuses never should be inserted in a live circuit. When it is necessary to remove a fuse, the circuit should be locked out. The fuse should be extracted with an insulated fuse puller. If the fuse is not protected by a switch, the supply end should be pulled out first. When the fuse is replaced, the supply end should be put in first.

A copper wire or other conductor must never be substituted for a fuse. A larger fuse never should be used to replace a blown fuse. Over fusing can cause overheated wiring and equipment, creating the very fire hazard that fuses are designed to prevent.

16.10.5 Circuit Breakers

A circuit breaker is a switch placed in a circuit so that it opens automatically if a certain temperature is reached or if too much current flows through the switch. A circuit breaker may operate instantly or be equipped with a timing device. There are two general types of circuit breakers:

1. Thermal - operating only on the basis of temperature rise.
2. Magnetic - operating only on the basis of the amount of current passing through the circuit.

16.10.6 NEC Standards

All circuits must be equipped with fuses or circuit breakers that will open if the actual current flow in a circuit exceeds the expected flow by 25 percent. For example, if a circuit is normally expected to carry a load of sixteen amperes, the fuse or circuit breaker must be rated at twenty amperes. The NEC further requires that the conductors in the circuit be able to carry the 25 percent excess load without overheating.

Welders require over-current protection set at not more than 200 percent (for arc welders) or 300 percent (for resistance welders) of the rated primary current of the welder. Overcurrent devices must be located where they are:

- readily accessible and capable of being reached quickly for operation, renewal, or inspections without requiring those to whom ready access is requisite to climb over or remove obstacles or to resort to portable ladders, and chairs.
- not exposed to physical damage
- not in the vicinity of easily ignitable material

Overcurrent devices frequently serve as disconnecting means. Therefore, they must be readily accessible when troubles occur. Furthermore, devices which are difficult to reach are apt to be neglected. In the case of overcurrent devices, neglect can lead to overheating and fires.

16.11 Detecting Overloads

Supervisors must make certain that circuits are not overloaded. Even if overcurrent protection devices function properly and prevent fire or heat hazards, there is still the inconvenience of resetting the breaker or replacing the fuse. Some signs of overloading include instances when:

- the fuse or circuit breaker frequently opens the circuit (if this happens, the machine causing the interruption should be identified)
- a ground fault circuit interrupter frequently interrupts the circuit
- an electrical machine is abnormally hot

- an extension cord becomes warm
- a cable bank, fuse box, or junction box becomes abnormally warm
- the insulation of a conductor is worn or frayed

If any of these conditions exist, a maintenance electrician should be called.

Before fuses are replaced or circuit breakers reset, an investigation should be made for the cause of the short circuit or overload. To ignore the warning signals which have been built into the system is foolish and shortsighted.

16.12 Maintaining Protective Devices

According to one researcher, the cost of the electrical system in any facility is only 7.5 percent of the total cost even though the facility is useless without electric power. Saving a few cents by cutting corners on maintenance is false economy and it undercuts the sound investment that has been made in designing and installing a safe system.

This chapter has reviewed the ways in which electrical protective devices are the safety valves of the system. Overcurrent protective equipment must be checked periodically to be certain that it works properly when troubles occur. The maintenance needs involve five steps:

1. Clean
2. Tighten
3. Lubricate
4. Inspect
5. Test

Fuses, relays, and circuit breakers must be cleaned periodically to eliminate dirt, dust, moisture, and contaminants which can make the system ineffective. All electrical connections must be tight. A loose connection generates excess heat, which can erode metal.

Lubrication is necessary to keep the mechanical joints or sliding parts in circuit breakers from becoming rusty over a period of time. Such lubrication must be done carefully so that improper lubricants are kept away from electrical contacts.

All electrical protective devices need to be checked to be certain that their interrupting rating and their setting is in accordance with the design. Such inspection will detect instances where the capacity of the device has been increased to keep the circuit from tripping. For example, if a 15-

ampere circuit breaker or fuse has been replaced by a 20- or 25-ampere device, an accident is just waiting to happen.

If it has a built-in test feature, the circuit breaker, fuse or relay can be subjected to simulated fault conditions to ensure that it operates in accordance with the manufacturer's specifications for the device being tested.

Such a maintenance program as the one described will ensure that devices have not been impaired by:

- dusty, oily, smoky, or corrosive atmospheres
- mechanical vibration
- excessive temperature
- tampering

16.13 Identifying Disconnecting Means

The electrical system for an industrial facility originates at a service entrance, from which feeders carry current to branch circuits. These in turn carry current to the outlets for lighting, machinery and equipment. It is important to identify the switches and circuit breakers which control the power to each particular circuit.

Each disconnecting means for motors and appliances, and each service, feeder or branch circuit at the point where it originates shall be legibly marked to indicate its purpose unless located and arranged so the purpose is evident. The marking shall be of sufficient durability to withstand the environment involved.

16.13.1 When Disconnection is Necessary

During the life of any electrical installation, there will be times when it is necessary to disconnect a branch circuit, a feeder or an entire service. Reasons for disconnections include:

- normal maintenance procedures, whether scheduled or unscheduled
- changes or additions to the system
- emergencies requiring quick action to disconnect power from a particular piece of equipment or portion of the system
- major emergencies such as fire or explosion requiring all power in the building be turned off quickly

If the appropriate disconnecting means is not obvious, mistakes may be made or vital time may be lost in tracing circuits to their source. If the circuit conductors, or the raceway containing them, are visible from the disconnecting means to the equipment involved, the purpose may be evident. However, a sign, label, tag, or nameplate on the disconnect is necessary in most cases to meet the needs of quick identification.

In many businesses, there is one person who is familiar with the electrical system and is usually the person responsible for maintaining electrical circuits and equipment. It is easy for the supervisor to assume that this key person will always be available to identify disconnecting means. But emergencies can and do occur when there is no time to consult the specialist. Harmful shocks and even electrocutions can result from the wrong switch being opened or closed in a panic situation. Fires can start if a faulted circuit is not disconnected quickly enough. Therefore, the supervisor must insist that switch-boards and control panels are adequately identified. See Chapter 18, Lockout/Tagout for more specific details for preventing energy to a system.

16.13.2 Steps in Identifying Disconnecting Means

Inadequate identification is a violation of the National Electrical Code and a poor safety practice. In order to avoid electrical hazards, the supervisor should ensure that the following steps be taken immediately.

1. Trace out all existing circuits, from service entrance to utilization equipment, and clearly mark each disconnecting device to indicate what circuit or what equipment it disconnects.
2. Ensure that exceptions to the above are permitted only where the circuit conductors or their raceway are clearly visible from the disconnecting means to the load.
3. Use labels or nameplates which are permanently legible.
4. Do not depend entirely on anyone's memory to identify circuits for proper labeling.
5. When numbers or letters are used to identify circuits and equipment, be sure that the system layout or key diagram is posted so that the key can be interpreted and the circuit located.

16.14 Marking Electrical Equipment

In addition to identifying the purpose of a disconnecting means, the NEC requires that equipment be marked with its appropriate ratings so that it can perform its function as intended. Observing the limiting conditions will prevent the operator from subjecting the equipment to conditions which will damage it and possibly cause injuries.

16.14.1 NEC Standards

According to NEC 110-21:

The manufacturer's name, trademark or other descriptive marking by which the organization responsible for the product may be identified shall be placed on all electric equipment. Other markings shall be provided giving voltage, current, wattage or other ratings .... The marking shall be of sufficient durability to withstand the environment involved.

A metalworking machine must have attached either to the control equipment enclosure or to the machine itself where plainly visible after installation a permanent nameplate listing supply voltage, phase, frequency, full-load currents, ampere rating of largest motor, short-circuit interrupting capacity of the machine overcurrent protective device if furnished, and diagram number. When the machine tool nameplate is marked "overcurrent protection provided at machine supply terminals," each set of supply conductors terminates in a single circuit breaker or set of fuses.

For AC transformer and DC rectifier arc welders, the nameplate must include name of manufacturer; frequency; number of phases; primary voltage; rated primary current; maximum open-circuit voltage; rated secondary current; basis of rating, such as the duty-cycle or time rating.

If equipment is connected to a voltage higher than its rating, violent failure may result. If the voltage is below its rating, the equipment may overheat and eventually fail. If alternating current equipment is energized with the wrong frequency or with direct current, violent failure may result. Any of these abuses can lead to burns and fires.

16.14.2 Use of the Nameplate

The supervisor must be certain that the nameplate on a machine or tool is not removed, covered by some part of the installation or obliterated by painting or other abuse.

When a nameplate gives an alternate or a maximum rating for the equipment, a specific marking should be added to the machine to indicate what voltage is actually being applied. For example, some machines can operate on either 115 or 230 volts, depending on internal conditions. The nameplate will not indicate which of these is the applied voltage so a supplementary marking such as a stencil or decal is necessary.

The manufacturer's name on the nameplate is useful if information needs to be sought or if replacement parts need to be ordered. The symbol or notice of testing or listing is also useful information to include on the name plate.

The supervisor will want to be certain that the proper electrical characteristics for the particular shop will be specified when new equipment is bought.

16.15 Special NEC Provisions

The industrial shop contains hazardous areas. These are places where flammable liquids and gases are stored and used. Equipment which can produce arcs, sparks or particles of hot metal should be either totally enclosed or so constructed that sparks and hot metal particles cannot escape. Included in this description are cutouts, switches, receptacles, charging panels, generators, motors, or other equipment having make-and-break or sliding contacts.

Special wiring is required in the auto shop because it is a hazardous location, containing vehicles which use volatile flammable liquids and gases for power and fuel.

The flexible cord by which electrical fixtures are suspended from the ceiling must be suitable for the type of service and approved for hard usage. Only flexible cord and connectors approved for extra-hard usage may be used for charging.

Connectors used for charging must be designed and installed so they will readily disconnect at any position of the charging cable. When a cord is suspended from overhead in order to connect a plug directly to the vehicle, it must be so arranged that the lowest point of sag is at least six inches above the floor.

Racks used for supporting battery cells and trays should be made of metal and treated to resist deteriorating action by the electrolyte. They must be provided with nonconducting members directly supporting the cells or with suitable insulating material other than paint or conducting members. Trays may be of wood or other nonconductive material treated to resist deterioration.

16.16 What the Worker Needs to Know

The following key safety points are ones in which the workers should be aware. This list is not exhaustive and may be modified to suit the needs of the particular shop situation.

1. Keep tools and cords away from heat, oil, and sharp edges that can damage electrical insulation.
2. Disconnect tools and extension cords by holding the plug, not by pulling on the cord. Be sure that the control switch on electrical equipment is in the "off" position before putting in a plug or pulling it out.
3. Never use a three prong grounded plug when the third prong broken off. Always plug in a three-prong plug into a properly installed three-prong socket.
4. Do not use electrical equipment in damp or wet areas.
5. Do not use electrical equipment on or near metal ladders capable of conducting electricity.
6. If tools or cords run hot, report the condition to the instructor. The insulation could be deteriorating. Never wrap a cloth around a tool too hot to hold. Sparks can ignite the cloth.
7. Immediately report damaged tools and equipments especially those that give off minor shocks. Immediately report any exposed live parts. Do not attempt to make repairs on your own.
8. Avoid using extension cords. When an extension cord must be used, choose one with the same ampere rating as the tool. Ensure that the insulation is intact and that all connections are tight and that the cord does not create a tripping hazard.
9. Use a Ground Fault Circuit Interrupter when using portable tools.
10. Do not overload circuits.

Summary

Electricity is essential to the everyday operations of the shop. When safety is viewed as an important component to the total shop program, the benefits of electricity can be enjoyed and its dangers recognized and controlled.

The factors affecting the severity of electric shock are voltage, resistance, duration of current flow, frequency of electrical waves, and the part of the body exposed. Protective measures used to prevent current from making contact with the human body are insolation, insulation, grounding, and overcurrent protection.

While fixed wiring is preferred over flexible cords, it cannot meet all of the needs of the facility. Whenever flexible cords are used, they should be installed, maintained, and stored according to the National Electric Code (NEC) standards.

Cord- and plug-connected equipment should be properly grounded to avoid electric shock. To ensure the safety of workers, proper grounding of such equipment is required by law. The characteristics of a properly grounded system are described by NEC. Grounding also protects against ground faults.

Overcurrent devices, such as the ground fault circuit interrupter will stop the flow of electricity in the event of excessive current flow from the ground, short circuit, or overload. Overcurrent protective equipment must be periodically inspected to ensure proper working condition.

It is important to clearly mark and identify the switches and circuit breakers which control the power to each particular circuit. Inadequate identification is a violation of the National Electric Code and an unsafe business practice. In addition, the NEC requires that electrical equipment be marked with its appropriate ratings.

Bibliography

Accident Prevention Manual for Industrial Operations, 7th ed., National Safety Council, Chicago, Illinois 60611. 1974.

National Electrical Code 1978, National Fire Protection Association, Boston, Massachusetts 02210. 1977.

U.S. Department of Labor, Occupational Safety and Health Administration, Ground-Fault Protection on Construction Sites, OSHA Bulletin 3007, 1977.

17

Controlling Occupational Exposure to Bloodborne Pathogens

ACCORDING TO the Occupational Safety and Health Administration (OSHA), health care workers are at risk of exposure to the hepatitis B virus (HBV) and human immunodeficiency virus (HIV). Traditionally, infection control measures to protect both patients and staff have been an important part of workers safety. Evidence compiled by the Center for Disease Control regarding the risk of diseases such as hepatitis B and AIDS as the result of occupational exposure indicates, however, that additional measures are needed to protect health care employees who are at risk.

The OSHA's standard for Occupational Exposure to Bloodborne Pathogens (Title 29 Code of Federal Regulations (CFR) 1910.1030)) protects employees who have occupational exposure to bloodborne pathogens. This standard describes how to determine who is covered and the ways to reduce workplace exposure to bloodborne pathogens.

Three commonly used terms are defined below.

- Occupational exposure - a reasonably anticipated skin, eye, mucous membrane, or parenteral contact with blood or other potentially infectious materials that may result from the performance of the employee's duties.
- Bloodborne pathogens - pathogenic microorganisms present in human blood that can cause disease.
- Other potentially infectious materials - certain human body fluids, such as saliva, and any body fluid visibly contaminated with blood.

17.1 Exposure Control Plan

The first step in reducing workplace exposure is to develop an exposure control plan. The plan must be updated at least annually and when alterations in procedures create new occupational exposure. The exposure control plan must be available to employees in written form. As required under the standard, a written exposure control plan requires the identification of job classifications with occupational exposure and an implementation schedule.

17.1.1 Identification of Job Classifications with Occupational Exposure

The exposure control plan must include a list of job classifications and, if necessary, specific job tasks where there is exposure to blood and other potentially infectious materials. An exposure determination must be based on the definition of occupational exposure and be made without regard to the use of personal protective clothing and equipment. The determination is made by reviewing job classifications within the practice setting, and then making a list divided into two groups.

The first group on the list includes job classifications in which all of the employees have occupational exposure, such as clinical dental hygienists. In instances where all the employees would fall under this first group, it is not necessary to list specific work tasks for that job classification.

The second group includes those job classifications in which some of the employees have occupational exposure. In this case, specific tasks and procedures or groups of tasks and procedures causing exposure must be listed for each job classification. An example would be a dental practice with two or more receptionists, where one of the receptionists might be assigned the task of filling in for the dental assistant.

17.1.2 Implementation Schedule

The exposure control plans must also include a schedule of how and when the provisions of the standard will be implemented. The schedule may simply be a calendar with brief notations describing the methods, or an annotated copy of the standard such as the infection control plan. The schedule must contain the timetables and methods for communicating hazards to employees, implementing preventative measures (Hepatitis B vaccination and universal precautions) and implementing control measures (engineering controls and work practice controls).

17.2 Communicating Hazards to Employees

Hazard training must be offered to all affected employees. The initial training for reassigned employees must be provided before tasks are assigned, at no cost to the employee, and during working hours. Training is also required for new employees at the time of initial assignment to tasks with occupational exposure or whenever job tasks change resulting in a change in occupational exposure.

Annual retraining for all affected employees must be provided. If employees have received training on bloodborne pathogens in the year preceding the standard, only training in those areas required by the standard and which was not included in the previous training needs to be provided. This training could be included with training on other aspects of office safety, such as infection control and chemical hazards.

Training sessions must be comprehensive in nature, yet appropriate for the educational level, literacy, and language of employees. Training sessions should provide an opportunity for interactive questions and answers. The trainer must be knowledgeable in the program components as they relate to the nature of the business. As a minimum, the training program must include the following:

1. An accessible copy of the regulatory text of the standard and an explanation of its content.
2. An explanation of the epidemiology and symptoms of bloodborne diseases.
3. An explanation of the modes of transmission of bloodborne pathogens.
4. An explanation of the employer's written exposure control plan and how to obtain a copy.
5. Information on how to recognize occupational exposure.
6. The methods to control occupational transmission of bloodborne pathogens.
7. How to select, use, remove, handle, decontaminate, and dispose of personal protective clothing and equipment.
8. Information on the hepatitis B vaccine and vaccination, the availability of vaccine, and the fact that the vaccination is available at no cost to the employee.
9. Information on emergencies involving blood and other potentially infectious materials.
10. An explanation of the reporting mechanisms for exposure incidents.

11. Information on the post-exposure evaluation and follow-up available by a health care professional.
12. An explanation of labels, signs, and other markings for contaminated materials, such as instruments and laundry.
13. A question and answer session open to any aspects of the training.

17.3 Preventative Measures

In addition to communicating hazards to employees and providing training to identify and control hazards, other preventive measures must be taken to ensure employee protection. Preventive measures such as hepatitis B vaccination, universal precautions, engineering controls, safe work practices, personal protective equipment, and housekeeping measures help reduce the risks of occupational exposure.

17.3.1 Hepatitis B Vaccination

Hepatitis B vaccination must be made available within 10 working days of initial assignment to every employee whose job classification or tasks result in occupational exposure. Hepatitis B vaccination and vaccine must be made available without cost to the employee, at a reasonable time and place for the employee, and by or under the supervision of a licensed health care professional. The health care professional must be a physician or nurse practitioner whose legal scope of practice allows them to perform the hepatitis B vaccination and post-exposure and follow-up required in the standard. The employer must provide the health care professional with a copy of the bloodborne pathogens standard. The health care professional will provide the employer with a written opinion stating whether hepatitis B vaccination is indicated for the employee or if the employee has received such vaccination.

Employers are not required to offer hepatitis B vaccination to employees who have previously completed the hepatitis B vaccination series, when immunity is confirmed through antibody testing, or if the vaccine is contraindicated for medical reasons.

The hepatitis B vaccination series must be administered according to the current guidelines of the U.S. Public Health Service, including recommendations made in the future for routine booster doses. For current information on the U.S. Public Health Service's recommendations on hepatitis B vaccination, contact the Centers for Disease Control and Prevention located in Atlanta, Georgia.

An employee who chooses not to accept the vaccine must sign the following statement of declination of hepatitis B vaccination. The statement can only be signed following appropriate training regarding hepatitis B. The training should include information about the efficacy, safety, benefits, and method of administration of hepatitis B vaccination and the fact that the vaccine and vaccination are provided free of charge to the employee. The statement is not a waiver, employees can request and receive the hepatitis B vaccination at a later date if they remain occupationally at risk for hepatitis B.

Statement of Declination of Hepatitis B Vaccination:

I understand that due to my occupational exposure to blood or other potentially infectious materials I may be at risk of acquiring hepatitis B virus (HBV) infection. I have been given the opportunity to be vaccinated with hepatitis B vaccine, at no charge to myself. However, I decline hepatitis B vaccination at this time. I understand that by declining this vaccine, I continue to be at risk of acquiring hepatitis B, a serious disease. If in the future I continue to have occupational exposure to blood or other potentially infectious materials and I want to be vaccinated with hepatitis B vaccine, I can receive the vaccination series at no charge to me.

Employee Signature _____
Date_____

17.3.2 Universal Precautions

The single most important measure to control transmission of HBV and HIV is to treat all human blood and other potentially infectious materials as if they were infectious for HBV and HIV. Application of this approach to infection control is referred to as Universal Precautions. Blood and saliva from all patients are considered potentially infectious materials. As defined in the standard, these fluids cause contamination due to the presence or the reasonably anticipated presence of blood or other potentially infectious materials on an item or surface.

17.4 Control Measures

Engineering and work practice controls are the primary methods used to control the transmission of HBV and HIV in the work place. Personal protective clothing and equipment are also necessary when occupational exposure to bloodborne pathogens remains even after instituting these controls. More information about personal protective clothing and equipment is provided later in this chapter.

17.4.1 Engineering Controls

Engineering controls are a primary method used to control the transmission of HBV and HIV in the work place. Engineering controls isolate or remove the hazard from employees. Rubber dams, high-speed evacuators, and special containers for contaminated sharp instruments are examples of engineering controls. Personal protective clothing and equipment are also necessary when occupational exposure to bloodborne pathogens remains even after instituting these controls. Engineering controls must be examined and maintained, or replaced, on a scheduled basis.

17.4.2 Work Practice Controls

Work practice controls reduce the likelihood of exposure by altering the manner in which the task is performed. All procedures must be performed in such a manner as to minimize splashing, spraying, spattering, and generating droplets of blood or other potentially infectious materials. This can be as simple as readjusting the position of the workstation. Work practice controls include the following:

1. Washing hands immediately, or as soon as feasible, after skin contact with blood or other potentially infectious materials occurs and after removing gloves or other personal protective equipment.
2. Flushing mucous membranes immediately or as soon as feasible if they are splashed with blood or other potentially infectious materials.
3. Prohibiting recapping, bending, or removing contaminated needles from syringes unless required by the medical procedure or no alternative is feasible in which case must be done by mechanical means, such as the use of forceps, or using a one-handed technique.

For example, recapping is permitted when administering multiple injections local anesthesia.

4. Eliminating the shearing and breaking of contaminated needles.

5. Discarding contaminated needles or disposable sharps into containers that are closable, puncture-resistant, leak-proof, colored red or labeled with the biohazard symbol shown in Figure 17.1. These containers must be easily accessible, maintained upright, and not allowed to overfill.

6. Placing contaminated, reusable sharp instruments in containers that are puncture-resistant, leak-proof, colored red or labeled with the biohazard symbol until properly processed. Reusable sharps must not be stored or processed in such a way that employees are required to reach by hand into the container to retrieve the instruments.

7. Prohibiting eating, drinking, smoking, applying cosmetics, and handling contact lenses in areas where there is occupational exposure.

8. Eliminating the storage of food and drink in refrigerators, cabinets or shelves, or on countertops where blood or other potentially infectious materials are present.

9. Storing, transporting, or shipping blood or other potentially infectious materials in containers that are closed, prevent leakage, colored red, or affixed with the biohazard label. Labeling requires a fluorescent orange or orange-red label with the biological symbol, along with the word "BIOHAZARD" in a contrasting color, affixed to the bag or container.

FIG. 17.1. The Biohazard Symbol.

17.5 Personal Protective Equipment

In addition to instituting engineering and work practice controls, the standard requires that appropriate personal protective equipment (PPE) also be used to reduce worker risk of exposure. PPE is specialized clothing or equipment worn by employees to protect themselves from exposure to blood or other potentially infectious materials. PPE do not allow blood or other potentially infectious materials to pass through to clothing, skin, or mucous membranes. This section includes information about PPE as it relates to reducing exposure to bloodborne pathogens. Additional information about PPE is provided in Chapter 13.

17.5.1 Responsibilities of the Employer

The employer has the following responsibilities for personal protective equipment:

- providing, maintaining, and replacing PPE at no cost to the employee
- ensuring accessibility to PPE in appropriate sizes
- providing hypoallergenic gloves, glove liners, powderless gloves or other similar alternatives if the employee has an allergy to the gloves usually provided at no cost to the employee
- ensuring employee use of PPE
- laundering and/or discarding PPE at no cost to the employee

17.5.2 General Requirements

General requirements for personal protective equipment include the following:

1. Gloves, clinic jackets, lab coats, and chin-length face shields, or the combination of masks with eye protection (such as glasses with solid side shields or goggles) must be worn whenever splashes, spray, spatter, or droplets of blood or other infectious materials may be generated.
2. Cotton or cotton/polyester clinic jackets or lab coats are usually satisfactory barriers for routine procedures.
3. When procedures are performed involving large quantities of blood, additional personal protective equipment is required.
4. The selection of appropriate personal protective equipment is based upon the quantity and type of exposure expected.
5. Face protection can be accomplished using a chin-length face shield or a combination of mask with eye protection.

6. Goggles or eyeglasses with solid side shields or face shields can provide adequate eye protection.
7. Clinic jackets, lab coats, gowns, and other protective clothing and equipment must be removed immediately or as soon as feasible when penetrated by blood or other infectious materials, and prior to leaving the work area.
8. Gloves must be worn when it is reasonably anticipated that an employee will have hand contact with blood or saliva during procedures; when performing vascular access procedures; or when handling instruments, materials, and surfaces that are contaminated.
9. Disposable gloves must be replaced upon the completion of the dental procedure, or if torn or punctured during the procedure.
10. Disposable gloves are not to be reused.
11. Utility gloves used for cleanup may be decontaminated for reuse but must be discarded if they are deteriorated or fail to function as a barrier.
12. Contaminated personal protective equipment must be placed in an appropriately designated area or container for storing, washing, decontaminating, or discarding.
13. Mechanical emergency respiratory devices and pocket masks are types of personal protective equipment designed to isolate a worker from contact with a victim's saliva during resuscitation. Unprotected mouth-to-mouth resuscitation should be avoided. The patient may expel saliva, blood, or other fluids during resuscitation.

17.6 Housekeeping

17.6.1 Equipment

The employer must ensure a clean and sanitary workplace. Requirements for maintaining clean and sanitary workplaces include:

1. Work surfaces, equipment, and other reusable items that have been contaminated by splashes, spills, or other contact with blood and other potentially infectious materials must be decontaminated with disinfectant upon completion of procedures.
2. If surfaces, equipment, and other items have been protected with coverings such as plastic wrap or foil, these coverings must be replaced when contaminated or at the end of the work-shift.

3. Reusable receptacles such as bins, pails, and cans that have a likelihood for becoming contaminated must be inspected and decontaminated on a regular basis and whenever visibly contaminated.

4. Broken glass that may be contaminated can be cleaned up with a brush or tongs but never picked up with hands, even if gloves are worn.

5. Equipment that has had contact with blood or other potentially infectious materials and serviced either on-site or shipped out of the facility for maintenance or other service, must either be decontaminated to the extent feasible or be labeled as a biohazard. If the equipment has been labeled as a biohazard, any parts which were not decontaminated must be clearly marked.

17.6.2 Waste

The removal of waste from the facility may be regulated by a combination of local, state, and federal laws. To comply with the federal bloodborne pathogens standard, special precautions are necessary when disposing of contaminated sharps and other regulated waste. Other regulated waste may include:

- liquid or semi-liquid blood or other potentially infectious materials
- items contaminated with blood or other potentially infectious materials that would release these substances in a liquid or semi-liquid state if compressed
- items caked with dried blood or other potentially infectious materials and are capable of releasing these materials during handling
- contaminated sharps
- pathological and microbiological wastes containing blood or other potentially infectious materials

Contaminated disposable sharps must be placed in containers that are closable, puncture resistant, leakproof, and are colored red or labeled. Other regulated waste must be contained in closable bags or containers that prevent leakage and are colored red or labeled. A secondary container is necessary for containers that are contaminated on the outside. The secondary container also must be closable, prevent leakage, and be color-coded or labeled. Table 17.1 lists the labeling requirements of waste containers.

TABLE 17.1
Labeling Requirements of Waste Containers

Item	No Label Needed*	Biohazard Label Required	Red Container Required
Regulated waste container (contaminated sharps containers)		X	X
Reusable contaminated sharps container (surgical instruments soaking in a tray)		X	X
Refrigerator/freezer holding blood or other potentially infectious material		X	
Containers used for storage, transport, or shipping of blood		X	X
Blood or blood products for clinical use	No labels required		
Individual specimen containers of blood or other potentially infectious materials remaining in facility	X	X	X
Contaminated equipment needing service (dialysis equipment, suction apparatus)		X- plus a label specifying where the contamination exists	
Specimens and regulated waste shipped from the primary facility to another facility for service or disposal			
Contaminated laundry	**	X	X
Contaminated laundry sent to another facility that does not use Universal Precautions		X	X

*No label needed provided Universal Precautions are used and specific use of container or item is known to all employees.
**Alternative labeling or color-coding is sufficient if it permits all employees to recognize the containers as requiring compliance with Universal Precautions.

17.6.3 Laundry

Contaminated laundry shall be handled as little as possible and with minimum agitation. Laundering contaminated articles, including employee clinic jackets and lab coats used as personal protective equipment, is the responsibility of the employer. This can be accomplished through the use of a washer and dryer in a designated area on-site, or the contaminated articles can be sent to a commercial laundry that processes contaminated laundry. The care and laundering of general work clothes such as uniforms used to provide a professional appearance and not used as personal protective equipment, are not the responsibility of the employer.

Contaminated laundry must be placed in bags or in red containers marked with the biohazard symbol. If the office uses Universal Precautions in handling all soiled laundry, alternative labeling is permitted, provided that all employees are appropriately trained and recognize that the bags contain contaminated laundry. If the laundry is sent off site for cleaning, it must be in bags or containers that are clearly marked with the biohazard symbol, unless the laundry facility utilizes Universal Precautions in the handling of all soiled laundry. If contaminated laundry is wet, the bags or containers must prevent leakage and soak-through. Gloves and other appropriate personal protective equipment must always be worn when handling contaminated laundry.

17.7 Post-Exposure Evaluation

The employer is responsible for establishing the procedure for evaluating exposure incidents. When evaluating an exposure incident, thorough assessment and confidentiality are critical issues. Employees should immediately report exposure incidents to their employer to initiate a timely follow-up process by a health care professional. Such a report initiates the procedure for a prompt request for evaluation of the source individual's HBV and HIV status. The employee who has had an exposure incident, referred to as the source individual, must be directed to a health care professional. The employer must provide the health care professional with a copy of the following items:

- the bloodborne pathogens standard
- a description of the employee's job duties as they relate to the incident
- a report of the specific exposure incident (accident report), including routes of exposure
- the results of the source individual's blood tests, if available
- relevant employee medical records, including their vaccination status

The health care professional will, if the employee consents, draw a baseline blood test to establish the employee's HIV and HBV status. The employee has the right to decline testing or to delay HIV testing for up to 90 days. During this time, the health care professional must preserve the employee's blood sample. The results of the source individual's blood tests are confidential and should be directed only to the attending health care professional. The source individual's blood test results must be made available as soon as possible to the exposed employee through consultation with the health care professional.

Following the post-exposure evaluation, the health care professional will provide a written opinion to the employer. This opinion is limited to a statement that the employee has been informed of the results of the evaluation and told of the need, if any, for further evaluation or treatment. All other findings are confidential.

The employer must provide a copy of the written opinion to the employee within 15 days of the evaluation. Requirements for the medical record and training records are discussed in the following section on recordkeeping.

17.8 Recordkeeping

There are two types of employee-related records required by the bloodborne pathogens standard, medical records and training records.

17.8.1 Medical Records

A medical record must be established for each employee with occupational exposure. This record is confidential and separate from other personnel records. This record may be kept on-site or may be retained by the health care professional that provides services to the employees. The medical record contains the hepatitis B vaccination status and includes the dates of the hepatitis B vaccination and the written opinion of the health care professional regarding the hepatitis B vaccination.

If an occupational exposure incident occurs, reports are added to the medical record to document the incident and the results of testing following the incident, as well as the written opinion of the health care professional. The medical record also must indicate which documents have been provided to the health care provider. Medical records must be maintained for 30 years past the last date of employment of the employee.

The confidentiality of medical records must be emphasized. No medical record or part of a medical record is to be disclosed except to the employee, those having the written consent of the employee, representatives of the Secretary of Labor, or as required or permitted by state or federal law.

17.8.2 Training Records

Training records document each training session and must be kept by the employer for 3 years. Training records must include the date of the training, a course content outline, the trainer's name and qualifications, and names and job titles of all persons attending the training sessions.

If the employer ceases to do business, medical and training records are transferred to the successor employer. If there is no successor employer, the employer must notify the Director of the National Institute for Occupational Safety and Health, U.S. Department of Health and Human Services for specific directions regarding disposition of the records at least 3 months prior to their intended disposal.

Upon request, both medical and training records must be made available to the Assistant Secretary of Labor, Occupational Safety and Health. Training records must also be available to employees or their representatives upon request. For further information about employee access to medical and exposure records, see Title 29 CFR 1910.20 (e), Access to Employee Exposure and Medical Records. Additional recordkeeping is required for employers with 11 or more employees (see Title 29 CFR 1904, Recordkeeping Guidelines for Occupational Injuries and Illnesses). For more information on AIDS, contact the Center for Disease Control National AIDS Clearinghouse at 1-800-458-5231.

Summary

An Exposure Control Plan is required for all businesses in which the employees who have occupational exposure to bloodborne pathogens. This plan, if thoughtfully and carefully developed, maintained, and implemented, can serve as an extremely important tool in reducing employees' risk of contracting a bloodborne disease. The plan should be revised and reviewed at least annually and whenever necessary to reflect new or modified tasks and procedures which affect occupational exposure and to reflect new or revised employee positions with occupational exposure.

Engineering and work practice controls are the primary methods used to control occupational exposure to bloodborne pathogens such as HBV and HIV. In addition to engineering and work practice controls, personal protective equipment may be necessary to reduce worker risk of exposure.

The employer is required to provide a clean and sanitary workplace. This includes properly maintaining work surfaces and equipment, ensuring the safe removal of regulated waste, and providing adequate laundering services. The employer is also responsible for developing and implementing a procedure for evaluating exposure incidents.

OSHA's bloodborne pathogens standard requires that employee medical and training records be created and maintained as described in the standard.

Bibliography

Access to Medical and Exposure Records-OSHA 3110.

Centers for Disease Control. Recommendations for the Prevention of HIV Transmission in Health Care Settings. MMWR, August 21, 1987, Vol. 36, No. 2S.

Chemical Hazard Communication-OSHA 3084.

How to Prepare for Workplace Emergencies-OSHA 3000 Occupational Exposure to Bloodborne Pathogens.

Occupational Exposure to Bloodborne Pathogens-OSHA 3127.

"Occupational Exposure to Bloodborne Pathogens" Title 29 CFR 1910.

18

Lockout/Tagout

LOCKOUT/TAGOUT is a method of keeping equipment from being set in motion and endangering workers. On September 1, 1989, OSHA issued a final rule on the Control of Hazardous Energy (Lockout/Tagout) in Volume 29 of the Code of Federal Regulations (29 CFR), Section 1910.147. This standard, which went into effect on January 2, 1990, helps safeguard employees from hazardous energy while they are performing servicing or maintenance on machines and equipment by identifying the practices and procedures necessary to shutdown and lockout or tagout machines and equipment. The standard also requires that employees receive training in their role in the lockout/tagout program and mandates that periodic inspections be conducted to maintain or enhance the energy control program.

In the early 1970's, OSHA adopted various lockout-related provisions of the then existing national consensus standards and Federal standards that were developed for specific types of equipment or industries. When the existing standards require lockout, the new rule supplements these existing standards by requiring the development and utilization of written procedures, the training of employees, and periodic inspections of the use of the procedures. OSHA has determined that lockout is a more reliable means of de-energizing equipment than tagout and that it should always be the preferred method used by employees. Except for limited situations, the use of lockout devices will provide a more secure and more effective means of protecting employees from the unexpected release of hazardous energy or startup of machines and equipment.

This rule requires that, in general, before servicing or maintenance is performed on machinery or equipment, the machinery or equipment must be turned off and disconnected from the energy source, and the energy-isolating device must be either locked or tagged out. OSHA estimates that adherence to the requirements of this standard can eliminate nearly 2 percent of all workplace deaths and can have a significant impact on worker safety and health in the U.S. The workers who actually service equipment such as craft workers, machine operators, and laborers face the greatest risk.

The following terms and phrases are used throughout this chapter:

1. Affected employee - An employee who performs the duties of his or her job in an area in which the energy control procedure is implemented and servicing or maintenance operations are performed. An affected employee does not perform servicing or maintenance on machines or equipment and, consequently, is not responsible for implementing the energy control procedure. An affected employee becomes an "authorized" employee whenever he or she performs servicing or maintenance functions on machines or equipment that must be locked or tagged.

2. Authorized employee - An employee who performs servicing or maintenance on machines and equipment. Lockout or tagout is used by these employees for their own protection.1.

3. Capable of being locked out - An energy-isolating device is considered capable of being locked out if it meets one of the following requirements:

   • it is designed with a hasp to which a lock can be attached
   • it is designed with any other integral part through which a lock can be affixed
   • it has a locking mechanism built into it
   • it can be locked without dismantling, rebuilding, or replacing the energy isolating device or permanently altering its energy control capability

4. Energized - Machines and equipment are energized when they are connected to an energy source or they contain residual or stored energy.

5. Energy-isolating device - Any mechanical device that physically prevents the transmission or release of energy. These include, but are not limited to, manually-operated electrical circuit breakers, disconnect switches, line valves, and blocks.

6. Energy source - Any source of electrical, mechanical, hydraulic, pneumatic, chemical, thermal, or other energy.

7. Energy control procedure - A written document that contains those items of information an authorized employee needs to know in order to safely control hazardous energy during servicing or maintenance of machines or equipment.

8. Energy control program - A program intended to prevent the unexpected energizing or the release of stored energy in machines or equipment. The program consists of energy control procedure(s), an employee training program, and periodic inspections.

9. Lockout - The placement of a lockout device on an energy-isolating device, in accordance with an established procedure, ensuring that the energy-isolating device and the equipment being controlled cannot be operated until the lockout device is removed.

10. Lockout device - Any device that uses positive means such as a lock, either key or combination type, to hold an energy-isolating device in a safe position, thereby preventing the energizing of machinery or equipment. When properly installed, a blank flange or bolted slip blind are considered equivalent to lockout devices.

11. Tagout - The placement of a tagout device on an energy-isolating device, in accordance with an established procedure, to indicate that the energy-isolating device and the equipment being controlled may not be operated until the tagout device is removed.

12. Tagout device - Any prominent warning device, such as a tag and a means of attachment, that can be securely fastened to an energy-isolating device in accordance with an established procedure. The tag indicates that the machine or equipment to which it is attached is not to be operated until the tagout device is removed in accordance with the energy control procedure.

18.1 Scope and Application

The lockout/tagout standard applies to general industry employment and covers the servicing and maintenance of machines and equipment in which the unexpected startup or the release of stored energy could cause injury to employees. If employees are performing servicing or maintenance tasks that do not expose them to the unexpected release of hazardous energy, the standard does not apply.

The standard establishes minimum performance requirements for the control of hazardous energy. The standard does not apply in the following situations:

- while servicing or maintaining cord and plug connected electrical equipment. (The hazards must be controlled by unplugging the equipment from the energy source; the plug must be under the exclusive control of the employee performing the servicing and/or maintenance.)
- during hot tap operations that involve transmission and distribution systems for gas, steam, water, or petroleum products when they are

performed on pressurized pipelines; when continuity of service is essential, and shutdown of the system is impractical; and employees are provided with an alternative type of protection that is equally effective.

The standard requires employers to establish procedures for isolating machines or equipment from the input of energy and affixing appropriate locks or tags to energy-isolating devices to prevent any unexpected energization, startup, or release of stored energy that would injure workers. When tags are used on energy-isolating devices capable of being locked out, the employer must provide additional means to assure a level of protection equivalent to that of locks. The standard also requires the training of employees and periodic inspections of the procedures to maintain or improve their effectiveness.

18.2 Normal Production Operations

OSHA recognizes that machines and equipment present many hazardous situations during normal production operations. These production hazards are covered by rules in other General Industry Standards, such as the requirements in Subpart O of Part 1910 for general machine guarding and guarding power transmission apparatus (1910.212 and 1910.219). In certain circumstances, however, some hazards encountered during normal production operations may be covered by the Lockout/tagout rule. The following sections illustrate some of these instances.

18.2.1 Servicing and/or Maintenance Operations

If a servicing activity such as lubricating, cleaning, or unjamming the production equipment-takes place during production, the employee performing the servicing may be subjected to hazards that are not encountered as part of the production operation itself. Workers engaged in these operations are covered by lockout/tagout when any of the following conditions occurs:

- The employee must either remove or bypass machine guards or other safety devices, resulting in exposure to hazards at the point of operation;

- The employee is required to place any part of his or her body in contact with the point of operation of the operational machine or piece of equipment; or
- The employee is required to place any part of his or her body into a danger zone associated with a machine operating cycle.

In the above situations, the equipment must be de-energized and locks or tags must be applied to the energy-isolation devices. In addition, when normal servicing tasks such as setting up equipment, and/or making significant adjustments to machines, do not occur during normal production operations, employees performing such tasks are required to lock out or tag out if they can be injured by unexpected energization of the equipment.

OSHA also recognizes that some servicing operations must be performed with the power on. Making many types of fine adjustments, such as centering the belt on conveyors, is one example. Certain aspects of troubleshooting, such as identifying the source of the problem as well as checking to ensure that it has been corrected, is another. OSHA requires the employer to provide effective protection for employees performing such operations. Although, in these cases, a power-on condition is essential either to accomplish the particular type of servicing or to verify that it was performed properly, lockout or tagout procedures are required when servicing or maintenance occurs with the power off.

18.2.2 Minor Servicing Tasks

Employees performing minor tool changes and adjustments and/or other minor servicing activities during normal production operations that are routine, repetitive, and integral to the use of the production equipment are not covered by the Lockout/tagout standard, provided the work is performed using alternative measures that give effective protection.

18.2.3 Energy Control Program

The lockout/tagout rule requires that the employer establish an energy control program that includes:

- documented energy control procedures
- an employee training program
- periodic inspections of the procedures

The standard requires employers to establish a program to ensure that machines and equipment are isolated and inoperative before any employee performs servicing or maintenance where the unexpected energization, start up, or release of stored energy could occur and cause injury.

The purpose of the energy control program is to ensure that, whenever the possibility of unexpected machine or equipment start-up exists or when the unexpected release of stored energy could occur and cause injury, the equipment is isolated from its energy source(s) and rendered inoperative prior to servicing or maintenance.

Employers have the flexibility to develop a program and procedures that meet the needs of their particular workplace and the particular types of machines and equipment being maintained or serviced.

18.2.4 Energy Control Procedure

This standard requires that energy control procedures be developed, documented, and used to control potentially hazardous energy sources whenever workers perform activities covered by the standard.

The written procedures must identify the information that authorized employees must know in order to control hazardous energy during servicing or maintenance. If this information is the same for various machines or equipment or if other means of logical grouping exists, then a single energy control procedure may be sufficient. If there are other conditions such as multiple energy sources, different connecting means, or a particular sequence that must be followed to shut down the machine or equipment, then the employer must develop separate energy control procedures to protect employees.

The energy control procedure must outline the scope, purpose, authorization, rules, and techniques that will be used to control hazardous energy sources as well as the means that will be used to enforce compliance. At a minimum, it includes, but is not be limited to, the following elements:

- a statement on how the procedure will be used
- the procedural steps needed to shut down, isolate, block, and secure machines or equipment
- the steps designating the safe placement, removal, and transfer of lockout/tagout devices and who has the responsibility for them
- the specific requirements for testing machines or equipment to determine and verify the effectiveness of locks, tags, and other energy control measures

The procedure must include the following steps:

- preparing for shutdown
- shutting down the machine(s) or equipment
- isolating the machine or equipment from the energy source(s)
- applying the lockout or tagout device(s) to the energy-isolating device(s)
- safely releasing all potentially hazardous stored or residual energy
- verifying the isolation of the machine(s) or equipment prior to the start of servicing or maintenance work

In addition, before lockout or tagout devices are removed and energy is restored to the machines or equipment, certain steps must be taken to reenergize equipment after servicing is completed, including:

- assuring that machines or equipment components are operationally intact
- notifying affected employees that lockout or tagout devices have been removed, and ensuring that all employees are safely positioned or removed from equipment
- assuring that lockout or tagout devices are removed from each energy-isolating device by the employee who applied the device (see sections 6 (e) and 6 (f) of 29 CFR 1910.147 for specific requirements of the standard)

18.2.5 Energy-Isolating Devices

The employer's primary tool for providing protection under the standard is the energy-isolating device, which is the mechanism that prevents the transmission or release of energy and to which all locks or tags are attached. This device guards against accidental machine or equipment start-up or the unexpected re-energization of equipment during servicing or maintenance. There are two types of energy-isolating devices: those capable of being locked and those that are not. The standard differentiates between the existence of these two conditions and the employer and employee responsibilities in each case.

When the energy-isolating device cannot be locked out, the employer must use tagout. Of course, the employer may choose to modify or replace the device to make it capable of being locked. When using tagout, the employer must comply with all tagout-related provisions of the standard and, in addition to the normal training required for all employees, must train his or her employees in the following limitations of tags:

- Tags are essentially warning devices affixed to energy-isolating devices and do not provide the physical restraint of a lock.
- When a tag is attached to an isolating means, it is not to be removed except by the person who applied it, and it is never to be bypassed, ignored, or otherwise defeated.
- Tags must be legible and understandable by all employees.
- Tags and their means of attachment must be made of materials that will withstand the environmental conditions encountered in the workplace.
- Tags may evoke a false sense of security. They are only one part of an overall energy control program.
- Tags must be securely attached to the energy-isolating devices so that they cannot be detached accidentally during use.

If the energy-isolating device is lockable, the employer shall use locks unless he or she can prove that the use of tags would provide protection at least as effective as locks and would assure full employee protection. Full employee protection includes complying with all tagout-related provisions plus implementing additional safety measures that can provide the level of safety equivalent to that obtained by using lockout. This might include removing and isolating a circuit element, blocking a controlling switch, opening an extra disconnecting device, or removing a valve handle to reduce the potential for any inadvertent energization.

Although OSHA acknowledges the existence of energy-isolating devices that cannot be locked out, the standard clearly states that whenever major replacement, repair, renovation, or modification of machines or equipment is performed and whenever new machines or equipment are installed, the employer must ensure that the energy-isolating devices for such machines or equipment are lockable. Such modifications and/or new purchases are most effectively and efficiently made as part of the normal equipment replacement cycle. All newly purchased equipment must be lockable.

18.3 Requirements for Lockout/Tagout Devices

When attached to an energy-isolating device, both lockout and tagout devices are tools that the employer can use in accordance with the requirements of the standard to help protect employees from hazardous energy. The lockout device provides protection by holding the energy-isolating device in the safe position, thus preventing the machine or equipment from becoming energized.

The tagout device accomplishes this by identifying the energy-isolating device as a source of potential danger; it indicates that the energy-isolating device and the equipment being controlled may not be operated until the tagout device is removed. Whichever devices are used, they must be singularly identified, must be the only devices used for controlling hazardous energy, and must meet the following requirements:

- Durable - Lockout and tagout devices must withstand the environment to which they are exposed for the maximum duration of the expected exposure. Tagout devices must be constructed and printed so that they do not deteriorate or become illegible, especially when used in corrosive (acid and alkali chemicals) or wet environments.
- Standardized - Both lockout and tagout devices must be standardized according to either color, shape, or size. Tagout devices must also be standardized according to print and format.
- Substantial - Lockout and tagout devices must be substantial enough to minimize early or accidental removal. Locks must be substantial to prevent removal except by excessive force of special tools such as bolt cutters or other metal cutting tools. Tag means of attachment must be nonreusable, attachable by hand, self-locking, and nonreleasable, with a minimum unlocking strength of no less than 50 pounds. The device for attaching the tag also must have the general design and basic characteristics equivalent to a one-piece nylon cable tie that will withstand all environments and conditions.
- Identifiable - Locks and tags must clearly identify the employee who applies them. Tags must also warn against hazardous conditions if the machine or equipment is energized and must include a legend such as the following: DO NOT START, DO NOT OPEN, DO NOT CLOSE, DO NOT ENERGIZE, DO NOT OPERATE.

18.4 Employee Training

The employer must provide effective initial training and retraining as necessary and must certify that such training has been given to all employees covered by the standard. The certification must contain each employee's name and dates of training.

For the purposes of the standard, there are three types of employees - authorized, affected, and other. The amount and kind of training that each employee receives is based upon:

- the relationship of that employee's job to the machine or equipment being locked or tagged out
- the degree of knowledge relevant to hazardous energy that he or she must possess

For example, the employer's training program for authorized employees (those who are charged with the responsibility for implementing the energy control procedures and performing the servicing or maintenance) must cover, at minimum, the following areas:

- details about the type and magnitude of the hazardous energy sources present in the workplace
- the methods and means necessary to isolate and control those energy sources (i.e., the elements of the energy control procedure(s)

By contrast, affected employees (usually the machine operators or users) and all other employees need only be able to:

- recognize when the control procedure is being implemented
- understand the purpose of the procedure and the importance of not attempting to start up or use the equipment that has been locked or tagged out

Because an "affected" employee is not one who is performing the servicing or maintenance, that employee's responsibilities under the energy control program are simple - whenever there is a lockout or tagout device in place on an energy-isolating device, the affected employee leaves it alone and does not attempt to operate the equipment.

Every training program must ensure that all employees understand the purpose, function, and restrictions of the energy control program and that authorized employees possess the knowledge and skills necessary for the safe application, use, and removal of energy controls. Training programs used for compliance with this standard, which is performance-oriented, should deal with the equipment, type(s) of energy, and hazard(s) specific to the workplace being covered.

Retraining must be provided whenever there is a change in job assignments, a change in machines, equipment or processes that present a new hazard, or a change in energy control procedures. Additional retraining must be conducted whenever a periodic inspection reveals, or whenever the employer has reason to believe, that there are deviations from or inadequacies in the employee's knowledge or use of the energy control procedure.

18.5 Periodic Inspections

Periodic inspections must be performed at least annually to assure that the energy control procedures (locks and tags) continue to be implemented properly and that the employees are familiar with their responsibilities under those procedures. In addition, the employer must certify that the periodic inspections have been performed. The certification must identify the machine or equipment on which the energy control procedure was used, the date of the inspection, the employees included in the inspection, and the name of the person performing the inspection.

For lockout procedures, the periodic inspection must include a review, between the inspector and each authorized employee, of that employee's responsibilities under the energy control procedure being inspected. When a tagout procedure is inspected, a review on the limitation of tags, in addition to the above requirements, must also be included with each affected and authorized employee.

18.6 Application of Controls and Lockout/Tagout Devices

The established procedure of applying energy controls includes specific elements and actions that must be implemented in sequence. These elements are briefly identified as follows:

- Prepare for shut down.
- Shut down the machine or equipment.
- Apply the lockout or tagout device.
- Render safe all stored or residual energy.
- Verify the isolation and de-energization of the machine or equipment.

18.7 Removal of Locks and Tags

Before lockout or tagout devices are removed and energy is restored to the machine or equipment, the authorized employee(s) must take the following actions or observe the following procedures:

1. Inspect the work area to ensure that non-essential items have been removed and that machine or equipment components are intact and capable of operating properly.
2. Check the area around the machine or equipment to ensure that all employees have been safely positioned or removed.
3. Notify affected employees immediately after removing locks or tags and before starting equipment or machines.
4. Make sure that locks or tags are removed only by those employees who attached them. (In the very few instances when this is not possible, the device may be removed under the direction of the employer provided that he or she strictly adheres to the specific procedures outlined in the standard.)

18.8 Additional Safety Requirements

Special circumstances exist when machines need to be tested or repositioned during servicing, outside (contractor) personnel are at the worksite, servicing or maintenance is performed by a group (rather than one specific person), or shifts or personnel changes occur.

18.9 Testing or positioning of machines

OSHA allows the temporary removal of locks or tags and the re-energization of the machine or equipment only when necessary under special conditions, for example, when power is needed for the testing or positioning of machines, equipment, or components. The re-energization must be conducted in accordance with the sequence of steps listed below:

1. Clear the machines or equipment of tools and materials.
2. Remove employees from the machines or equipment area.
3. Remove the lockout or tagout devices as specified in the standard.
4. Energize and proceed with testing or positioning.

5. De-energize all systems, isolate the machine or equipment from the energy source and reapply lockout or tagout devices as specified.

Summary

Lockout/tagout is a technique for keeping equipment from being set in motion and endangering workers. The OSHA standard establishes minimum performance requirements for the control of hazardous energy and for the placement and removal of lockout and tagout devices.

The OSHA standard also requires the establishment of and energy control program with documented energy control procedures, and employee training program, and periodic inspections of the procedures.

Bibliography

OSHA (1997) "Control of Hazardous Energy (Lockout/Tagout)", U.S. Department of Labor, Washington, D.C.

OSHA (1999) "A Guide to the Control of Hazardous Energy (Lockout/Tagout)". N.C. Department of Labor, Raleigh, N.C.

19

Behavior Change Methods

WHEN DEVELOPING a behavior change program for an employee, it is useful to first consider whether the problem reflects a lack of skill, a lack of motivation, or a combination of these two. A worker may operate a machine improperly either because she does not know the proper way to operate it or because she may not be sufficiently motivated to operate it correctly. Distinguishing between these sources of faulty performance is important because fundamentally different behavior control procedures are used to correct the problem depending on its source. Training is used to solve skill problems and changes in consequences is used to solve motivation problems.

19.1 Identifying Skill and Motivation Problems

Businesses often fail to make the distinction between skill and motivation problems and consequently employ ineffective solutions. Although there are cases in which motivation is provided when training is needed, the more common mistake is to provide training when motivation is needed. Examples of the latter include providing training to cause workers to spend more of their time attending to assigned tasks and less time in non-productive pursuits. The workers already knew how to do their jobs, therefore, they had no need for training. An example of providing motivation when training is needed is the management of new sales people in businesses such as life insurance. Although the new sales staff posses the necessary motivation, many do not have the skills to sell insurance. Their efforts arc not reinforced by making sales and they gradually give up. Frequently, a simple investigation can reveal whether training or motivation is needed to solve the problem.

Several procedures are useful to distinguish between skill and motivation problems. None of these procedures is fool-proof and each should be used cautiously. Several procedures should be considered before any definitive broad-scale solutions are attempted.

19.1.1 Identifying Lack of Skill

First, ask is a person sometimes behaves as desired. If a person sometimes behaves correctly, the chances are that the skills are present but, on occasion, there is insufficient motivation. This rule can prove faulty in cases in which the worker may have the skills to perform correctly in some circumstances but not others. The performance inconsistencies may result because of changes in the circumstances of the task.

Second, ask if just about everyone in a particular job knows how to perform the job correctly. If they do, one can assume that the appropriate skills are common knowledge and that the particular worker also knows how to behave but simply isn't motivated to do so. This assumption can be a incorrect when the one worker has been inadequately trained. It is also useful to find out if the worker expends reasonable energy in doing assigned jobs. If there is good evidence that the worker works hard but has particular performance deficits, especially those that consistently involve the same set of skills, the problem is most likely a lack of skill.

A straightforward test of whether a poor performance involves a lack of skill or a lack of motivation consists of asking or having a direct supervisor ask the person in question to correctly perform the task. If the person can do as asked, it is evident that the person has the necessary skills. Failures to perform on other occasions are likely the results of insufficient motivation.

19.1.2 Identifying Lack of Motivation

Problems with lack of motivation are evident when the worker in question could perform correctly if extremely motivated. The supervisor must discover whether or not the worker, if highly motivated, would be able to behave as desired. Hypothetically, if the worker were offered $100 to do as requested, would he be able to provide the desired performance? If the answer to this hypothetical question is yes, it is safe to assume that the worker has the necessary skills.

Answers to these questions may, of course, be inconclusive. The problem may involve both a lack of skill and a lack of motivation. In this case, both training and improved motivation must be provided. Management may ask whether it is better to provide training when motivation is needed or motivation when training is in order. There is no correct answer to this question, however, the motivated person will seek out knowledge and training and solve his problems without formal assistance. The skilled but unmotivated worker is without useful recourse until some productive motivation occurs.

19.2 Dealing with Motivation Problems

The following steps are recommended for dealing with motivation problems:

1. Specify the objectives. Specifying the objectives simply means knowing exactly what it is you plan to try to accomplish. Know which behaviors you want people to engage in and which ones you want eliminated. Know what results you hope to achieve.
2. Consider sharing the objectives. Sharing the objectives can be controversial. One theory suggests a good way to manage behaviors is to involve workers in goal setting. This does not mean dictating goals to them, or deciding which goals would be useful for them and then trying to sell them these goals. Rather, the goal-setting theory says that people should participate in the selection and discovery of the goals that are appropriate for them. However, the rule is to "consider" sharing the objective, which means that judgment may be needed to determine those instances in which you may better achieve your goals if they are not shared. This includes instances in which a program would successfully change a worker's motivation while their prior knowledge of the goals would be offensive to them.
3. Analyze the existing consequences for and against the desired performance. If someone is engaging in an undesired behavior, then the undesirable behavior is being reinforced and the reinforcement for the desired behavior is absent or weak in comparison to reinforcement for other behavior. Perhaps, the desired behavior is even being negatively reinforced in some way. You will find it useful to discover the forms of reinforcement that are operating. In the case of unsafe behavior, it is likely that there are production pressures that cause people to hurry or take short cuts that expose them to hazards. Behaviors that best reduce exposures to hazards may require more effort. Peer groups may

reinforce macho, tough or indifferent behavior. The only reinforcer or punisher that would encourage safe behavior may be the long-range and improbable chance that the worker will be harmed. The sources of reinforcement and punishment must be understood in order to control them.

4. Provide reinforcement for the desired behavior or results of behavior. Under almost all circumstances it will be useful to provide reinforcement of the desired behavior or performance. This is accomplished by understanding the reinforcers in effect for the workers, selecting a effective reinforcer(s), and arranging for those reinforcers to occur following the desired behavior or performance.

5. Eliminate existing reinforcements for undesired behaviors. This often involves changing peer group pressure from reinforcing hazardous behaviors to reinforcing safe behavior. It also includes eliminating pressure from the management to take hazardous short-cuts. In many cases, the major impediments to safe behavior are the effort and bother necessary to perform jobs safely. These impediments may be impossible to eliminate. Accordingly, the solution will consist of making the reinforcement for behaving safely sufficiently powerful to offset the natural consequences that promote hazardous behavior.

6. Adjust the consequences as required. Once some experience at adjusting consequences is acquired, the alert manager will find more ways that useful consequences can be put to work. One should not expect to be perfectly satisfied with their first attempt at improving motivation.

19.3 Using Social Reinforcement

Social reinforcement is the behavior of one person that reinforces another person. The best known example is praise or recognition. It is often the case that one person can reinforce a particular response of another person simply by praising the other person dependent on the occurrence of the response. Management should spend a few minutes each day looking for things that people are doing well and praising them when they are observed doing them well.

Social reinforcement should be timely, honest, sincere, of good variety, and descriptive. A praise or compliment should not be delayed. The sooner after the response the praise occurs, the better. A praise must be honest.

The person must have done a good job and the desired behavior must have occurred. A inaccurate praise reflects poorly on the person doing the praising and adds the risk that the unwanted behavior will reoccur. An effective praiser is discriminating, taking the time to notice what people are doing.

Praise should be sincere. An effective praiser has an appreciation for the welfare of other people and wants to help them. The efforts of an insincere praiser are wasted. While the insincere praise may be effective in the short-term, it will not be effective in the long-term.

The praise should not consist of the same sentiment over and over again. If the same sentiment is used repeatedly, the praise will be considered insincere. Workers will be bored by the repetitive praises. "You are doing a good job, John." "Good job, John." "That's a good job, John." "Good job, John." Want to hear it again? Probably neither would John. If you are going to praise, take time to learn a lot of ways to praise and ways to say nice things to and about people.

Because it is difficult to offer praises immediately after the desired behaviors have occurred, the praise should include a description of the desired behavior. For example, a description praise would be to say "Good job, John, you got your crew to remove the dangerous piles of lumber almost as soon as we found them." This praise avoids misunderstandings, naturally adds to the variety of the praise, and demonstrates to the persons being praised that you know what is going on.

19.3.1 Performance Feedback

Performance feedback occurs when specific information about performance is provided to the performer. Performance feedback functions differently in different settings. However, under certain conditions, performance feedback can constitute reinforcement, punishment, or a combination of the two. It might serve as an antecedent stimulus to prompt a worker to behave in a certain way. It might serve as a prompt for a supervisor to provide social reinforcement or punishment to a worker for the good or poor performances reflected in the feedback. In any case, performance feedback can produce improvements in a variety of kinds of performances.

Performance feedback should be timely, accurate, provided in graphic form for individuals and/or specific groups of workers, and be provided in relation to realistic expectations and organizational objectives. It is recommend that feedback on most forms of performance be provided on a daily or weekly basis. This means that the feedback will occur often and immediately following the behavior. If feedback is to be provided as infrequently as once per month, it is important to accompany the feedback with some tangible reinforcers.

If performance feedback is to function as reinforcement or punishment, or as a prompt for a supervisor to deliver reinforcement or punishment, it is essential that the feedback be accurate. If the feedback is inaccurate, the resultant feedback or punishment will result in problems discussed above.

Graphic feedback has considerable advantage over most other forms of feedback. It provides quantitative information about the performance and functional pictures about performances in relation to standards. It also provides information about trends in performance which may allow for the reinforcement of behaviors that are trending towards goals long before the goals are reached.

In most instances, performance feedback should be specific to individual performance and should be directed at individuals. It may be desirable to publicly display graphic feedback about individual performance, particularly if that performance is exemplary or improving impressively. In such cases, it is hoped that the public display will bring social reinforcement to the individual. Public display of poor performance should be used with caution because of the chances of bringing unnecessary punishment on a person, perhaps forcing the person to justify his performance by attacking company policy or job standards.

Graphic displays of the aggregate of the performances of a group may be considered in instances where it is desirable to generate group pressures for performance. Again, the possibilities of the kinds of reactions likely to be generated by the graphic display should be considered.

Standards or goals, if they are realistic and fair, can be plotted on graphs indicating performance. Such standards can be developed with any desired degree of input by the person(s) for whom the standards are intended. The standards provide a level against which current performance can be compared and a means of estimating the rate at which improvements towards the standard are being made.

19.4 Using Token Reinforcement

Tangible reinforcers are items which are reinforcing for the person who receives them. A token reinforcer is one which has little intrinsic value but which can be exchanged for many things of value. Money is the best known example of a token reinforcer. Trading stamps are token reinforcers, as are points, that are arbitrarily awarded to allow for score-keeping in contests and games. In contrast to the backup reinforcers for which they can be exchanged, token reinforcers can be easily handled and can be highly generalized if they are exchangeable for a variety of backup reinforcers. Therefore, they are likely to be valuable to the recipient regardless of the reinforcers for which the recipient is, at the moment, deprived.

Token reinforcement should be timely, accurate, constructed to include many backup reinforcers, and descriptive. Token reinforcement should be delivered frequently and with little delay following the occurrence of desired performance. As with all reinforcements, its delivery must be contingent on the occurrence of the desired behavior. If undesired behaviors are reinforced, they will continue to occur.

The greater the variety of backup reinforcers for which the tokens can be exchanged, the better the chances that the tokens will continue to be reinforcing for an individual. It also increased the likelihood that there will be a desired backup reinforcer for every individual receiving tokens. It will be useful to consult the workers who will receive the tokens to determine the kinds of things that will be reinforcing for them.

Money is a highly generalized token reinforcer. Money may not be as useful as more specific reinforcers to promote a particular behavior. If money is used as a reinforcer it is likely to be added to the family resources and doled out for food, sneakers, gasoline, and utility bills. However, a token which can be traded only for more particularized backups such as golf clubs, vacation trips, meals at favorite restaurants, tickets to sporting events, or personal prizes, may be more effective to reinforce desired behaviors.

19.5 Using Punishment

It is important to understand some of the complications that can result from the use of punishment. For several decades, psychologists thought that punishment was not effective in reducing the frequency of unwanted behavior. One interpretation of the effects of punishment was that it could be used to temporarily suppress behavior but it was not useful to eliminate behavior. If punishment is weak, and if the unwanted behavior is being maintained by uncontrolled reinforcement, the punishment may only suppress the behavior temporarily. However, it is also clear that under many circumstances, strong punishment can lessen unwanted behavior for very long periods of time, perhaps indefinitely. Whether it is useful to use punishment to weaken behavior is a much more complicated question.

In circumstances in which an unwanted behavior is occurring, it is safe to assume that there is some source of reinforcement for the unwanted behavior. If there were no reinforcement, it would not be occurring. Further, if the wanted behavior is not occurring or is weak in comparison to the unwanted behavior, there must be insufficient reinforcement for it. These considerations set the stage for a discussion of the problems that often accompany the use of punishment.

First, punishment can suppress the unwanted behavior without an increase in the frequency of the wanted behavior. This occurs when the reinforcement for the wanted behavior is inadequate. In contrast, reinforcement would more likely strengthen the wanted behavior at the expense of the unwanted behavior.

Punishment may have to be used indefinitely to maintain the suppression of unwanted behavior. In such circumstances, expect that an instance of punishment will produce suppression of the unwanted behavior for a while but that the unwanted behavior will eventually begin to recover and will have to be punished again to keep it suppressed.

Punishment may produce undesired emotional responses ranging from passive responses such as crying to aggressive responses such as physical retaliation. Punishment can also prompt subversion. People will start looking for ways to exhibit the undesired behavior and avoid the punishment. This case can occur when the reinforcement for the unwanted behavior is not eliminated and the reinforcement for the wanted behavior remains weak.

The likelihood of subversion can increase dramatically if several people cooperate to protect each other. Such cooperation is likely if all involved are subject to punishment and all are inadequately reinforced for positive behavior. In such instances, it is likely that there will be social reinforcement from the peer group to cooperate in subverting the punishment. This peer reinforcement is in addition to the other sources of reinforcement for the unwanted behavior.

If punishment is severe, and it may have to be severe if it is to off-set the effects of the reinforcement that is maintaining the unwanted behavior, the effects may generalize to behaviors other than the target behaviors. In such instances, useful as well as harmful behaviors are suppressed.

The person who administers the majority of punishment, particularly severe punishment, may come to be regarded as a punishing stimulus and people will avoid him. This can be a problem particularly for a supervisor who should be playing a key role in cooperative endeavors but who is being consistently avoided because of his frequent association with punishment.

There are many problems associated with the use of punishment and it may not be a particularly useful way to motivate behavior in a positive manner. Even though there are serious problems associated with the use of punishment, there may be circumstances when punishment must be considered. Clear examples would be instances where a person, despite proper training and warning, behaves in a way that seriously threatens their own or another person's life or well-being.

When punishment is indicated, the following conditions should be met:

1. The rules and the consequences for breaking them should be stated in advance.
2. People should know the proper way to behave.
3. There should be many more opportunities for reinforcement for correct behaviors than punishment for inappropriate behaviors.
4. Punishment should be reserved for particularly serious behaviors.
5. The punishment should occur as immediately as possible following the behavior.
6. The duration of the punishment should be short.
7. The punishment should be directed towards the behavior, not the person.
8. The punishment should be accompanied by a very brief description of the behavior being punished.
9. Punishment should be administered consistently and fairly.

Several of these conditions are aimed at avoiding the problems that can accompany the use of punishment, others at maximizing the chances of producing the desired effect. Having rules and consequences stated in advance is aimed at minimizing unfortunate reactions to punishment and at insuring that the person knows the correct behavior, so they are not punished because of a lack of skill or knowledge.

Using more reinforcement for correct behaviors than punishment for incorrect behaviors and reserving punishment for particularly serious behaviors has several purposes. This practice will help:

- ensure there is reinforcement for the desired behavior
- keep supervisors from being avoided by the workers
- minimize attempts at subverting punishment
- clarify the management's intent to generate behaviors that will help people rather than to arbitrarily exercise power

The importance of ensuring that consequences occur immediately following the behavior has previously been discussed. The duration of punishment should be short to bring quick closure to the event and allow the individuals involved to give their attention instead to finding ways to promote desired behaviors.

Ensuring the punishment is directed towards the behavior and not the person will help make it clear that a good performance is expected from the person. It will also help to minimize the likelihood that the person will have to fight back or attempt to subvert the supervisor's efforts. Briefly describing the behavior being punished will help to ensure that you suppress the intended behavior.

If a behavior is deserving of punishment once, it is deserving of punishment every time it occurs. Punishment between individuals should be consistent unless differences in skills or knowledge are evident. If the same unwanted behavior occurs over and over, one can assume that there is a lack of skills, a reinforcement for the unwanted behavior is present, or there is insufficient reinforcement for the desired behavior.

19.6 Information-Presenting, Education, and Training

The terms education and training are used in so many imprecise ways a terminological distinction is in order. First, information-presenting is simply the presenting of information, by any means by the trainer or educator. Information-presenting specifies something about the behavior of the instructor but nothing about changes of the behavior of the person who is the object of the exercise. This person may sleep through the presentation and there may be absolutely no change in behavior.

With education there are procedural guarantees that the person who is the object of the work has some particular behavior. Generally, the guarantees provide only that the person has some written or spoken verbal behavior about the topic of interest. The guarantees usually take the form of tests that are used to sample the verbal repertoire of the learner. Universities typically give tests that require written verbal responses, and less frequently, spoken verbal responses. Universities provide some sort of guarantee that a person can talk or write about certain things. However, the nature of what is learned is seldom specified with much precision. The point is that universities and other educational institutions seldom check to see if learners acquire real-world skills. They simply sample what a person says or writes about real-world phenomena.

Training includes some procedural guarantee that the learner learns the specific real-world behavior of interest. Again, the guarantee is usually provided by a test or checkout. However, with training, the test probes the learner's repertoire of on-the-job skills rather than verbal knowledge about on-the-job skills.

Working with these distinctions, it is clear that most of the "training" that occurs in business and industry is actually information-presenting. Certain information is presented by the trainer or teacher and there are no guarantees that the process has any effect on the behavior of the learner. When business and industry "training" provides tests that might probe the repertoires of the learners, they usually sample verbal skills of what the learners can write or say about the phenomena of interest. We rarely know whether business and industry "training" has effects on the behaviors of interest because the effort rarely includes tests for those behaviors.

Do not assume that information-presenting and education have no effects on work behavior. Either process may have some effects. The point is that as they are ordinarily conducted, we do not know what those effects, if there are any, are. Furthermore, in the absence of on-the-job tests of skills, there are no mechanisms for improving the effects information-presenting and education can have on real work skills.

Suppose that we always want to provide training rather than just information or education, when a person is assumed to have a skill deficit. How can that be accomplished reasonably? It will be useful to distinguish three kinds of training which we will call corrective feedback, simple training, and complex training. Corrective feedback is used to briefly correct incorrect performance. Simple training is used to develop a skill having only a few components, perhaps using a fire extinguisher or cleaning up a spill. Complex training is useful when the performance includes a large number of components such as would be the case with learning to safely fly an airplane or to minimize exposures to toxic chemicals in a complex job.

19.7 Using Corrective Feedback

Corrective feedback is information about how to perform that is provided to someone who is performing incorrectly. It is used to prompt a correct response the next time it is appropriate. Corrective feedback is used as an alternative to punishment in circumstances in which you assume that a person is well motivated but has failed to perform correctly because of a skill deficit. The skill deficit is so simple or involves so few responses that more formalized training is unnecessary. Useful corrective feedback is:

- timely
- non-punitive
- descriptive of the incorrect behavior
- explanatory of why the performances were incorrect
- descriptive of the correct behavior
- done in such a way to include practice of the correct alternative, an observation of the practiced performance, further feedback and reinforcement if possible.

An example of the first several steps of corrective feedback might be, "You just walked across the puddle of spilled lubricant. That is dangerous because you could slip, fall down, and hurt yourself. The proper thing to do is throw some absorbent onto the spill and gradually take up the absorbent from the edge of the puddle so you never have to walk on it. Make sure you have a dry, non-slippery surface on which to stand as you work."

Independently of intent, many people have difficulty providing feedback in the face of an incorrect performance, unless the feedback is in the form of disciplinary action. Remember, corrective feedback is used when the performer behaved incorrectly because of a lack of skill or knowledge about correct performance. Presumably the person is motivated and trying to perform correctly. In this case, punishment is inappropriate.

What is appropriate is to tell the person that he has performed incorrectly and describe the details of the incorrect performance. The person will then know exactly what was not done correctly. It is also useful to provide an explanation of why their way of doing something is not correct. That explanation will help them to learn the principles that are involved in the performance and perhaps to better understand the importance of performing the task correctly.

The feedback should specify the correct way to behave. A demonstration is often a useful way to communicate how you want someone to behave. Again, a brief explanation of why the suggested way is correct may help the person to learn the principles involved in the performance.

Remember that you are actually training someone when you provide corrective feedback. Notice that good training provides the learner an opportunity to practice and subsequently demonstrate the correct performance. The demonstration of course, provides an opportunity for the trainer to make sure the skill is correctly performed and to provide feedback to confirm that it is correct or to change it if it is not.

19.8 Conducting Simple Training

The steps in carrying out simple training are to:

1. Specify the objectives.
2. Communicate the objectives to the learner.
3. Provide for practice of the objective behaviors.
4. Test for the objective behaviors.
5. Provide feedback.
6. Provide reinforcement.
7. Recycle to mastery.
8. Provide for maintenance of the skills.
9. Evaluate and modify the training as indicated.

This list of procedures allows for very effective and straightforward training that is more easily accomplished than the length of the list may imply.

The goals of the training should be clear before beginning the training. Specifying the objectives involves determining the exact behaviors which will be involved in the training. There are good methods to determine behavioral objectives for training. These include watching or videotaping an expert doing the job is the standard method and then carrying out the job yourself and asking the expert to criticize your performance is a check on your understanding.

If goals of the training are clear, the next logical step is to show or tell the learner exactly what they are. Much of the research and controversy over how to train is centered around questions about what are the best methods for communicating objectives. For the moment, assume that showing someone is a good method. You can show someone with a live demonstration. If you are going to be training people repetitively, it may be useful to videotape or film someone doing the job. If the other procedures are used correctly, the exact method by which you communicate the objectives is probably not particularly important, as long as they are communicated accurately.

To insure that someone can perform as desired, they should be provided an opportunity to practice of the desired skill. By doing so, you not only increase their chances of learning, you set the stage for ensuring that they learn.

Feedback and reinforcement provide for the strengthening of a correct performance or the correction of one that is in error.

The procedures involved in testing and recycling to mastery provide the guarantee that the learner has the desired skills.

Providing for maintenance can involve either making sure that there are reasonable consequences for the skill as they are used on the job, and in the case of infrequently used skills, it might include refresher testing and training.

On-the-job tests provide evidence of the effectiveness of training as well as evidence of the skills learned. Therefore, data from the tests can be used as a basis for evaluating training. Such an evaluation may determine whether or not training is effective and if it should be changed in particular ways.

These training procedures actually make good common sense when you begin to translate the general forms of the methods into particulars. Suppose you want to train people to operate a soda acid fire extinguisher (this does not include the judgments involved in knowing when such a fire extinguisher should be used). This includes a simple collection of behavioral objectives. A person should lift the extinguisher from its wall hanger, carry it to the fire, grasp it by the top and bottom, turn it upside down, and direct the stream from the extinguisher at the base of the flames.

You might best communicate these objectives by showing the learner the specific steps in correct sequence. After your demonstration, you might ask the learner to practice the steps as you watch. While the learner is practicing, provide corrective feedback for any mistakes and confirmation if the performance is correct. Probably nothing more than social reinforcement, "That's good!" or "That's the way to do it, good job!" is necessary or appropriate. You might provide an on-the-job simulated test to see if the learner can perform as specified. If there is a failure, you simply repeat your demonstration and allow for more practice, testing, feedback, and reinforcement until a perfect performance results.

Maintenance of the skill depends on the opportunity to use the acquired skills and on the reinforcement dependent on their occurrence. If evaluation of the training is important, it can be carried out by data collected on the performances during the on-the-job tests.

19.9 Conducting Complex Training

Slightly more complex training is required if the skills to be learned are greater in number. The list of training methods includes:

1. Specify the objectives.
2. Sequence the objectives.

3. Break collections of objectives into short lessons.
4. Communicate the objectives to the learner.
5. Provide practice (memory) guides.
6. Provide for practice on the objective behaviors.
7. Provide frequent tests.
8. Accumulate test criteria.
9. Provide feedback.
10. Provide reinforcement.
11. Recycle to mastery.
12. Provide for maintenance of the skills.
13. Evaluate and modify the training as indicated.

Five of these procedures are additions to those specified for simple training and are necessary to accommodate the greater number of behaviors involved in the desired skills.

Sequencing to determine the order in which the objectives will be presented is particularly important as the list of objectives becomes long. Several criteria are deserving of consideration as the objectives are sequenced. First, skills that are prerequisites for other behaviors should be taught before the other behaviors. If skill in A is necessary before B can be learned, then A should be taught before B. Generally, it is a good idea to present easier to learn skills before the more difficult ones. This allows the cooperation and confidence of the learner to build before greater demands are imposed. This usually, but not always, includes teaching simpler skills before more complex skills.

Research generally suggests that people have much shorter attention spans than most training programs and most forms of education accommodate. Instead of having a few long training sessions, a better approach, especially as the list of objectives becomes long. is to have several short training sessions to communicate objectives to the learner. A functional rule is to keep the number of objectives presented in one lesson as small as is necessary so that the lessons will be from five to fifteen minutes long. This stays close to attention spans of adult learners and allows for manageable numbers of objectives during practice.

A practice or memory guide is a list of objectives involved in a lesson. The guide is given to learners to help them remember the objectives from the time they are communicated until they can be practiced.

Frequent tests allow the learner to get feedback and, hopefully, reinforcement before errors become permanently incorporated into repertoires. Frequent tests also minimize the number of new objectives for which the learner is responsible at any one test.

If a learner acquires skills A, B, C, D, and E and then quits practicing them while learning skills F, G, H, I, and J, there is a tendency for the first learned skills to be lost as the new ones are learned. The solution to this is to have earlier learned skills reviewed in later lessons and tests. A particularly effective form of review is to incorporate criteria for passing earlier tests into later tests. This accumulation of lesson criteria is designed to insure that the skills are maintained over the period of learning. Otherwise, the methods for training complex behaviors are like those for training simpler collections of behaviors.

19.10 Applications to Safety and Health

These types of methods can be combined and incorporated into programs to promote improved performance in a very large number of different areas. There have been programs to control tardiness, to promote attendance at work, to promote productivity, to control waste, to reduce fuel consumption, and to improve quality of production. Safety and health represents one particular kind of application.

19.10.1 Case Study Example: Bus Drivers

A bus company had an ongoing safety training program, but the drivers still had many accidents. All evidence suggested a large percentage of these accidents could be avoided. The drivers were removed from supervision for long periods of time. The Safety Consultant recommended that the company adopt an incentive program to reinforce drivers for avoiding accidents and injuries.

A token system, unique to the bus company, was constructed. Individual drivers were given points if they went for pay periods without accidents. They received additional points if other drivers on the same teams also avoided accidents. The points could be exchanged for such things as tickets to ball games, coupons for purchases at fast food restaurants, and even bus passes.

The safety records of the drivers who were first allowed to work with the token reinforcement system quickly exceeded that of control drivers who were not on the program. When the control drivers were placed on the reinforcement program, their safety records also improved. The return on investment for this program was about three to one.

Summary

Behavior can be changed to improve the safety and health of workers in business and industry. Behavioral methods seem to be quite general in the sense that all work behaviors are subject to the principles presented. Therefore, whenever control of a particular behavior is important to safety and health, appropriate methods can be developed with a good chance that results will be successful. Places of work appear to have a large variety of reinforcers that can be used to strengthen desired behaviors. Although typical business practice involves presenting these to workers in-dependently of useful behaviors, they can be incorporated into productive programs in straightforward and effective ways.

Bibliography

Blanchard. K. and Johnson. S. (1981) The One Minute Manager. Berkley Books: New York.

Haynes, R. S. (1980) A behavioral approach to the reduction of accidents of city bus operators. Doctoral dissertation, University of Kansas, Lawrence, Kansas.

Hopkins, B. L., Conard, B.C., and Duellman, D. L. (1979) Behavior management for occupational safety and health. U.S. DHEW, National Institute for Occupational Safety and Health, Cincinnati Ohio.

Komaki, J., Barwick, K. K., and Scott, L. R. A behavioral approach to occupational safety. Journal of Applied Psychology 34, 1978, 63, 434-445.

Sulzer-Azaroff. B. (1982) Behavioral approaches to occupational health and safety. In L. W. Frederiksen (ed.) Handbook of Organizational Behavior Management, John Wiley and Sons: New York.

20

Violence in the Workplace

AN AVERAGE of 20 workers are murdered each week in the United States. The majority of these murders are robbery-related crimes. In addition, an estimated 1 million workers are assaulted annually in U.S. workplaces. Most of these assaults occur in service settings such as hospitals, nursing homes, and social service agencies. Factors that place workers at risk for violence in the workplace include interacting with the public, exchanging money, delivering services or goods, working late at night or during early morning hours, working alone, guarding valuables or property, and dealing with violent people or volatile situations.

20.1 Violence in the Workplace

Violence in the workplace has received considerable attention in the press and among safety and health professionals. Much of the reason for this attention is the magnitude of this problem in U.S. workplaces. Unfortunately, sensational acts of coworker violence (which form only a small part of the problem) are often emphasized by the media to the exclusion of the almost daily killings of taxicab drivers, convenience store clerks and other retail workers, security guards, and police officers.

These deaths often go virtually unnoticed, yet their numbers are staggering: 1,071 workplace homicides occurred in 1994. These homicides included 179 supervisors or proprietors in retail sales, 105 cashiers, 86 taxicab drivers, 49 managers in restaurants or hotels, 70 police officers or detectives, and 76 security guards. An additional 1 million workers were assaulted each year. These figures indicate that an average of 20 workers are murdered and 18,000 are assaulted each week while at work or on duty. Death or injury should not be an inevitable result of one's chosen occupation, nor should these staggering figures be accepted as a cost of doing business in our society.

Although no definitive strategy will ever be appropriate for all workplaces, we must begin to change the way work is done in certain settings to minimize or remove the risk of work-place violence. We must also change the way we think about workplace violence by shifting the emphasis from reactionary approaches to prevention, and by embracing workplace violence as an occupational safety and health issue.

20.1.1 Defining Workplace Violence

Defining workplace violence has generated considerable discussion. Some would include in the definition any language or actions that make one person uncomfortable in the work-place; others would include threats and harassment; and all would include any bodily injury inflicted by one person on another. Thus the spectrum of workplace violence ranges from offensive language to homicide, and a reasonable working definition of workplace violence is as follows: violent acts, including physical assaults and threats of assault, directed toward persons at work or on duty. Most studies to date have focused primarily on physical injuries, since they are clearly defined and easily measured. This chapter examines multiple sources and acknowledges differences in definitions and coverage to learn as much as possible from these varied efforts.

20.1.2 The Study of Workplace Violence

The circumstances of workplace violence also vary and may include robbery-associated violence; violence by disgruntled clients, customers, patients, and inmates; violence by coworkers, employees, or employers; and domestic violence that finds its way into the workplace. These circumstances all appear to be related to the level of violence in communities and in society in general. Several reasons exist for focusing specifically on workplace violence:

1. Violence is a substantial contributor to death and injury on the job. NIOSH data indicate that homicide has become the second leading cause of occupational injury death, exceeded only by motor-vehicle-related deaths. Estimates of nonfatal workplace assaults vary dramatically, but a reasonable estimate from the National Crime Victimization Survey is that approximately 1 million people are assaulted while at work or on duty each year; this figure represents

15% of the acts of violence experienced by U.S. residents aged 12 or older.

2. The circumstances of workplace violence differ significantly from those of all homicides. For example, 75% of all workplace homicides in 1993 were robbery-related; but in the general population, only 9% of homicides were robbery-related, and only 19% were committed in conjunction with any kind of felony (robbery, rape, and arson). Furthermore, 47% of all murder victims in 1993 were related to or acquainted with their assailants, whereas the majority of workplace homicides (because they are robbery-related) are believed to occur among persons not known to one another. Only 17% of female victims of workplace homicides were killed by a spouse or former spouse, whereas 29% of the female homicide victims in the general population were killed by a husband, ex-husband, boyfriend, or ex-boyfriend.

3. Workplace violence is not distributed randomly across all workplaces but is clustered in particular occupational settings. More than half (56%) of workplace homicides occurred in retail trade and service industries. Homicide is the leading cause of death in these industries as well as in finance, insurance, and real estate. Eighty-five percent of nonfatal assaults in the workplace occur in service and retail trade industries. As the U.S. economy continues to shift toward the service sectors, fatal and nonfatal workplace violence will be an increasingly important occupational safety and health issue.

4. The risk of workplace violence is associated with specific workplace factors such as dealing with the public, the exchange of money, and the delivery of services or goods. Consequently, great potential exists for workplace-specific prevention efforts such as bullet-resistant barriers and enclosures in taxicabs, convenience stores, gas stations, emergency departments, and other areas where workers come in direct contact with the public; locked drop safes and other cash-handling procedures in retail establishments; and threat assessment policies in all types of workplaces.

Long-term efforts to reduce the level of violence in U.S. society must address a variety of social issues such as education, poverty, and environmental justice. However, short-term efforts must address the pervasive nature of violence in our society and the need to protect workers. We cannot wait to address workplace violence as a social issue alone but must take immediate action to address it as a serious occupational safety issue.

20.2 Homicide in the Workplace

Data from the National Traumatic Occupational Fatalities (NTOF) Surveillance System indicate that 9,937 workplace homicides occurred during the 13-year period from 1980 through 1992, with an average workplace homicide rate of 0.70/100,000 workers (Table 20.1). Over the course of the 1980s, workplace homicides decreased; but in the 1990s, the numbers began to increase, surpassing machine-related deaths and approaching the number of workplace motor-vehicle-related deaths (Figure 20.1).

TABLE 20.1
Workplace homicides in the United States, 1980-92*

Year	Number	Rate**
1980	929	.96
1981	944	.94
1982	859	.86
1983	721	.72
1984	660	.63
1985	751	.70
1986	672	.61
1987	649	.58
1988	699	.61
1989	696	.59
1990	725	.61
1991	875	.75
1992	757	.64
Total	9,937	.70

*Data not available for New York City and Connecticut.
**Per 100,000 workers.

20.2.1 Sex

The majority (80%) of workplace homicides during 1980-92 occurred among male workers. The leading cause of occupational injury death varied by sex, with homicides accounting for 11% of all occupational injury deaths among male workers and 42% among female workers [NIOSH 1995]. The majority of female homicide victims were employed in retail trade (46%) and service (22%) industries (Table 20.2). A large number of male homicide victims were employed not only in retail trade (36%) and service (16%) industries but in public administration (11%) and transportation/ communication/ public utilities (11%) (Table 20.2). Although homicide is the leading cause of occupational injury death among female workers, male workers have more than three times the risk of work-related homicide (Table 20.3).

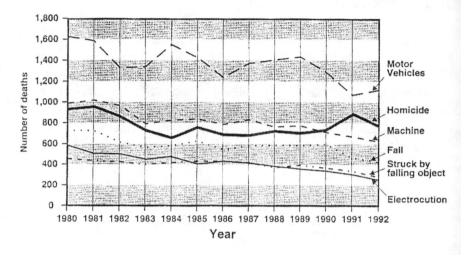

NOTE: Data were not available for New York City and Connecticut.
FIG. 20.1 Leading Causes of Occupational Injury Deaths - United States, 1980-92.

TABLE 20.2
Workplace Homicides by Industry and Sex – United States 1980-92*

Industry	Homicides (% of Total)**	
	Male Workers	Female Workers
Retail Trade	36.1	45.5
Services	16.0	22.2
Public Administration	10.5	2.9
Transportation/communication/public utilities	10.6	3.8
Manufacturing	7.0	4.9
Construction	4.1	0.6
Agriculture/forestry/fishing	2.7	0.6
Finance/insurance/real estate	2.4	6.8
Wholesale Trade	1.7	1.1
Mining	0.6	0.1
Not Classified	8.5	11.7

*Data for New York City and Connecticut were not available for 1992.
**Percentages add to more than 100% because of rounding.

20.2.2 Age

The largest number of work-place homicides occurred among workers aged 25 to 34, whereas the rate of workplace homicide increased with age (Table 20.3). The highest rates of workplace homicide occurred among workers aged 65 and older; the rates for these workers were more than twice those for workers aged 55-64 (Table 20.3). This pattern held true for both male and female workers.

TABLE 20.3
Workplace homicides by age group and sex--United States, 1980-92*,**

Age Group	Male Workers		Female Workers		All Workers	
	Number	Rate	Number	Rate	Number	Rate
16-19	242	0.55	102	0.25	344	0.41
20-24	796	.87	285	.35	1,081	.62
25-34	2,020	.89	591	.33	2,611	.65
35 -44.	1,841	.99	423	.28	2,265	.68
45--54	1,344	1.04	293	.29	1,637	.71
55--64	1,055	1.22	191	.31	1,246	.84
65+	620	2.59	115	.71	735	1.83
Total***	7,935	--	2,001	--	9,937	--
Average	--	1.01	--	.32	--	.70

*Data from New York City and Connecticut were not available for 1992.
**Rates are per 100,000 workers.
***Totals include victims for whom age data were missing (17 male workers) and one worker whose sex was not reported.

20.2.3 Method of Homicide

Between 1980 and 1992, 76% of work-related homicides were committed with firearms, and another 12% resulted from wounds inflicted by cutting or piercing instruments (Table 20.4). During this period, the number of firearm-related homicides declined then gradually increased, with the number of firearm-related workplace homicides in 1991 exceeding that in 1980 (Figure 20.2). Firearms accounted for an increasing percentage of the total workplace homicides over the 13-year period: 74% in 1980 and 84% in 1991. Firearms were used in 79% of the workplace homicides in 1992.

TABLE 20.4
Workplace homicides by method--United States, 1980-92*

Method	Number	% of Total
Firearm	7,590	76.4
Cutting or piercing instrument	1,231	12.4
Strangulation	185	1.9
All other methods	931	9.4

*Data for New York City and Connecticut were not available for 1992.

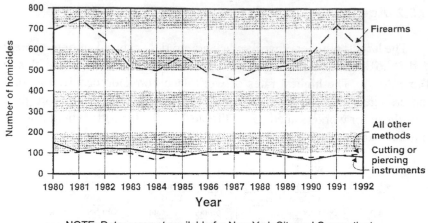

NOTE: Data were not available for New York City and Connecticut.
FIG. 20.2 Work-related Homicides by Method and Year.

20.2.4 Industry and Occupation

During the 13-year period 1980-92, the greatest number of deaths occurred in the retail trade (3,774) and service (1,713) industries, whereas the highest rates per 100,000 workers occurred in retail trades (1.6), public administration (1.3), and transportation/communication/ public utilities (0.94) (Table 20.5).

TABLE 20.5
Workplace homicides by industry--United States, 1980-92*

Industry	Number	% of Total	Rate**
Retail trade	3,774	38.0	1.60
Public administration	889	8.9	1.30
Transportation/communication/public utilities	917	9.2	.94
Agriculture/forestry/fishing	222	2.2	.50
Mining	45	0.5	.40
Service	1,713	17.2	.38
Construction	335	3.4	.37
Finance/insurance/real estate	327	3.3	.35
Wholesale trade	155	1.6	.27
Manufacturing	650	6.5	.24
Not classified	910	9.1	--

*Data for New York City and Connecticut were not available for 1992.
**Per 100,000 workers.

At the more detailed levels of industry (Table 20.8), the largest number of deaths occurred in grocery stores (N=330), eating and drinking places (N=262), taxicab services (N=138), and justice/ public order establishments

(N=137). Taxicab services had the highest rate of work-related homicide during the 3-year period 1990-92 (41.4/100,000). This rate was nearly 60 times the national average rate of work-related homicides (0.70/100,000). This figure was followed by rates for liquor stores (7.5), detective/ protective services (7.0), gas service stations (4.8), and jewelry stores (4.7) (Table 20.6).

TABLE 20.6
Workplace homicides in high-risk industries--United States, 1980-89 and 1990-92*,**

Industry	1980-89		1990-92	
	Number	Rate	Number	Rate
Taxicab services	287	26.9	138	41.4
Liquor stores	115	8.0	30	7.5
Gas service stations	304	5.6	68	4.8
Detective/protective services	152	5.0	86	7.0
Justice/public order establishment	640	3.4	137	2.2
Grocery stores	806	3.2	330	3.8
Jewelry stores	56	3.2	26	4.7
Hotels/motels	153	1.5	33	0.8
Barber shops	14	1.5	4	***
Eating/drinking places	734	1.5	262	1.5

*Data for New York City and Connecticut were not available for 1992.
**Rates are per 100,000 workers.
***Rate was not calculated because of the instability of rates based on small numbers.

When detailed occupations were analyzed for 1990-92 (Table 20.7), the highest homicide rates were found for taxicab drivers/chauffeurs (22.7), sheriffs/bailiffs (10.7), police and detectives-public service (6.1), gas station/ garage workers (5.9), and security guards (5.5).

20.3 Nonfatal Assaults in the Workplace

20.3.1 Estimated Magnitude of the Problem

Limited information is available in the criminal justice and public health literature regarding the nature and magnitude of nonfatal workplace violence. The risk of workplace victimization was related more to the task performed than to the demographic characteristics of the person performing the job. Factors related to an increased risk for workplace victimization included routine face-to-face contact with large numbers of people, the handling of money, and jobs that required routine travel or that did not have a single worksite. The delivery of passengers or goods and dealing with the public were the factors associated with an increased risk for work-place assault.

TABLE 20.7
Workplace homicides in high-risk* occupations--United States, 1983-89 and 1990-92**,***

Occupation	1983-89		1990-92	
	Number	Rate	Number	Rate
Taxicab driver/chauffeur	197	15.1	140	22.7
Sheriff/bailiff	73	10.9	36	10.7
Police and detective--public service	267	9.0	86	6.1
Hotel clerk	29	5.1	6	2.0
Gas station/garage worker	83	4.5	37	5.9
Security guard	160	3.6	115	5.5
Stock handler/bagger	189	3.1	95	3.5
Supervisor/proprietor, sales	662	2.8	372	3.3
Supervisor, police and detective	12	2.2	0	§
Barber	14	2.2	4	§
Bartender	49	2.1	20	2.3
Correctional institution officer	19	1.5	3	§
Salesperson, motor vehicle and boat	21	1.1	17	2.0
Salesperson, other commodities	98	1.0	73	1.7
Sales counter clerk	13	1.2	18	3.1
Fire fighter	18	1.4	8	1.3
Logging occupation	4	§	6	2.3
Butcher/meatcutter	11	.6	12	1.5

*High-risk occupations have workplace homicide rates that are twice the average rate during one or both time periods.
**Data for New York City and Connecticut were not available for 1992.
***Rates are per 100,000 workers.
§Rate was not calculated because of the instability of rates based on small numbers.

An annual survey of approximately 250,000 private establishments indicate that 22,400 workplace assaults occurred in 1992, representing 1% of all cases involving days away from work.

Unlike homicides, nonfatal workplace assaults are distributed almost equally between men (44%) and women (56%). The majority of the nonfatal assaults reported occurred in the service (64%) and retail trade (21%) industries. Of those in services, 27% occurred in nursing homes, 13% in social services, and 11% in hospitals. In retail trade, 6% occurred in grocery stores, and another 5% occurred in eating and drinking places (Table 20.8). The source of injury in 45% of the cases was a health care patient (Figure 20.3), with another 31% described as other person and 6% as coworker or former coworker.

Nearly half (47%) of the workplace assaults were described as incidents involving hitting, kicking, or beating; there were also cases of squeezing, pinching, scratching, biting, stabbing, and shooting, as well as rapes and threats of violence (Table 20.9). The median days away from work as the result of an assault was 5, but this figure varied by type of assault (Table 20.9). Another study analyzed workplace victimizations by type of work setting and found that 61% occurred in private companies, 30% occurred among government employees, and 8% of the victims were self-employed.

TABLE 20.8
Violent acts resulting in days away from work in 1992, by industry

Industry	Violent acts resulting in days away from work (% of total)
Services	64
Nursing homes	27
Social services	13
Hospitals	11
Other services	13
Retail trades	21
Grocery stores	6
Eating and drinking places	5
Other retail	10
Transportation/communication/public utilities	4
Finance/insurance/real estate	4
Other	4
Manufacturing	3

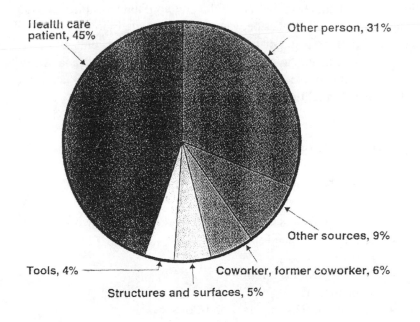

Fig. 20.3 Violent Acts Resulting in Days Away From Work, by Source of Injury-United States, 1992.

TABLE 20.9
Violent acts resulting in days away from work--private industry, 1992

Type of Violent Act	No. of Cases	Median Days Away From Work
Hitting, kicking, beating	10,425	5
Squeezing, pinching, scratching, twisting	2,457	4
Biting	901	3
Stabbing	598	28
Shooting	560	30
All other specified acts (e.g., rape, threats)	5,157	5

20.3.2 Discussion

Nonfatal assaults in the workplace clearly affect many workers and employers. Although groups at high risk for workplace homicide and nonfatal workplace assaults share similar characteristics such as interaction with the public and the handling of money, there are also clear differences. For example, groups such as health care workers are not at elevated risk of workplace homicide, but they are at greatly increased risk of nonfatal assaults. Some of the distinctions between fatal and nonfatal workplace assaults can be attributed to differences between robbery-related violence and violence resulting from the anger or frustration of customers, clients, or coworkers, with robbery-related violence being more likely to result in a fatal outcome. The premeditated use of firearms to facilitate robberies is also likely to influence the lethality of assaults in the workplace.

20.4 Risk Factors and Prevention Strategies

20.4.1 Risk Factors

A number of factors may increase a worker's risk for workplace assault. These factors include the following:

- contact with the public
- exchange of money
- delivery of passengers, goods, or services
- having a mobile workplace such as a taxicab or police cruiser
- working with unstable or volatile persons in health care, social service, or criminal justice settings

- working alone or in small numbers
- working late at night or during early morning hours
- working in high-crime areas
- guarding valuable property or possessions
- working in community-based settings

20.4.2 Prevention Strategies

20.4.2.1 Environmental Designs

Commonly implemented cash-handling policies in retail settings include procedures such as using locked drop safes, carrying small amounts of cash, and posting signs and printing notices that limited cash is available. It may also be useful to explore the feasibility of cashless transactions in taxicabs and retail settings through the use of machines that accommodate automatic teller account cards or debit cards. These approaches could be used in any setting where cash is currently exchanged between workers and customers.

Physical separation of workers from customers, clients, and the general public through the use of bullet-resistant barriers or enclosures has been proposed for retail settings such as gas stations and convenience stores, hospital emergency departments, and social service agency claims areas. The height and depth of counters (with or without bullet-resistant barriers) are also important considerations in protecting workers, since they introduce physical distance between workers and potential attackers. Consideration must nonetheless be given to the continued ease of conducting business; a safety device that increases frustration for workers or for customers, clients, or patients may be self-defeating.

Visibility and lighting are also important environmental design considerations. Making high-risk areas visible to more people and installing good external lighting should decrease the risk of workplace assaults.

Access to and egress from the workplace are also important areas to assess. The number of entrances and exits, the ease with which non-employees can gain access to work areas because doors are unlocked, and the number of areas where potential attackers can hide are issues that should be addressed. This issue has implications for the design of buildings and parking areas, landscaping, and the placement of garbage areas, outdoor refrigeration areas, and other storage facilities that workers must use during a work shift.

Numerous security devices may reduce the risk for assaults against workers and facilitate the identification and apprehension of perpetrators.

These include closed-circuit cameras, alarms, two-way mirrors, card-key access systems, panic-bar doors locked from the outside only, and trouble lights or geographic locating devices in taxicabs and other mobile workplaces.

Personal protective equipment such as body armor has been used effectively by public safety personnel to mitigate the effects of workplace violence. For example, the lives of more than 1,800 police officers have been saved by such vests.

20.4.2.2 Administrative Controls

Staffing plans and work practices such as escorting patients and prohibiting unsupervised movement within and between areas are effective work practices. Increasing the number of staff on duty may also be appropriate in any number of service and retail settings. The use of security guards or receptionists to screen persons entering the workplace and controlling access to actual work areas has been suggested.

Work practices and staffing patterns during the opening and closing of establishments and during money drops and pickups should be care-fully reviewed for the increased risk of assault they pose to workers. These practices include having workers take out garbage, dispose of grease, store food or other items in external storage areas, and transport or store money.

Policies and procedures for assessing and reporting threats allow employers to track and assess threats and violent incidents in the workplace. Such policies clearly indicate a zero tolerance of workplace violence and provide mechanisms by which incidents can be reported and handled. In addition, such information allows employers to assess whether prevention strategies are appropriate and effective. These policies should also include guidance on recognizing the potential for violence, methods for defusing or deescalating potentially violent situations, and instruction about the use of security devices and protective equipment. Procedures for obtaining medical care and psychological support following violent incidents should also be addressed. Training and education efforts are clearly needed to accompany such policies.

20.4.2.3 Behavioral Strategies

Training employees in nonviolent response and conflict resolution has been suggested to reduce the risk that volatile situations will escalate to physical violence. Also critical is training that addresses hazards associated with specific tasks or worksites and relevant prevention strategies. Training should not be regarded as the sole prevention strategy but as a component in a comprehensive approach to reducing workplace violence. To increase vigilance and compliance with stated violence prevention policies, training should emphasize the appropriate use and maintenance of protective equipment, adherence to administrative controls, and increased knowledge and awareness of the risk of work-place violence.

20.5 Developing and Implementing a Workplace Violence Prevention Program and Policy

The first priority in developing a workplace violence prevention policy is to establish a system for documenting violent incidents in the workplace. Such data are essential for assessing the nature and magnitude of workplace violence in a given workplace and quantifying risk. These data can be used to assess the need for action to reduce or mitigate the risks for work-place violence and implement a reasonable intervention strategy.

Implementation of the reporting system, a workplace violence prevention policy, and specific prevention strategies should be publicized company-wide, and appropriate training sessions should be scheduled. The demonstrated commitment of management is crucial to the success of the program. The success and appropriateness of intervention strategies can be monitored and adjusted with continued data collection.

A written workplace violence policy should clearly indicate a zero tolerance of violence at work, whether the violence originates inside or outside the workplace. Note that when violence or the threat of violence occurs among coworkers, firing the perpetrator may or may not be the most appropriate way to reduce the risk for additional or future violence. The employer may want to retain some control over the perpetrator and require or provide counseling or other care, if appropriate. The violence prevention policy should explicitly state the consequences of making threats or committing acts of violence in the workplace.

A workplace violence prevention program should also include procedures and responsibilities to be taken in the event of a violent incident in the workplace. This program should who is responsible for the immediate care of the victim(s), re-establishing work areas and processes, and organizing and carrying out stress debriefing sessions with victims, their coworkers, and perhaps the families of victims and coworkers. Employee assistance programs, human resource professionals, and local mental health and emergency service personnel can offer assistance in developing these strategies.

20.5.1 Responding to an Immediate Threat of Workplace Violence

For a situation that poses an immediate threat of workplace violence, all legal, human resource, employee assistance, community mental health, and law enforcement resources should be used to develop a response. The risk of injury to all workers should be minimized.

If a threat has been made that refers to particular times and places, or if the potential offender is knowledgeable about workplace procedures and time frames, patterns may need to be shifted. For example, a person who has leveled a threat against a worker may indicate, "I know where you park and what time you get off work!" In such a case it may be advisable to change or even stagger departure times and implement a buddy system or an escort by security guard for leaving the building and getting to parking areas. The threat should not be ignored in the hope that it will resolve itself or out of fear of triggering an outburst from the person who has lodged the threat. If someone poses a danger to himself or others, appropriate authorities should be notified and action should be taken.

20.5.2 Dealing with the Consequences of Workplace Violence

Much discussion has also centered around the role of stress in workplace violence. The most important thing to remember is that stress can be both a cause and an effect of workplace violence. That is, high levels of stress may lead to violence in the workplace, but a violent incident in the workplace will most certainly lead to stress, perhaps even to post-traumatic stress disorder.

Employers should be sensitive to the effects of workplace violence and provide an environment that promotes open communication; they should also have in place an established procedure for reporting and responding to violence. Appropriate referrals to employee assistance programs or other local mental health services may be appropriate for stress debriefing sessions after critical incidents.

Summary

Clearly, violence is pervasive in U.S. workplaces, accounting for 1,071 homicides in 1994 and approximately a million nonfatal assaults each year. The circumstances of workplace homicides differ substantially from those portrayed by the media and from homicides in the general population. For the most part, workplace homicides are not the result of disgruntled workers who take out their frustrations on coworkers or supervisors, or of intimate partners and other relatives who kill loved ones in the course of a dispute; rather, they are mostly robbery-related crimes. Although no single intervention strategy is appropriate for all workplaces and no definitive strategies can be recommended at this time, immediate action should be taken to reduce the toll of work-place homicide on our Nation's workforce.

What are the hazards?

An average of 20 workers are murdered each week in the United States. The of these murders are robbery-related crimes. In addition, an estimated 1 million workers are assaulted annually in U.S. workplaces. Most of these assaults occur in service settings such as hospitals, nursing homes, and social service agencies. Factors workers at risk for violence in the workplace include interacting with the public, exchanging money, delivering services or goods, working late at night or during early morning hours, working alone, guarding valuables or property, and dealing with violent people or volatile situations.

How are workers exposed or put at risk?

Anyone can become the victim of a workplace assault, but the risks are much greater in certain industries and occupations. For workplace homicides, taxicab drivers have the highest risk of any occupational group; for nonfatal workplace assaults, health care, community services, and retail settings are at increased risk.

What recommendations has the Federal government made to protect worker health?

A number of environmental, administrative, and behavioral strategies have the potential for reducing the risk of workplace violence. No single strategy is appropriate for all workplaces, but all workers and employers should assess the risk of violence in their workplaces and take appropriate action to reduce those risks. Collecting information about all incidents of workplace violence helps determine whether prevention are necessary, appropriate, and effective.

Where can I get more information?

The references and related reading fist at the end of this chapter provide a useful inventory of published reports and literature. A number of unions, employer groups, and professionals in occupations safety and health, human resources, and employee assistance have also developed materials regarding workplace violence. Any resource should be evaluated in light of the violence experience in specific workplaces.

Bibliography

BLS (1995). National census of fatal occupational injuries, 1994. Washington, DC: U.S. Department of Labor, BLS News, USDL-95-288.

Chelimsky E., Jordan F.C., Russell LS., Strack J.R. (1979). Security and the small business retailer. Washington, DC: National Institute of Law Enforcement and Criminal Justice, Law Enforcement Assistance Administration, Department of Justice.

Jenkins E.L., Kisner S.M., Fodbroke D.E., Layne L.A., Stout N.A., Castillo D.N. et. al. (1993). Fatal injuries to workers in the United States, 1980-1989: a decade of surveillance; national profile. Washington, DC: US Government Printing Office, DHHS (NIOSH) Publication No.93-108.

Kraus J.F (1987). Homicide while at work: persons, industries, and occupations at high risk. American Journal of Public Health.

NIOSH (1996). Current Intelligence Bulletin number 57, "Violence in the workplace-risk factors".

Publication DHHS, 96-100.

Shepherd E., ed. (1994). Violence in health care; a practical guide to copping with violence and caring of victims. New York, NY: Oxford University Press.

Toscano G, Weber W (1995). Violence in the workplace. Compensation and working conditions. Washington, DC: U.S. Department of Labor, Bureau of Labor Statistics.

21

Asbestos Safety

ASBESTOS IS the generic term for a group of naturally occurring, fibrous minerals with high tensile strength, flexibility and resistance to thermal, chemical, and electrical conditions.

In the construction industry, asbestos is found in older installed products such as shingles, floor tiles, cement pipe and sheet, roofing felts, insulation, ceiling tiles, fire-resistant drywall, and acoustical products. Very few asbestos-containing products are currently being installed. Consequently, most worker exposures occur during the removal of asbestos and the renovation and maintenance of older buildings and structures containing asbestos.

Asbestos fibers enter the body by the inhalation or ingestion of airborne particles that become embedded in the tissues of the respiratory or digestive systems. Exposure to asbestos can cause disabling or fatal diseases such as asbestosis (an emphysema like condition), lung cancer, mesothelioma (a cancerous tumor that spreads rapidly in the cells of membranes covering the lungs and body organs), and gastrointestinal cancer. The symptoms of these diseases generally do not appear for 20 or more years after initial exposure.

OSHA began regulating workplace asbestos exposure in 1970, adopting a permissible exposure limit (PEL) to regulate worker exposures. Over the years, more information on the adverse health effects of asbestos exposure has become available, prompting the agency to revise the asbestos standard several times to better protect workers. On August 10, 1994, OSHA estimates that about 42 additional cancer deaths per year will be avoided in all industries.

21.1 Work Classification

OSHA's revised standard establishes a new classification system for asbestos construction work. The revised standards clearly spells out mandatory, simple, technological work practices to follow to reduce worker exposures. Four classes of construction activity are matched with increasingly stringent control requirements. Table 21.1, at the end of this chapter, contains a list of provisions broken down by work classification.

Class I asbestos work—the most potentially hazardous class of asbestos jobs—involves the removal of thermal system insulation and sprayed-on or troweled-on surfacing asbestos-containing materials or presumed asbestos-containing materials. Thermal system insulation includes asbestos-containing materials applied to pipes, boilers, tanks, ducts, or other structural components to prevent heat loss or gain. Surfacing materials include decorative plaster on ceilings, acoustical asbestos-containing materials on decking, or fireproofing on structural members.

Class II work includes the removal of other types of asbestos-containing materials that are not thermal system insulation such as resilient flooring and roofing materials containing asbestos. Examples of Class H work include removal of floor or ceiling tiles, siding, roofing, or transite panels.

Class III asbestos work includes repair and maintenance operations where asbestos-containing or presumed asbestos-containing materials are disturbed.

Class IV operations include custodial activities where employees clean up asbestos-containing waste and debris. This includes dusting contaminated surfaces, vacuuming contaminated carpets, mopping floors, and cleaning up asbestos-containing or presumed asbestos-containing materials from thermal system insulation.

21.2 Scope and Application

The asbestos standard for the construction industry regulates asbestos exposure for the following activities:

- demolishing or salvaging structures where asbestos is present
- removing or encapsulating asbestos-containing materials
- constructing, altering, repairing, maintaining, or renovating asbestos-containing structures or substrates
- installing asbestos-containing products
- cleaning up asbestos spills/emergencies
- transporting, disposing, storing, containing, and housekeeping involving asbestos or asbestos-containing products on a construction site

21.3 Provisions of the Standard

OSHA sets out several provisions employers must follow to comply with the asbestos standard. The agency has established strict exposure limits and requirements for exposure assessment, medical surveillance, record keeping, "competent persons," regulated areas, and hazard communication.

21.3.1 Permissible Exposure Limit (PEL)

Employers must ensure that no employee is exposed to an airborne concentration of asbestos in excess of 0.1 f/cc as an 8-hour time-weighted average (TWA).

OSHA also established a short-term exposure limit (STEL) for asbestos. Employers must ensure that no employee is exposed to an airborne concentration of asbestos in excess of 1 f/cc as averaged over a sampling period of 30 minutes.

21.3.2 Exposure Assessments and Monitoring

Employers must assess all asbestos operations for their potential to generate airborne fibers. Employers must use exposure-monitoring data to assess employee exposures.

21.3.2.1 Initial Exposure Assessments

The designated "competent person" must assess exposures immediately before or as the operation begins to determine expected exposures. The assessment must be done in time to comply with all standard requirements triggered by exposure data or the lack of a negative exposure assessment and to provide the necessary information to ensure all control systems are appropriate and work properly.

The initial exposure assessment must be based on the following:

- The results of employee exposure monitoring.
- All observations, information, or calculations indicating employee exposure to asbestos, including any previous monitoring.
- The presumption that employees performing Class I asbestos work

are exposed in excess of the PEL and STEL until exposure monitoring proves they are not.

21.3.2.2 Negative Exposure Assessments

For any specific asbestos job that trained employees perform, employers may show that exposure will be below the PEL by performing an assessment and confirming it by the following:

- "objective data" demonstrating an asbestos-containing material or activities involving it cannot release airborne fibers in excess of the PEL and STEL
- "historical data" from prior monitoring for similar asbestos jobs performed within 12 months of the current job and obtained during work operations conducted under similar conditions
- employees' training and experience were no more extensive for previous jobs than training for current employees
- data show a high degree of certainty that employee exposures will not exceed the PEL and STEL under current conditions
- current initial exposure monitoring used breathing zone air samples representing the 8-hour TWA and 30-minute short-term exposures for each employee in those operations most likely to result in exposures over the PEL for the entire asbestos job

21.3.2.3 Exposure Monitoring

Employee exposure measurements must be made from breathing zone air samples representing the 8-hour TWA and 30-minute short-term exposures for each employee.

Employers must take one or more samples representing full-shift exposure to determine the 8-hour TWA exposure in each work area. To determine short-term employee exposures, employers must take one or more samples representing 30-minute exposures for the operations most likely to expose employees above the STEL in each work area.

Employers must allow affected employees and their designated representatives to observe any employee exposure monitoring. When observation requires entry into a regulated area, the employer must provide and require the use of protective clothing and equipment.

21.3.2.4 Periodic Monitoring

For Class I and II jobs, employers must monitor daily each employee working in a regulated area, unless a negative exposure assessment for the entire operation already exists and nothing has changed. When all employees use supplied-air respirators operated in positive-pressure mode, however, employers may discontinue daily monitoring. Note that for employees performing Class I work using control methods not recommended in the standard, employers must continue daily monitoring, even when employees use supplied-air respirators.

For operations other than Class I and II, employers must monitor all work where exposures can possibly exceed the PEL often enough to validate the exposure prediction.

If periodic monitoring shows employee exposures below the PEL and STEL, the employer may discontinue monitoring for the represented employees.

21.3.2.5 Additional Monitoring

Changes in processes, control equipment, level of personnel experience, or work practices that could result in exposures above the PEL or STEL, regardless of a previous negative exposure assessment for a specific job, require additional monitoring.

21.3.3 Medical Surveillance

Employers must provide a medical surveillance program for all employees who:

- for a combined total of 30 or more days per year, engage in Class I, II, or III work or are exposed at or above the PEL or STEL
- wear negative-pressure respirators

A licensed physician must perform or supervise all medical exams and procedures, provided at no cost to employees and at a reasonable time. Employers must make medical exams and consultations available to employees:

- prior to employee assignment to an area where negative-pressure

respirators are worn
- within 10 working days after the 30th day of exposure for employees assigned to an area where exposure is at or above the PEL for 30 or more days per year
- at least annually thereafter
- when the examining physician suggests them more frequently

If the employee was examined within the past 12 months and that exam meets the criteria of the standard, however, another medical exam is not required.

Medical exams must include the following:

- a medical and work history
- completion of a standardized questionnaire with the initial exam and an abbreviated standardized questionnaire with annual exams
- a physical exam focusing on the pulmonary and gastrointestinal systems
- any other exams or tests suggested by the examining physician

Employers must provide the examining physician:

- a copy of OSHA's asbestos standard and its appendices
- a description of the affected employee's duties relating to exposure
- the employee's representative exposure level or anticipated exposure level
- a description of any personal protective equipment and respiratory equipment used
- information from previous medical exams not otherwise available

It is the employer's responsibility to obtain the physician's written opinion, containing results of the medical exam and:

- any medical conditions of the employee that increase health risks from asbestos exposure
- any recommended limitations on the employee or protective equipment used
- a statement that the employee has been informed of the results of the medical exam and any medical conditions resulting from asbestos exposure
- a statement that the employee has been informed of the increased risk of lung cancer from the combined effect of smoking and asbestos exposure

The physician must not reveal in the written opinion specific findings or diagnoses unrelated to occupational exposure to asbestos. The employer must provide a copy of the physician's written opinion to the affected employee within 30 days after receipt.

21.3.4 Record Keeping

21.3.4.1 Objective Data Records

Where employers use objective data to demonstrate that products made from or containing asbestos cannot release fibers in concentrations at or above the PEL or STEL, they must keep an accurate record for as long as it is relied on and include:

- the exempt product
- the source of the objective data
- the testing protocol, test results, and analysis of the material for release of asbestos
- a description of the exempt operation and support data
- other data relevant to operations, materials, processes, or employee exposures

21.3.4.2 Monitoring Records

Employers must keep records of all employee exposure monitoring for at least 30 years, including:

- the date of measurement
- the operation involving asbestos exposure that was monitored
- sampling and analytical methods used and evidence of their accuracy
- the number, duration, and results of samples taken
- the type of protective devices worn
- the name, social security number, and exposures of the represented employees

Employers must make exposure records available when requested to affected employees, former employees, their designated representatives, and/or OSHA's Assistant Secretary.

21.3.4.3 Medical Surveillance Records

Employers must keep all medical surveillance records for the duration of the employee's employment plus 30 years, including:

- the employee's name and social security number
- the employee's medical exam results, including the medical history, questionnaires, responses, test results, and physician's recommendations
- the physician's written opinions
- any employee medical complaints related to asbestos exposure
- a copy of the information provided to the examining physician

Employee medical surveillance records must be available to the subject employee, anyone having specific written consent of that employee, and/or OSHA's Assistant Secretary.

21.3.4.4 Other Record Keeping Requirements

Employers must maintain all employee training records for 1 year beyond the last date of employment for each employee.

Where data demonstrates presumed asbestos-containing materials do not contain asbestos, building owners or employers must keep the records for as long as they rely on them. Building owners must maintain written notifications on the identification, location, and quantity of any asbestos-containing or presumed asbestos-containing materials for the duration of ownership and transfer the records to successive owners.

When an employer ceases to do business without a successor to keep the records, the employer must notify the Director of the National Institute for Occupational Safety and Health (NIOSH) at least 90 days prior to their disposal and transmit them as requested.

21.3.5 "Competent Person" Requirements

On all construction sites with asbestos operations, employers must name a "competent person" qualified and authorized to ensure worker safety and health. Under these requirements for safety and health prevention programs, the "competent person" must frequently inspect job sites, materials, and equipment.

In addition, for Class I jobs the "competent person" must inspect onsite at least once during each work shift and upon employee request. For Class II and III jobs, the "competent person" must inspect often enough to assess changing conditions and upon employee request.

At work sites where employees perform Class I or II asbestos work, the "competent person" must supervise

- the setup and ensure the integrity of regulated areas, enclosures, or other containments by onsite inspection
- setup procedures to control entry to and exit from the enclosure or area
- all employee exposure monitoring, ensuring it is properly conducted
- use of required protective clothing and equipment by employees working within the enclosure and/or using glove bags
- proper setup, removal, and performance of engineering controls, work practices, and personal protective equipment through onsite inspections
- employee use of hygiene facilities and required decontamination procedures
- notification requirements

The "competent person" must attend a comprehensive training course for contractors and supervisors certified by the U.S. Environmental Protection Agency (EPA) or a state-approved training provider or a course that is equivalent in length and content. For Class III and IV asbestos work, training must include a course equivalent in length and content to the 16-hour "Operations and Maintenance" course developed by EPA for maintenance and custodial workers.

21.3.6 Regulated Areas

A regulated area is a marked off site where employees work with asbestos, including any adjoining area(s) where debris and waste from asbestos work accumulates or where airborne concentrations of asbestos exceed or can possibly exceed the PEL.

All Class I, II, and III asbestos work or any other operations where airborne asbestos exceeds the PEL must be done within regulated areas. Authorized personnel only may enter. The designated "competent person" supervises all asbestos work performed in the area.

Employers must mark off the regulated area in any manner that minimizes the number of persons within the area and protects persons outside the area from exposure to airborne asbestos. Critical barriers or negative-pressure enclosures may mark off the regulated area.

Posted warning signs demarcating the area must be easily readable and understandable. The signs must bear the following information:

- DANGER
- ASBESTOS
- CANCER AND LUNG DISEASE HAZARD
- AUTHORIZED PERSONNEL ONLY
- RESPIRATORY AND PROTECTIVE CLOTHING ARE REQUIRED IN THIS AREA

Employers must supply a respirator to all persons entering regulated areas. Employees must not eat, drink, smoke, chew tobacco or gum, or apply cosmetics in regulated areas. An employer performing work in a regulated area must inform other employers onsite of the nature of the work, regulated area requirements, and measures taken to protect onsite employees.

The contractor creating or controlling the source of asbestos contamination must abate the hazards. All employers with employees working near regulated areas must assess each day the enclosure's integrity or the effectiveness of control methods to prevent airborne asbestos from migrating.

A general contractor on a construction project must oversee all asbestos work, even though he or she may not be the designated "competent person." As supervisor of the entire project, the general contractor determines whether asbestos contractors comply with the standard and ensures they correct any problems.

21.3.7 Communication of Hazards

21.3.7.1 Notification Requirements

The communication of asbestos hazards is vital to prevent further overexposure. Most asbestos-related construction involves previously installed building materials. Building owners often are the only or best source of information concerning them. The owners and employers of potentially exposed employees have specific duties under the standard.

Before beginning work, building owners must identify at the work site all thermal system insulation, sprayed or troweled-on surfacing materials in buildings, and resilient flooring material installed before 1981. Building owners also must notify, in writing, the following persons of the presence, locations, and quantity of asbestos-containing or presumed asbestos-containing materials:

- prospective employers applying or bidding for work in or adjacent to areas containing asbestos
- the owner's employees who work in or nearby these areas
- other employers on multi-employer work sites with employees
- working in or adjacent to the area
- tenants who will occupy the areas containing such materials

All employers discovering asbestos-containing materials on a work site must notify the building owner and other employers onsite within 24 hours of its presence, location, and quantity. Employers also must inform building owners and employees working in nearby areas of the precautions taken to confine airborne asbestos. Within 10 days of project completion, employers must inform building owners and other employers onsite of the current locations and quantity of remaining asbestos-containing materials and any final monitoring results.

At any time, employers or building owners may demonstrate that a presumed asbestos-containing material does not contain asbestos by inspecting the material (conducted according to the requirements of the Asbestos Hazard Response Act (AHERA)(40 CFR 763, Subpart E)) and by performing tests to prove asbestos is not present.

Employers do not have to inform employees of asbestos-free building materials present; however, employers must retain the information, data, and analysis supporting the determination. See the record keeping requirements section of this chapter for more specific information.

21.3.7.2 Signs

At the entrance to mechanical rooms or areas containing thermal system insulation and surfacing asbestos-containing materials, the building owner must post signs identifying the material present, its specific location, and appropriate work practices that ensure it is not disturbed.

Employers must post warning signs in regulated areas to inform employees of the dangers and necessary protective steps to take before entering.

21.3.7.3 Labels

Employers must, when possible, attach warning labels to all products and containers of asbestos, including waste containers, and all installed asbestos products. Labels must be printed in large, bold letters on a contrasting background and used in accordance with OSHA's Hazard Communication Standard (29 CFR 1910.1200). All labels must contain a warning statement against breathing asbestos fibers and contain the following legend:

- DANGER
- CONTAINS ASBESTOS FIBERS
- AVOID CREATING DUST
- CANCER AND LUNG DISEASE HAZARD

Labels are not required where:

1. Asbestos is present in concentrations less than 1 percent by weight.
2. A bonding agent, coating, or binder has altered asbestos fibers, prohibiting the release of airborne asbestos over the PEL or STEL during reasonable use, handling, storage, disposal, processing, or transportation.

When building owners or employers identify previously installed asbestos or presumed asbestos-containing materials, labels or signs must be attached or posted to inform employees which materials contain asbestos. Attached labels must be clearly noticeable and readable.

21.3.7.4 Employee Information and Training

General Training Requirements. Employers must, at no cost to employees, provide a training program for all employees installing and handling asbestos-containing products and for employees performing Class I through IV asbestos operations. Employees must receive training prior to or at initial assignment and at least annually thereafter.

Training courses must be easily understandable for employees and must inform them of:

- ways to recognize asbestos
- the adverse health effects of asbestos exposure

- the relationship between smoking and asbestos in causing lung cancer
- operations that could result in asbestos exposure and the importance of protective controls to minimize exposure
- the purpose, proper use, fitting instruction, and limitations of respirators
- the appropriate work practices for performing asbestos jobs
- medical surveillance program requirements
- the contents of the standard
- the names, addresses, and phone numbers of public health organizations that provide information and materials or conduct smoking cessation programs
- sign and label requirements and the meaning of legends on them

The employer also must provide, at no cost to employees, written materials relating to employee training and self-help smoking cessation programs.

Additional Training Based on Work Class. For Class I and II operations, training must be equivalent in curriculum, method, and length to the EPA Model Accreditation Plan (MAP) asbestos abatement worker training (40 CFR 763, Subpart E). For employees performing Class II operations involving one generic category of building materials containing asbestos (e.g., roofing, flooring, or siding materials or transite panels), training may be covered in an 8-hour course that includes "hands-on" experience.

For Class III operations, training must be equivalent in curriculum and method to the 16-hour "Operations and Maintenance" course developed by EPA for maintenance and custodial workers whose work disturbs asbestos-containing materials. The course must include "hands-on" training on proper respirator use and work practices.

For Class IV operations, training must be equivalent in curriculum and method to EPA awareness training. Training must focus on the locations of asbestos-containing or presumed asbestos-containing materials and the ways to recognize damage and deterioration and avoid exposure. The course must be at least 2 hours in length.

21.4 Methods of Compliance

21.4.1 Control Measures

For all covered work, employers must use the following control methods to comply with the PEL and STEL:

- local exhaust ventilation equipped with a high efficiency particulate air (HEPA) filter dust collection systems
- enclosure or isolation of processes producing asbestos dust
- ventilation of the regulated area to move contaminated air away from the employees' breathing zone and toward a filtration or collection device equipped with a HEPA filter
- feasible engineering and work practice controls to reduce exposure to the lowest possible levels, supplemented by respirators to reach the PEL or STEL or lower

Employers must use the following engineering controls and work practices for all operations regardless of exposure levels:

- vacuum cleaners equipped with HEPA filters to collect all asbestos-containing or presumed asbestos-containing debris and dust
- wet methods or wetting agents to control employee exposures, except when infeasible (e.g., due to the creation of electrical hazards, equipment malfunction, and slipping hazards)
- prompt cleanup and disposal in leak-tight containers of asbestos-contaminated wastes and debris

The following work practices and engineering controls are prohibited for all asbestos-related work or work that disturbs asbestos or presumed asbestos-containing materials, regardless of measured exposure levels or the results of initial exposure assessments:

- high-speed abrasive disc saws not equipped with a point-of-cut ventilator or enclosure with HEPA-filtered exhaust air
- compressed air to remove asbestos or asbestos-containing materials unless the compressed air is used with an enclosed ventilation system
- dry sweeping, shoveling, or other dry cleanup of dust and debris
- employee rotation to reduce exposure

In addition, OSHA's asbestos standard established specific requirements for each class of asbestos work in construction.

Class I Work. A designated "competent person" must supervise all Class I work, including installing and operating the control system. Employers must place critical barriers over all openings to regulated areas or use another barrier or isolation method to prevent airborne asbestos from migrating for:

- all Class I jobs removing more than 25 linear or 10 square feet of thermal system insulation or surfacing material
- all other Class I jobs without a negative exposure assessment
- where employees are working in areas adjacent to a Class I regulated area

Otherwise, employers must perform perimeter area surveillance during each work shift. No asbestos dust should be visible. Perimeter monitoring must show that clearance levels are met or that perimeter area levels are no greater than background levels.

For all Class I jobs:

- HVAC systems must be isolated in the regulated area by sealing with a double layer of 6 mil plastic or the equivalent
- impermeable drop cloths must be placed on surfaces beneath all removal activity
- all objects within the regulated area must be covered with secured impermeable drop cloths or plastic sheeting
- for jobs without a negative exposure assessment or where exposure monitoring shows the PEL is exceeded, employers must ventilate the regulated area to move the contaminated air away from the employee breathing zone and toward a HEPA-filtration or collection device

In addition, employees performing Class I work must use one or more of the following control methods:

- negative-pressure enclosure systems must be used where the configuration of the work area makes it impossible to erect the enclosure
- glove bag systems can be used to remove asbestos-containing or presumed asbestos-containing materials from straight runs of piping
- negative-pressure glove bag systems can be used to remove asbestos or presumed asbestos-containing materials from piping
- negative-pressure glove box systems can be used to remove asbestos or presumed asbestos-containing materials from pipe runs
- water spray process systems may be used to remove asbestos or presumed asbestos-containing materials from cold-line piping if employees carrying out the process have completed a 40-hour training course on its use in addition to training required for all

employees performing Class I work
- a small walk-in enclosure that accommodates no more than 2 people (mini-enclosure) may be used if the disturbance or removal can be completely contained by the enclosure

Employers may use different or modified engineering and work practice controls if the following provisions are met:

1. The control method encloses, contains, or isolates the process or source of airborne asbestos dust, or captures and redirects the dust before it enters into the employees' breathing zone.
2. A certified industrial hygienist or licensed professional engineer qualified as a project designer evaluates the work area, the projected work practices, and the engineering controls and certifies, in writing, that based on evaluations and data the planned control method adequately reduces direct and indirect employee exposure to or below the PEL under worst-case conditions. The planned control method also must prevent asbestos contamination outside the regulated area, as measured by sampling meeting the requirements of EPA's "Asbestos in Schools" rule or perimeter monitoring.
3. Before using alternative methods to remove more than 25 linear or 10 square feet of thermal system insulation or surfacing material, employers must send a copy of the evaluation and certification to the OSHA National Office, Office of Technical Support, Room N3653, 200 Constitution Avenue, NW, Washington, DC 20210.

Class II Work. The "competent person" must supervise all Class II work. Employers must use critical barriers over all openings to the regulated area or another barrier or isolation method to prevent airborne asbestos from migrating for:

- all indoor Class II jobs without a negative exposure assessment
- where changing conditions indicate exposure above the PEL
- where asbestos-containing materials are not removed substantially intact

Otherwise, employers must perform perimeter area monitoring to verify that the barrier works properly. Impermeable drop cloths must be placed on all surfaces beneath removal activities.

All Class II asbestos work can use the same work practices and requirements as Class I asbestos jobs. Alternatively, Class II work can be performed more easily using simple work practices set out in the standard for specific jobs.

For removing vinyl and asphalt flooring materials containing asbestos or installed in buildings constructed before 1981 and not verified as asbestos-free, employers must ensure that employees:

- do not sand flooring or its backing
- do not rip up resilient sheeting
- do not dry sweep
- do not use mechanical chipping unless performed in a negative-pressure enclosure
- use vacuums equipped with HEPA filters to clean floors
- use wet methods when removing resilient sheeting by cutting
- use wet methods to scrape residual adhesives and/or backing
- remove tiles intact, unless impossible
- may omit wetting when tiles are heated and removed intact
- assume resilient flooring material, including associated mastic and backing, are asbestos-containing, unless an industrial hygienist determines it asbestos-free

To remove asbestos-containing roofing materials, employers must ensure that employees:

- remove them intact
- use wet methods when possible
- continuously mist cutting machines during use, unless the "competent person" determines misting to be unsafe
- immediately HEPA-vacuum all loose dust along the cut
- lower as soon as possible or by the end of the work shift any unwrapped or unbagged roofing material to the ground via a covered, dust-tight chute, crane, or hoist
- transfer unwrapped materials to a closed receptacle to prevent dispersing the dust when lowered
- isolate roof-level heating and ventilation air intake sources or shut down the ventilation system

When removing cementitious asbestos-containing siding and shingles or transite panels, employers must ensure that employees:

- do not cut, abrade, or break siding, shingles, or transite panels unless methods less likely to result in asbestos fiber release cannot be used

- spray each panel or shingle with amended water before removing
- lower to the ground any unwrapped or unbagged panels or shingles via a covered dust-tight chute, crane, or hoist, or place them in an impervious waste bag or wrap them in plastic sheeting, as soon as possible or by the end of the work shift
- cut nails with fiat, sharp instruments

When removing asbestos-containing gaskets, employers must ensure that employees:

- remove gaskets within glove bags if they are visibly deteriorated and unlikely to be removed intact
- thoroughly wet the gaskets with amended water prior to removing
- immediately place the wet gaskets in a disposal container, and scrape using wet methods to remove residue

For removal of any other Class II asbestos-containing material, employers must ensure employees:

- do not cut, abrade, or break the material unless infeasible
- thoroughly wet the material with amended water before and during removal
- remove the material intact, if possible
- immediately bag or wrap removed asbestos-containing materials or keep them wet until transferred to a closed receptacle at the end of the work shift

Employers may use different or modified engineering and work practice controls if:

1. They can demonstrate by employee exposure data during the use of such methods and under similar conditions that employee exposure will not exceed the PEL under any anticipated circumstances.
2. The "competent person" evaluates the work area, the projected work practices, and the engineering controls and certifies, in writing, that they will reduce all employee exposure to below the PEL under expected conditions. The evaluation must be based on exposure data for conditions closely resembling those of the current job and for employees with equivalent training and experience.

Class III Work. Employers must use wet methods and local exhaust ventilation, when feasible, during Class III work. Where drilling, cutting, abrading, sanding, chipping, breaking, or sawing thermal system insulation or surfacing materials occurs, employers must use impermeable drop cloths as well as mini-enclosures, glove bag systems, or other effective isolation methods. Where no negative exposure assessment exists or monitoring shows the PEL is exceeded, employers must contain the area with impermeable drop cloths and plastic barriers or other isolation methods and ensure that employees wear respirators.

Class IV Work. Employees conducting Class IV asbestos work must have attended an asbestos awareness training program. Employees must use wet methods and HEPA vacuums to promptly clean asbestos-containing or presumed asbestos-containing debris. When cleaning debris and waste in regulated areas, employees must wear respirators. In areas where thermal system insulation or surfacing material is present, employees must assume that all waste and debris contain asbestos.

21.4.2 Respiratory Protection

Respirators must be used during all:

- class I asbestos jobs
- class H work where an asbestos-containing material is not removed substantially intact
- class H and III work not using wet methods
- class H and III work without a negative exposure assessment
- class III jobs where thermal system insulation or surfacing asbestos-containing or presumed asbestos-containing material is cut, abraded, or broken
- class IV work within a regulated area where respirators are required
- work where employees are exposed above the PEL or STEL
- emergencies

Employers must provide respirators at no cost to employees, selecting the appropriate type from among those approved by the Mine Safety and Health Administration (MSHA) and NIOSH.

For all employees performing Class I work in regulated areas and for jobs without a negative exposure assessment, employers must provide full-facepiece supplied-air respirators operated in pressure-demand mode and equipped with an auxiliary positive-pressure, self-contained breathing apparatus.

Employers must provide half-mask purifying respirators, other than disposable respirators, equipped with high-efficiency filters for Class II and III asbestos jobs without a negative exposure assessment and for Class III jobs where work disturbs thermal system insulation or surfacing asbestos-containing or presumed asbestos-containing materials.

Employers must institute a respiratory program in accordance with Respiratory Protection, 29 CFR 1910.134. Employers must permit employees using filter respirators to change the filter elements when breathing resistance increases; employers must maintain an adequate supply of filters for this purpose. Employers must permit employees wearing respirators to leave work areas to wash their faces and respirator face-pieces as necessary to prevent skin irritation.

Employers must ensure that the respirators issued have the least possible facepiece leakage and fit properly. For employees wearing negative-pressure respirators, employers must perform either quantitative or qualitative face fit tests with the initial fitting and at least every 6 months following. The qualitative fit tests can be used only for fit testing of half-mask respirators where they are permitted or for full-facepiece air-purifying respirators where they are worn at levels where half-facepiece air-purifying respirators are permitted. Employers must conduct qualitative and quantitative fit tests and use the tests to select face pieces that provide the required protection.

Employers must not assign any employee, who based on the most recent physical exam and the examining physician's recommendations would be unable to function normally, to tasks requiring respirator use. Employers must assign such employees to other jobs or give them the opportunity to transfer to different positions in the same geographical area and with the same seniority, status, pay rate, and job benefits as before transferring, if such positions are available.

21.4.3 Protective Clothing

Employers must provide and require the use of protective clothing such as coveralls or similar whole-body clothing, head coverings, gloves, and foot coverings for:

- any employee exposed to airborne asbestos exceeding the PEL or STEL
- work without a negative exposure assessment

- any employee performing Class I work involving the removal of over 25 linear or 10 square feet of thermal system insulation or surfacing asbestos-containing or presumed asbestos-containing materials

Employers must launder contaminated clothing to prevent the release of airborne asbestos in excess of the PEL or STEL. Any employer who gives contaminated clothing to another person for laundering must inform him or her of the contamination.

Employers must transport contaminated clothing in sealed, impermeable bags or other closed impermeable containers bearing appropriate labels. The "competent person" must examine employee work suits at least once per work shift for rips or tears. Rips or tears found while the employee is working must be mended or replaced immediately.

21.4.4 Hygiene Facilities

21.4.4.1 Decontamination Requirements for Class I Asbestos Work

For employees performing Class I asbestos jobs involving over 25 linear or 10 square feet of thermal system insulation or surfacing asbestos-containing or presumed asbestos-containing materials, employers must create a decontamination area adjacent to and connected with the regulated area. Employees must enter and exit the regulated area through the decontamination area.

The decontamination area must be composed of an equipment room, shower area, and clean room in series. The equipment room must be supplied with impermeable, labeled bags and containers to store and dispose of contaminated protective equipment. Shower facilities must be adjacent to both the equipment and clean rooms, unless work is performed outdoors or this arrangement is impractical. If so, employers must ensure that employees remove asbestos contamination from their work suits in the equipment room using a HEPA vacuum before proceeding to a shower non-adjacent to the work area; or remove their contaminated work suits in the equipment room, don clean work suits, and proceed to a shower non-adjacent to the work area. Figure 21.1 is a sample shop layout meeting these requirements.

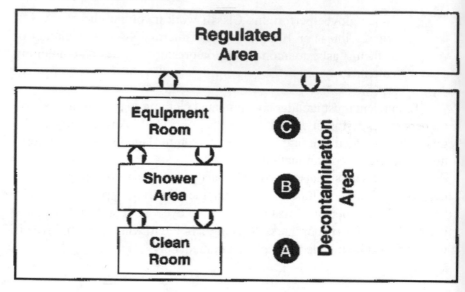

The clean room must have a locker or appropriate storage container for each employee unless work is performed outdoors or this arrangement is not possible. In such a case, employees may clean protective clothing with a portable HEPA vacuum before leaving the regulated area. Employees then must change into "street clothing" in clean change areas.

Before entering the regulated area, employees must enter the decontamination area through the clean room, remove and deposit street clothing within a provided locker,; and put on protective clothing and respiratory protection before leaving the clean area. To enter the regulated area, employees must pass through the equipment room.

Before exiting the regulated area, employees must remove all gross contamination and debris and then remove their protective clothing in the equipment room, depositing the clothing in labeled, impermeable bags or containers. Employees must shower before entering the clean room to change into "street clothing."

When employees consume food or beverages at the Class I work site, employers must provide lunch areas with airborne asbestos levels below the PEL and STEL.

21.4.4.2 Decontamination Requirements for Other Class I and Class II and III Asbestos Work Without a Negative Exposure

Assessment and Where Exposures Exceed the PEL

Employers must establish an equipment area adjacent to the regulated area for the decontamination of employees and their equipment. The area must be covered by an impermeable drop cloth on the floor or horizontal work surface and must be large enough to accommodate equipment cleaning and personal protective equipment removal without spreading contamination beyond the area. Before removing work clothing, employees must clean it with a HEPA vacuum. All equipment and the surfaces of containers filled with asbestos-containing materials must be cleaned prior to removal. Employers must ensure employees enter and exit the regulated area through the equipment area.

21.4.4.3 Decontamination Requirements for Class IV Work

Employers must ensure employees performing Class IV work within a regulated area comply with the hygiene practices required of employees performing work with higher classifications in that regulated area. Otherwise, employees cleaning up thermal system insulation or asbestos-containing debris must use decontamination facilities required for Class II and III work where exposure exceeds the PEL or no negative exposure assessment exists.

21.4.4.4 Smoking

Employers must ensure that employees performing any class of asbestos work do not smoke in any work area with asbestos exposure.

21.4.5 Housekeeping

Asbestos waste, scrap, debris, bags, containers, equipment, and contaminated clothing consigned for disposal must be collected and disposed of in sealed, labeled, impermeable bags or other closed, labeled impermeable containers. Employees must use HEPA-filtered vacuuming equipment and must empty it so as to minimize asbestos entry into the workplace.

All vinyl and asphalt flooring material must remain intact unless the building owner demonstrates that the flooring does not contain asbestos. Sanding flooring material is prohibited. Employees stripping finishes must use wet methods and low abrasion pads at speeds lower than 300 revolutions per minute. Burnishing or dry buffing may be done only on flooring with enough finish that the pad cannot contact the flooring material. Employees must not dust, sweep, or vacuum without a HEPA filter in an area containing thermal system insulation or surfacing material or visibly deteriorated asbestos-containing materials. Employees must promptly clean and dispose of dust and debris in leak-tight containers.

TABLE 21.1
Quick Reference of Provisions by Work Class

	Class 1	Class 2	Class 3	Class 4
Definition	Removal of thermal systems insulation (TSI) and surfacing materials (SM)	Removal or all other asbestos not TSI or SM	Maintenance and repair operations disturbing asbestos-containing materials	Housekeeping and custodial operations (including construction site cleanup)
Regulated Areas	Required (signs required)	Required (signs required)	Required (signs required)	Required (signs required)
"Competent Person"	Required onsite • Inspect each work-shift • Contractors and supervisors training required	Required onsite •Inspect often •Contractors and supervisors training required	Required onsite • Inspect often • Contractors and supervisors training required	Required onsite • Inspect often • Contractors and supervisors training required
Air Monitoring	• Initial if no negative exposure assessment (NEA) • Daily if no NEA • Terminate if < permissible exposure limit (PEL) • Additional if conditions change	• Initial if no NEA • Daily if no NEA • Terminate if < PEL • Additional if conditions change	• Initial if no NEA • Periodic to accurately predict if > PEL • Terminate if < PEL • Additional if conditions change	
Medical Surveillance	Required if • Wearing negative-pressure respirator • > PEL • 30 days exposure per year	Required if • Wearing negative-pressure respirator • > PEL • 30 days exposure/year	Required if • Wearing negative-pressure respirator • > PEL • 30 days exposure/year	Required if • Wearing negative-pressure respirator • > PEL

TABLE 21.1 (Continued)

Definition	Class 1	Class 2	Class 3	Class 4
Respirators	Mandatory for all Class 1 jobs	Mandatory if • non-intact removal • no NEA • > PEL • dry removal (except for roofing) • in emergencies	Half-mask air-purifying respirator minimum if • no NEA • TSI or SM disturbed • > PEL Mandatory if • dry removal (except for roofing) • in emergencies	Mandatory • in regulated area where required • if > PEL • in emergencies
Protective Clothing and Equipment	Required for all jobs if • > 25 linear or 10 square feet of TSI or SM removal • no NEA • > PEL	Required for all jobs if • no NEA • > PEL	Required for all jobs if • no NEA • > PEL	Required for all jobs if • no NEA • > PEL
Training	Equivalent to Asbestos Hazard Response Act (AHERA) worker course	Equivalent to AHERA worker course or specific work practices in removing one ACM only	Equivalent to AHERA operation and maintenance course	Equivalent to AHERA Awareness Training
Decontamination Procedures	Full decon unit required of > 25 linear or 10 square feet TSI or SM removal • connected shower/clean room required • vacuum, change, shower elsewhere • detailed procedures Lunch areas required If < 25 linear or 10 square feet TSI or SM removal or > PEL or no NEA • equipment room/area required	If > PEL or no NEA • equipment room/area required • dropcloths required • area must accommodate cleanup • must decontaminate all personal protective equipment • must enter regulated area through equipment room/decon area No smoking in work area	If > PEL or no NEA • equipment room/area required • dropcloths required • area must accommodate cleanup • must decontaminate all personal protective equipment • must enter regulated area through equipment room/decon area If NEA, must vacuum No smoking in work area	If > PEL or no NEA • equipment room/area required • dropcloths required • area must accommodate cleanup • must decontaminate all personal protective equipment • must enter regulated area through equipment room/decon area No smoking in work area

TABLE 21.1 (Continued)

Definition	Class 1	Class 2	Class 3	Class 4
Decontamination Procedures	• dropcloths required • area must accommodate cleanup • must decontaminate all personal protective equipment • must enter regulated area through equipment room/decon area No smoking in work area			
Required Work Practices and Engineering Controls	• wet methods • HEPA vacuum • Prompt cleanup/disposal	• wet methods • HEPA vacuum • Prompt cleanup/disposal	• wet methods • HEPA vacuum • Prompt cleanup/disposal	• wet methods • HEPA vacuum • Prompt cleanup/disposal
Required Work Practices and Engineering Controls to Comply with Permissible Exposure Limit (PEL)	• HEPA local exhaust • Enclosure • Directed ventilation • Other work practices • Supplement with respirators	• HEPA local exhaust • Enclosure • Directed ventilation • Other work practices • Supplement with respirators	• HEPA local exhaust • Enclosure • Directed ventilation • Other work practices • Supplement with respirators	• HEPA local exhaust • Enclosure • Directed ventilation • Other work practices • Supplement with respirators
Prohibited Work Practices and Engineering Controls	• High speed abrasive disc saws without HEPA • Compressed air without capture device • Dry sweeping/shoveling • Employee rotation	• High speed abrasive disc saws without HEPA • Compressed air without capture device • Dry sweeping/shoveling • Employee rotation	• High speed abrasive disc saws without HEPA • Compressed air without capture device • Dry sweeping/shoveling • Employee rotation	• High speed abrasive disc saws without HEPA • Compressed air without capture device • Dry sweeping/shoveling • Employee rotation

TABLE 21.1 (Continued)

Definition	Class 1	Class 2	Class 3	Class 4
Controls and Work Practices	• Critical barriers/isolation methods required if - .>25 linear or 10 square feet of TSI or SM removal - < 25 linear or 10 square feet TSI or SM removal only if no NEA or adjacent workers • HVAC isolation required • Dropcloths required • Directed ventilation required if no NEA or > PEL Also, one or more of the following controls must be used: • Negative-pressure enclosure • Glove bag for straight runs of pipe • Negative-pressure glove bag for pipe runs • Negative-pressure glove box for pipe runs • Water spray process • Mini-enclosure	For indoor work only: • Critical barriers/isolation methods required if - No NEA - Likely > PEL - Non-intact removal • Dropcloths required If > PEL, must use: • Local HEPA exhaust • Process isolation • Directed ventilation • Additional feasible control supplemented with respirators For removal of vinyl and asphalt flooring materials: • No standing • HEPA vacuum • Wet methods • No dry sweeping • Chipping done in negative-pressure enclosure • Intact removal, if possible • Dry heat removal allowed • Assume contains asbestos without an analysis	• Critical barriers required - if no NEA - > PEL via monitoring • dropcloths required • local HEPA exhaust required Enclosure or isolation of operation required if: • TSI or SM is drilled, cut, abraded, sanded, sawed, or chipped	• See Required Work Practices and Engineering Controls

TABLE 21.1 (Continued)

Definition	Class 1	Class 2	Class 3	Class 4
		For removal of built-up roofing materials or asbestos-cement shingles: • Intact removal, if possible • Wet methods, if feasible • Cutting machine misting • HEPA-vacuum debris • Lower by day's end • Control dust of unbagged material • Roof vent system protected For removal or cementitious siding, shingles, or transite panels: • Intact removal, if possible • Wet methods • Lower via dust-tight chute by day's end • Cut nail heads For removal of gaskets: • Use glove bags if not intact • Wet removal • Prompt disposal • Wet scraping		

TABLE 21.1 (Continued)

Definition	Class 1	Class 2	Class 3	Class 4
		Additional requirements: • Wet methods • Intact removal, if possible • Cutting, abrading, or breaking prohibited		

Summary

Asbestos has been very useful in many products from ceiling and floor tiles to car brakes and clutches. But now we have learned that it is hazardous to ones health to inhale asbestos fibers. Asbestos related health problems are very serious, but they can take more than 20 years to show up.

The most serious health problem is lung cancer. The risk of cancer is almost 10 times greater for those who work with asbestos and smoke. The most likely problem is asbestosis, a lung disease that produces breathing and heart problems. Asbestosis is acquired from an accumulation of asbestos fiber in the lungs and it can lead to cancer. The symptoms of asbestosis include shortness of breath (in advanced stages, even when you are resting), coughing, noising breathing, fatigue, and a vague feeling of sickness.

Another lung disease associated with asbestos exposure is mesothelioma, a cancer of the chest lining that's nearly always fatal. Symptoms include shortness of breath, chest pain, and stomach pains.

It is important to follow safety procedures when working with asbestos. For this reason, OSHA has made annual medical exams for exposed workers a required part of asbestos regulations. These exams are part of what OSHA calls a "medical surveillance program," that is designed to help keep workers healthy.

Permissible Exposure Limits. The standard sets a Permissible Exposure Limit (PEL) of 0.1 fiber per cubic centimeter of air over an 8-hour day. The limit is equivalent to the maximum average concentration of asbestos that is safe to work around in an average day. If a worker is exposed over that amount the employer needs to take the following action to protect its workers by monitoring the air, employee training, and medical checkups.

Reducing Risk to Exposure. Small business can use engineering controls and methods to reduce the amount of fibers getting into the air workers breathe. The use of the following equipment can also make it safer to work with asbestos:

- automatic bag opening equipment
- local exhaust systems that collect air in closed containers
- dust collection and cleaning systems
- hoods to cover operations that create or release fibers
- tools with exhaust systems or wet sprays
- shrouds for tool grinders

Bibliography

OSHA (1995) "Asbestos Standard for the Construction Industry". U.S. Department of Labor, Washington, D.C.

22

Construction Safety

CONSTRUCTION WORK has a high injury rate when compared to most other types of work. OSHA has continued to inspect construction sites, finding safety problems with excavation, concrete placement, and temporary electric service. This chapter provides guidelines to improve safety and reduce injury during construction activities.

22.1 Concrete and Masonry Construction

The Occupational Safety and Health Administration's standard for concrete and masonry construction, Subpart Q, Concrete and Masonry Construction, Title 29 of the Code of Federal Regulations (CFR), Part 1926.700 through 706, sets forth requirements with which employers must comply to protect construction workers from accidents and injuries resulting from the premature removal of form work, the failure to brace masonry walls, the failure to support precast panels, the inadvertent operation of equipment, and the failure to guard reinforcing steel.

Subpart Q prescribes performance-oriented requirements designed to help protect all construction hazards associated with concrete and operations at construction, demolition, alteration or repair work sites. Other relevant provisions in both general industry and construction standards (29CFR Parts 1910 and 1926) also apply to these operations.

Subpart Q is divided into the following major groups each of which is discussed in more detail in the following paragraphs:

- Scope, application, and definitions (29 CFR 1926.700)
- General requirements (29 CFR 1926.702)
- Equipment and tools (29 CFR 1926.702)
- Cast-in-place concrete (20 CFR 1926.703)
- Precast concrete (29 CFR 1926.704)
- Lift-slab construction (29 CFR 1926.705)

- Masonry construction (29 CFR 1926.706)

Some of the terms used in this chapter are defined below.

- Bull Float. A tool used to spread out and smooth concrete. Formwork. The total system of support for freshly placed or partially cured concrete, including the mold or sheeting (form) that is in contact with the concrete as well as all supporting members including shores, reshores, hardware, braces, and related hardware.
- Jacking Operation. Lifting vertically a slab (or group of slabs) from one location to another, for example, from the casting location to a temporary (parked) location, or from a temporary location to another temporary location, or to the final location in the structure—during a lift-slab construction operation.
- Lift Slab. A method of concrete construction in which floor and roof slabs are cast on or at ground level and, using jacks, are lifted into position.
- Limited Access Zone. An area alongside a masonry wall, that is under construction, and that is clearly demarcated to limit access by employees.
- Precast Concrete. Concrete members (such as walls, panels, slabs, columns, and beams) that have been formed, cast, and cured prior to final placement in a structure.
- Reshoring. The construction operation in which shoring equipment (also called reshores or reshoring equipment) is placed, as the original forms and shores are removed in order to support partially cured concrete and construction loads.
- Shore. A supporting member that resists a compressive force imposed by a load.
- Tremie. A pipe through which concrete may be deposited under water.
- Vertical Slip Forms. Forms that are jacked vertically during the placement of concrete.

22.1.1 Construction Loads

Employers must not place construction loads on a concrete structure or portion of a concrete structure unless the employer determines, based on information received from a person who is qualified in structural design, that the structure or portion of the structure is capable of supporting the intended loads.

22.1.2 Reinforcing Steel

All protruding reinforcing steel, onto and into which employees could fall, must be guarded to eliminate the hazard of impalement.

22.1.3 Post-Tensioning Operations

Employees (except those essential to the post-tensioning operations) must not be permitted to be behind the jack during tensioning operations. Signs and barriers must be erected to limit employee access to the post-tensioning area during tensioning operations.

22.1.4 Concrete Buckets

Employees must not be permitted to ride concrete buckets.

22.1.5 Working Under Loads

Employees must not be permitted to work under concrete buckets while the buckets are being elevated or lowered into position. To the extent practicable, elevated concrete buckets must be routed so that no employee or the fewest employees possible are exposed to the hazards associated with falling concrete buckets.

22.1.6 Personal Protective Equipment

Employees must not be permitted to apply a cement, sand, and water mixture through a pneumatic hose unless they are wearing protective head and face equipment.

22.1.7 Equipment and Tools

The standard also includes requirements for the following equipment and operations:

- bulk cement storage
- concrete mixers
- power concrete trowels
- concrete buggies
- concrete pumping systems
- concrete buckets
- tremies
- bull floats
- masonry saws
- lockout/tagout procedures

22.2 General Requirements for Cast-in-place Concrete

22.2.1 General Requirements for Formwork

Formwork must be designed, fabricated, erected, supported, braced, and maintained so that it will be capable of supporting without failure all vertical and lateral loads that might be applied to the formwork. As indicated in the Appendix to the standard, formwork that is designed, fabricated, erected, supported, braced, and maintained in conformance with Sections 6 and 7 of the American National Standard for Construction and Demolition Operations - Concrete and Masonry Work (ANSI) A10.9-1983, also meets the requirements of this paragraph.

22.2.2 Drawings or Plans

Drawings and plans, including all revisions for the jack layout, formwork (including shoring equipment), working decks, and scaffolds, must be available at the jobsite.

22.2.3 Shoring and Reshoring

All shoring equipment (including equipment used in reshoring operations) must be inspected prior to erection to determine that the equipment meets the requirements specified in the formwork drawings. Damaged shoring equipment must not be used for shoring. Erected shoring equipment must be inspected immediately prior to, during, and immediately after concrete placement. Shoring equipment that is found to be damaged or weakened after erection must be immediately reinforced.

The sills for shoring must be sound, rigid, and capable of carrying the maximum intended load. All base plates, shore heads, extension devices, and adjustment screws must be in firm contact and secured, when necessary, with the foundation and the form.

Eccentric loads on shore heads must be prohibited unless these members have been designed for such loading.

If single-post shores are used one on top of another (tiered), then additional shoring requirements must be met. The shores must be as follows:

- designed by a qualified designer and the erected shoring must be inspected by an engineer qualified in structural design
- vertically aligned
- spliced to prevent misalignment
- adequately braced in two mutually perpendicular directions at the splice level. Each tier also must be diagonally braced in the same two directions.

Adjustment of single-post shores to raise formwork must not be made after the placement of concrete.

Reshoring must be erected, as the original forms and shores are removed, whenever the concrete is required to support loads in excess of its capacity.

22.2.4 Vertical Slip Forms

The steel rods or pipes on which jacks climb or by which the forms are lifted must be specifically designed for that purpose and adequately braced where not encased in concrete. Forms must be designed to prevent excessive distortion of the structure during the jacking operation. Jacks and vertical supports must be positioned in such a manner that the loads do not exceed the rated capacity of the jacks.

The jacks or other lifting devices must be provided with mechanical dogs or other automatic holding devices to support the slip forms whenever failure of the power supply or lifting mechanisms occurs. The form structure must be maintained within all design tolerances specified for plumbness during the jacking operation. The predetermined safe rate of lift must not be exceeded. All vertical slip forms must be provided with scaffolds or work platforms where employees are required to work or pass.

22.2.5 Reinforcing Steel

Reinforcing steel for walls, piers, columns, and similar vertical structures must be adequately supported to prevent overturning and collapse. Employers must take measures to prevent unrolled wire mesh from recoiling. Such measures may include, but are not limited to, securing each end of the roll or turning over the roll.

22.2.6 Removal of Formwork

Forms and shores (except those that are used for slabs on grade and slip forms) must not be removed until the employer determines that the concrete has gained sufficient strength to support its weight and superimposed loads. Such determination must be based on compliance with one of the following conditions:

- The plans and specifications stipulate conditions for removal of forms and shores, and such conditions have been followed.
- The concrete has been properly tested with an appropriate American Society for Testing and Materials (ASTM) standard test method designed to indicate the concrete compressive strength, and the test results indicate that the concrete has gained sufficient strength to support its weight and superimposed loads.

Reshoring must not be removed until the concrete being supported has attained adequate strength to support its weight and all loads placed upon it.

22.2.7 Precast Concrete

Precast concrete wall units, structural framing, and tilt-up wall panels must be adequately supported to prevent overturning and to prevent collapse until permanent connections are completed.

Lifting inserts that are embedded or otherwise attached to tilt-up wall panels must be capable of supporting at least two times the maximum intended load applied or transmitted to them; lifting inserts for other precast members must be capable of supporting four times the load. Lifting hardware shall be capable of supporting at least five times the maximum intended load applied or transmitted to the lifting hardware. Only essential employees are permitted under precast concrete that is being lifted or tilted into position.

22.2.8 Lift-Slab Operations

Lift-slab operations must be designed and planned by a registered professional engineer who has experience in lift-slab construction. Such plans and designs must be implemented by the employer and must include detailed instructions and sketches indicating the prescribed method of erection. The plans and designs must also include provisions for ensuring lateral stability of the building/structure during construction.

Jacking equipment must be marked with the manufacturer's rated capacity and must be capable of supporting at least two and one-half times the load being lifted during jacking operations and the equipment must not be overloaded. For the purpose of this provision, jacking equipment includes any load bearing component that is used to carry out the lifting operation(s). Such equipment includes, but is not limited to, the following: threaded rods, lifting attachments, lifting nuts, hook-up collars, T-caps, shearheads, columns, and footings.

Jacks/lifting units must be designed and installed so that they will neither lift nor continue to lift when loaded in excess of their rated capacity. Jacks/ lifting units must have a safety device which will cause the jacks/lifting units to support the load at any position in the event of their malfunction or loss of ability to continue to lift.

No employee, except those essential to the jacking operation, shall be permitted in the building while any jacking operation is taking place unless the building has been reinforced sufficiently to ensure its integrity during erection. The phrase "reinforced sufficiently to ensure its integrity" as used in this paragraph means that a registered professional engineer, independent of the engineer who designed and planned the lifting operation, has determined from the plans that if there is a loss of support at any jack location, that loss will be confined to that location and the structure as a whole will remain stable.

Under no circumstances shall any employee who is not essential to the jacking operation be permitted immediately beneath a slab while it is being lifted.

22.2.9 Masonry Construction

Whenever a masonry wall is being constructed, employers must establish a limited access zone prior to the start of construction. The limited access zone must be:

- equal to the height of the wall to be constructed plus 4 feet (1.2 meters) and shall run the entire length of the wall
- on the side of the wall that will be unscaffolded
- restricted to entry only by employees actively engaged in constructing the wall
- kept in place until the wall is adequately supported to prevent overturning and collapse unless the height of the wall is more than 8 feet (2.4 meters) and unsupported, in which case it must be braced. The bracing must remain in place until permanent supporting elements of the structure are in place.

22.3 Ground Fault Protection on Construction Sites

This section is intended to help employers and employees responsible for electrical equipment to provide protection against 120-volt electrical hazards on the construction site, the most common being ground fault electrical shock, through the use of ground fault circuit interrupters (GFCIs) or through an assured equipment grounding conductor program.

To help cope with the electrical hazards at construction sites, the Occupational Safety and Health Administration (OSHA) issued a revision of OSHA safety and health regulation, 29 Code of Federal Regulations Part 1926, Subpart K (Electrical Standards for Construction). This revision was published in the Federal Register on July 11, 1986, and contains the requirements for the GFCI and the assured equipment grounding conductor program. Following the OSHA rules will help reduce the number of injuries and accidents from electrical hazards. Work disruptions should be minor, and the necessary inspections and maintenance should require little time.

22.3.1 Insulation and Grounding

Insulation and grounding are two recognized means of preventing injury during electrical equipment operation. Conductor insulation may be provided by placing nonconductive material such as plastic around the conductor. Grounding may be achieved through the use of a direct connection to a known ground such as a metal cold water pipe.

Consider, for example, the metal housing or enclosure around a motor or the metal box in which electrical switches, circuit breakers, and controls are placed. Such enclosures protect the equipment from dirt and moisture and prevent accidental contact with exposed wiring. However, there is a hazard associated with housings and enclosures. A malfunction within the equipment such as deteriorated insulation may create an electrical shock hazard. Many metal enclosures are connected to a ground to eliminate the hazard.

If a "hot" wire contacts a grounded enclosure, a ground fault results which normally will trip a circuit breaker or blow a fuse. Metal enclosures and containers are usually grounded by connecting them with a wire going to ground. This wire is called an equipment grounding conductor. Most portable electric tools and appliances are grounded by this means. There is one disadvantage to grounding: a break in the grounding system may occur without the user's knowledge.

Insulation may be damaged by hard usage on the job or simply by aging. If this damage causes the conductors to become ex-posed, the hazards of shocks, burns, and fire will exist. Double insulation may be used as additional protection on the live parts of a tool, but double insulation does not provide protection against defective cords and plugs or against heavy moisture conditions.

The use of a ground-fault circuit interrupter (GFCI) is one method used to overcome grounding and insulation deficiencies.

22.3.2 The Ground-Fault Circuit Interrupter

The ground-fault circuit interrupter (GFCI) is a fast-acting circuit breaker which senses small imbalances in the circuit caused by current leakage to ground and, in a fraction of a second, shuts off the electricity. The GFCI continually matches the amount of current going to an electrical device against the amount of current returning from the device along the electrical path. Whenever the amount "going" differs from the amount "returning" by approximately 5 milliamps, the GFCI interrupts the electric power within as little as 1/40 of a second.

However, the GFCI will not protect the employee from line-to-line contact hazards (such as a person holding two "hot" wires or a hot and a neutral wire in each hand). It does provide protection against the most common form of electrical shock hazard—the ground fault. It also provides protection against fires, overheating, and destruction of insulation on wiring.

22.3.3 Hazards Associated with Portable Tools

With the wide use of portable tools on construction sites, the use of flexible cords often becomes necessary. Hazards are created when cords, cord connectors, receptacles, and cord- and plug-connected equipment are improperly used and maintained.

Generally, flexible cords are more vulnerable to damage than is fixed wiring. Flexible cords must be connected to devices and to fittings so as to prevent tension at joints and terminal screws.

Because a cord is exposed, flexible, and unsecured, joints and terminals become more vulnerable. Flexible cord conductors are finely stranded for flexibility, but the strands of one conductor may loosen from under terminal screws and touch another conductor, especially if the cord is subjected to stress or strain.

A flexible cord may be damaged by activities on the job, by door or window edges, by staples or fastenings, by abrasion from adjacent materials, or simply by aging. If the electrical conductors become exposed, there is a danger of shocks, burns, or fire. A frequent hazard on a construction site is a cord assembly with improperly connected terminals.

When a cord connector is wet, hazardous leakage can occur to the equipment grounding conductor and to humans who pick up that connector if they also provide a path to ground. Such leakage is not limited to the face of the connector but also develops at any wetted portion of it.

When the leakage current of tools is below 1 ampere, and the grounding conductor has a low resistance, no shock should be perceived. However, should the resistance of the equipment grounding conductor increase, the current through the body also will increase. Thus, if the resistance of the equipment grounding conductor is significantly greater than 1 ohm, tools with even small leakages become hazardous.

22.3.4 Preventing and Eliminating Hazards

GFCIs can be used successfully to reduce electrical hazards on construction sites. Tripping of GFCIs, interruption of current flow, is sometimes caused by wet connectors and tools. It is a good practice to limit exposure of connectors and tools to excessive moisture by using watertight or sealable connectors. Providing more GFCIs or shorter circuits can prevent tripping caused by the cumulative leakage from several tools or by leakages from extremely long circuits.

22.3.5 Employer's Responsibility

OSHA ground-fault protection rules and regulations have been determined necessary and appropriate for employee safety and health. Therefore, it is the employer's responsibility to provide either: (a) ground-fault circuit interrupters on construction sites for receptacle outlets in use and not part of the permanent wiring of the building or structure; or (b) a scheduled and recorded assured equipment grounding conductor program on construction sites, covering all cord sets, receptacles which are not part of the permanent wiring of the building or structure, and equipment connected by cord and plug which are available for use or used by employees.

22.3.6 The Use of Ground-Fault Circuit Interrupters

The employer is required to provide approved ground-fault circuit interrupters for all 120-volt, single-phase, 15- and 20-ampere receptacle outlets on construction sites that are not a part of the permanent wiring of the building or structure and that are in use by employees. If a receptacle or receptacles are installed as part of the permanent wiring of the building or structure and they are used for temporary electric power, GFCI protection shall be provided. Receptacles on the ends of extension cords are not part of the permanent wiring and, therefore, must be protected by GFCIs whether or not the extension cord is plugged into permanent wiring. These GFCIs monitor the current-to-the-load for leakage to ground.

When this leakage exceeds 5 mA +/- 1 mA, the GFCI interrupts the current. They are rated to trip quickly enough to prevent electrocution. This protection is required in addition to, not as a substitute for, the grounding requirements of OSHA safety and health rules and regulations, 29 CFR 1926. The requirements which the employer must meet, if he or she chooses the GFCI option, are stated in 29 CFR 1926.404(b)(1)(ii).

22.3.7 Assured Equipment Grounding Conductor Program

The assured equipment grounding conductor program covers all cord sets, receptacles which are not a part of the permanent wiring of the building or structure, and equipment connected by cord and plug which are available for use or used by employees. The requirements which the program must meet are stated in 29 CFR 1926.404(b)(1)(iii) and Table 22.1, but employers

TABLE 22.1
Wiring and Design Protection

Safety and Health Regulations Part 1926 Subpart K (Partial)
§ 1926.404 wiring design and protection.

(b) Branch circuits-(1) Ground-fault protection-(i) General.
The employer shall use either ground-fault circuit interrupters as specified in paragraph (b)(1)(ii) of this section or an assured equipment grounding conductor program as specified in paragraph (b)(1)(iii) of this section to protect employees on construction sites. These requirements are in addition to any other requirements for equipment grounding conductors.

(ii) Ground-fault circuit interrupters.
All 120-volt, single-phase, 15- and 20-ampere receptacle outlets on construction sites, which are not a part of the permanent wiring of the building or structure and which are in use by employees, shall have approved ground-fault circuit interrupters for personnel protection. Receptacles on a two-wire, single-phase portable or vehicle- mounted generator rated not more than 5kW, where the circuit conductors of the generator are insulated from the generator frame and all other grounded surfaces, need not be protected with ground-fault circuit interrupters.

(iii) Assured equipment grounding conductor program.
The employer shall establish and implement an assured equipment grounding conductor program on construction sites covering all cord sets, receptacles which are not a part of the building or structure, and equipment connected by cord and plug which are available for use or used by employees. This program shall comply with the following minimum requirements:

(A) A written description of the program, including the specific procedures adopted by the employer, shall be available at the jobsite for inspection and copying by the Assistant Secretary and any affected employee.

(B) The employer shall designate one or more competent persons (as defined in § 1926.32(f)) to implement the program.

(C) Each cord set, attachment cap, plug, and receptacle of cord sets, and any equipment connected by cord and plug, except cord sets and receptacles which are fixed and not exposed to damage, shall be visually inspected before each day's use for external defects, such as deformed or missing pins or insulation damage, and for indications of possible internal damage. Equipment found damaged or defective shall not be used until repaired.

(D) The following tests shall be performed on all cord sets, receptacles which are not a part of the permanent wiring of the building or structure, and cord- and plug-connected equipment required to be grounded:
   (1) All equipment grounding conductors shall be tested for continuity and shall be electrically continuous.
   (2) Each receptacle and attachment cap or plug shall be tested for correct attachment of the equipment grounding conductor. The equipment grounding conductor shall be connected to its proper terminal.

(E) All required tests shall be performed:
   (1) Before first use;
   (2) Before equipment is returned to service following any repairs;
   (3) Before equipment is used after any incident which can be reasonably suspected to have caused damage (for ex-ample, when a cord set is run over); and
   (4) At intervals not to exceed 3 months, except that cord sets and receptacles which are fixed and not exposed to damage shall be tested at intervals not exceeding 6 months.

(F) The employer shall not make available or permit the use by employees of any equipment which has not met the requirements of this paragraph (b)(1)(iii) of this section.

(G) Tests performed as required in this paragraph shall be re-corded. This test record shall identify each receptacle, cord set, and cord- and plug-connected equipment that passed the test and shall indicate the last date it was tested or the interval for which it was tested. This record shall be kept by means of logs, color coding, or other effective means and shall be maintained until replaced by a more current record. The record shall be made available on the jobsite for inspection by the Assistant Secretary and any affected employee.

may provide additional tests or procedures. OSHA requires that a written description of the employer's assured equipment grounding conductor program, including the specific procedures adopted, be kept at the jobsite. This program should outline the employer's specific procedures for the required equipment inspections, tests, and test schedule.

The required tests must be recorded and the record maintained until replaced by a more current record. The written program description and the recorded tests must be made available, at the jobsite, to OSHA and to any affected employee upon request. The employer is required to designate one or more competent persons to implement the program.

Electrical equipment noted in the assured equipment grounding conductor program must be visually inspected for damage or defects before each day's use. Any damaged or defective equipment must not be used by the employee until repaired.

Two tests are required by OSHA. One is a continuity test to ensure that the equipment grounding conductor is electrically continuous. It must be performed on all cord sets, receptacles which are not part of the permanent wiring of the building or structure, and on cord- and plug-connected equipment which is required to be grounded. This test may be performed using a simple continuity tester, such as a lamp and battery, a bell and battery, an ohmmeter, or a receptacle tester.

The other test must be performed on receptacles and plugs to ensure that the equipment grounding conductor is connected to its proper terminal. This test can be performed with the same equipment used in the first test.

These tests are required before first use, after any repairs, after damage is suspected to have occurred, and at 3-month intervals. Cord sets and receptacles which are essentially fixed and not exposed to damage must be tested at regular intervals not to exceed 6 months. Any equipment which fails to pass the required tests shall not be made available or used by employees.

22.4 Trenching and Excavation Safety

A trench is a narrow excavation, less than 15' wide, deeper than its width. An excavation is a cavity or depression cut or dug into the earth's surface. Trenches and excavations are among the most hazardous construction work sites. There is no room for risk. Any condition that is ignored or overlooked can cause an accident. If often takes only a few seconds for an accident to happen.

Trenching and excavation safety can protect workers from cave-ins and other disasters when trenching or excavating. It involves knowing the types of hazards that workers face during this kind of work and protecting workers by using common sense and safe working practices. Accidents usually happen when shortcuts are taken. Some common short cuts include:

- Inadequate shoring: The failure to support trench or excavation walls properly in order to save time or cut costs or because the trench will only be open for a short time.
- Lack of knowledge: Not knowing the hazards involved and the precautions necessary to minimize the risks.
- Poor judgment: Poor judgment of soil, weather, climate conditions and so forth, all which determine the sloping, strength of the support that is needed.

22.4.1 Know the Hazards

The most serious hazard is cave-ins which can crush or suffocate workers. Other hazards may include:

- Obstacles such as carelessly placed tools and equipment, or excavated material which can cause injuries due to slips, trips, or falls.
- Poisonous gases can quickly overcome workers with little or no warning.
- Lack of oxygen can pose a serious risk in trenches or excavations. Always test oxygen levels before entering any area where entry and exit are limited and where normal air is in short supply such as in a confined space.
- Combustible vapors and gases which can build up in confined spaces leading to a danger of fire and/or explosion.

22.4.2 Causes of Cave-ins

Cave-ins occur when an unsupported wall is weakened or undermined by too much weight, pressure, or an unstable bottom. One of the danger signs is surface cracking in the ground nearby. These cracks usually occur near the edge of the trench or excavation within half the width of the trench.

Overhangs and bulges also present hazards. An overhang at the top or a bulge in a wall can cause soil to topple or slide into the trench or excavation. If any of these conditions occur, work should be stopped immediately and the problem reported to the management.

22.4.3 Know When to Take Precautions

Weather and climate may call for stronger shoring, extra bracing, or steeper sloping. Water flow from rain or melting snow, ground water, storm drains, damaged water lines or nearby streams, can loosen soil and increase pressure on walls. Frozen ground can thaw, weakening the walls of trench or excavation. Very dry weather can also be dangerous because it tends to loosen soil. Long tem excavations may need extra weather and climate protection. Sides and faces should be covered with plastic or sprayed with anti-moisture chemical to reduce danger.

22.4.4 Other Factors Requiring Precautions

Soils with high slit or sand content are very unstable unless properly shored or sloped. Even if the excavation is less than five feet deep. Wet or back-filled soil is also unstable and needs wall support. Without proper support, hard rock can cave in if faults are split during cutting or digging cutting of digging operations.

Vibrations can loosen soil and cause walls to collapse unless proper shoring or sloping is used. Sources of vibration include:

- moving machinery
- vehicle, railway or air traffic
- blasting operations
- machines being used in nearby buildings such punch presses and forging hammers

Heavy equipment, pipes, or timbers can exert great pressure on trench walls. Keep such equipment as far back as possible from the excavation site. If heavy loads need to be near the site, provide extra wall support-braces, sheet-piling, or shoring. Buildings, trees, curbs, and utility poles can also put stress on walls or cause soil to move. For maximum protection, shoring, bracing, or underpinning may be required.

Always store excavated material at least two feet from the edge of a trench or excavation. Do not let it accumulate near wall sides.

22.4.5 Preventing Wall Collapse

Use proper techniques and equipment to prevent collapse. Use sheeting made of timber or steel, bracing, and jacks (screw or hydraulic) to support trench walls. The correct type of sheeting should be determined by qualified and authorized personnel. In addition, the following guidelines should be followed:

- Use only materials in good condition.
- Do not use shoring to support heavy equipment.
- Use heavier shoring materials for trenches to be left open over a long period of time.
- Use special care removing shoring. Remove from the bottom up, using lifting tackle or jacks.
- Shoring should be inspected and maintained daily.
- To prevent cave-ins, the side of a trench or excavation should be sloped so that soil will not slide.
- Use a trench shield. A trench shield is a heavy frame, often made of steel, to which steel plates are bolted or welded. It can be moved along the trench and is especially useful for sewer and pipeline work because it allows prompt backfilling.

The angle of repose is the steepest angle at which trench or excavation walls will lie without sliding. The more unstable the soil, the flatter the angle should be. For example, 90° for solid rock, 45° for average soil, and 26° for loose sand.

22.4.6 Site Protection

Site protection is needed for workers to protect themselves from rocks or other objects kicked or thrown into the trench. Site protection is also needed for the protection of pedestrians, vehicles, and children especially around trenches filled with water. Safety measures include:

- fences
- covers for manhole opening

- barricades
- warning signs
- guardrails
- flags
- security guards

It is also important to provide a lighting system if necessary, so safety can be maintained at night too.

22.5 Materials Handling and Storing Safety

Handling and storing materials involves diverse operations such as hoisting tons of steel with a crane, driving a truck loaded with concrete blocks, manually carrying bags or materials, and stacking drums, barrels, kegs, lumber, or loose bricks.

The efficient handling and storing of materials is vital to industry. These operations provide continuous flow of raw materials, parts, and assemblies through the workplace, and ensure that materials are available when needed. Yet, the improper handling and storing of materials can cause costly injuries.

22.5.1 Potential Hazards

Workers frequently cite the weight and bulkiness of objects being lifted as major contributing factors to their injuries. In 1990, for example, 400,000 workplace accidents resulted in back injuries. Workers also frequently cited body movement as contributing to their injuries. Bending, followed by twisting and turning, were the more commonly cited movements that caused back injuries. Back injuries accounted for more than 20 percent of all occupational illnesses, according to data from the National Safety Council. By 1994, the U.S. Bureau of Labor Statistics reported there were 613,251 over-exertion cases with lost-workdays. The majority of those cases were due to lifting (367,424), pushing/pulling (93,325), and carrying (68,992). Those cases represent 27 percent of all lost-workday cases.

In addition, workers can be injured by falling objects, improperly stacked materials, or by various types of equipment. When manually moving materials, workers should be aware of potential injuries, including the following:

- strains and sprains from improperly lifting loads or from carrying loads that are either too large or too heavy
- fractures and bruises caused by being struck by materials or by being caught in pinch points
- cuts and bruises caused by falling materials that have been improperly stored or by incorrectly cutting ties or other securing devices

22.5.2 Safety and Health Program Guidelines

To have an effective materials handling and storing safety and health program, managers must take an active role in its development. First-line supervisors must be convinced of the importance of controlling hazards associated with materials handling and storing and must be held accountable for employee training. An ongoing safety and health program should be used to motivate employees to continue using necessary protective gear and observing proper job procedures.

OSHA's recommended *Safety and Health Program Management Guidelines* issued in 1989 can provide a blueprint for employers seeking guidance on how to effectively manage and protect worker safety and health. The four main elements of an effective occupational safety and health program are

- management commitment and employee involvement
- worksite analysis
- hazard prevention and control
- safety and health training

These elements encompass principles such as establishing and communicating clear goals of a safety and health management program; conducting worksite examinations to identify existing hazards and the conditions under which changes might occur; effectively designing the jobsite or job to prevent hazards; and providing essential training to address the safety and health responsibilities of both management and employees.

Instituting these practices, along with providing the correct materials handling equipment, can add a large measure of worker safety and health in the area of materials handling and storing.

22.5.3 Moving, Handling, and Storing Materials

When manually moving materials, employees should seek help when a load is so bulky it cannot be properly grasped or lifted, when they cannot see around or over it, or when they cannot safely handle the load.

When placing blocks under a raised load, an employee should ensure that the load is not released until his or her hands are removed from under the load. Blocking materials and timbers should be large and strong enough to support the load safely. Materials with evidence of cracks, rounded corners, splintered pieces, or dry rot should not be used for blocking.

Handles or holders should be attached to loads to reduce the chances of getting fingers pinched or smashed. Workers also should use appropriate protective equipment. For loads with sharp or rough edges, wear gloves or other hand and forearm protection. In addition, to avoid injuries to the eyes, use eye protection. When the loads are heavy or bulky, the mover also should wear steel-toed safety shoes or boots to prevent foot injuries if he or she slips or accidentally drops a load.

When mechanically moving materials, avoid overloading the equipment by letting the weight, size, and shape of the material being moved dictate the type of equipment used for transporting it. All materials handling equipment has rated capacities that determine the maximum weight the equipment can safely handle and the conditions under which it can handle that weight. The equipment-rated capacity must be displayed on each piece of equipment and must not be exceeded except for load testing.

When picking up items with a powered industrial truck, the load must be centered on the forks and as close to the mast as possible to minimize the potential for the truck tipping or the load falling. Never overload a lift truck since it would be hard to control and could easily tip over. Do not place extra weight on the rear of a counterbalanced forklift to allow an overload. The load must be at the lowest position for traveling and the truck manufacturer's operational requirements must be followed. All stacked loads must be correctly piled and cross-tiered, where possible. Precautions also should be taken when stacking and storing material.

Stored materials must not create a hazard. Storage areas must be kept free from accumulated materials that cause tripping, fires, or explosions, or that may contribute to the harboring of rats and other pests. Stored materials inside buildings must not be placed within 6 feet of hoist ways or inside floor openings nor within 10 feet of an exterior wall. When stacking and piling materials, it is important to be aware of such factors as the materials' height and weight, how accessible the stored materials are to the user, and the condition of the containers where the materials are being stored.

Non-compatible material must be separated in storage. Employees who work on stored materials in silos, hoppers, or tanks, must be equipped with lifelines and safety belts.

All bound material should be stacked, placed on racks, blocked, interlocked, or otherwise secured to prevent it from sliding, falling, or collapsing. A load greater than that approved by a building official may not be placed on any floor of a building or other structure. Where applicable, load limits approved by the building inspector should be conspicuously posted in all storage areas.

When stacking materials, height limitations should be observed. For example, lumber must be stacked no more than 16 feet high if it is handled manually; 20 feet is the maximum stacking height if a forklift is used. For quick reference, walls or posts may be painted with stripes to indicate maximum stacking heights.

Used lumber must have all nails removed before stacking. Lumber must be stacked and leveled on solidly supported bracing. The stacks must be stable and self-supporting. Stacks of loose bricks should not be more than 7 feet in height. When these stacks reach a height of 4 feet, they should be tapered back 2 inches for every foot of height above the 4-foot level. When masonry blocks are stacked higher than 6 feet, the stacks should be tapered back one-half block for each tier above the 6-foot level.

Bags and bundles must be stacked in interlocking rows to remain secure. Bagged material must be stacked by stepping back the layers and cross-keying the bags at least every ten layers. To remove bags from the stack, start from the top row first. Baled paper and rags stored inside a building must not be closer than 18 inches to the walls, partitions, or sprinkler heads. Boxed materials must be banded or held in place using cross-ties or shrink plastic fiber.

Drums, barrels, and kegs must be stacked symmetrically. if stored on their sides, the bottom tiers must be blocked to keep them from rolling. When stacked on end, put planks, sheets of plywood dunnage, or pallets between each tier to make a firm, flat, stacking surface. When stacking materials two or more tiers high, the bottom tier must be chocked on each side to prevent shifting in either direction.

When stacking, consider the need for availability of the material. Material that cannot be stacked due to size, shape, or fragility can be safely stored on shelves or in bins. Structural steel, bar stock, poles, and other cylindrical materials, unless in racks, must be stacked and blocked to prevent spreading or tilting. Pipes and bars should not be stored in racks that face main aisles; this could create a hazard to passersby when removing supplies.

To reduce potential accidents associated with workplace equipment, employees need to be trained in the proper use and limitations of the equipment they operate. This includes knowing how to effectively use equipment such as conveyors, cranes, and slings.

22.5.4 Conveyors

When using conveyors, workers' hands may be caught in nip points where the conveyor medium runs near the frame or over support members or rollers; workers may be struck by material falling off the conveyor; or they may become caught on or in the conveyor, being drawn into the conveyor path as a result.

To reduce the severity of an injury, an emergency button or pull cord designed to stop the conveyor must be installed at the employee's work station. Continuously accessible conveyor belts should have an emergency stop cable that extends the entire length of the conveyor belt so that the cable can be accessed from any location along the belt. The emergency stop switch must be designed to be reset before the conveyor can be restarted. Before restarting a conveyor that has stopped due to an overload, appropriate personnel must inspect the conveyor and clear the stoppage before restarting. Employees must never ride on a materials handling conveyor.

Where a conveyor passes over work areas or aisles, guards must be provided to keep employees from being struck by falling material. If the crossover is low enough for workers to run into it, the guard must be either marked with a warning sign or painted a bright color to protect employees.

Screw conveyors must be completely covered except at loading and discharging points. At those points, guards must protect employees against contacting the moving screw; the guards are movable, and they must be interlocked to prevent conveyor movement when not in place.

22.5.5 Cranes

Employers must permit only thoroughly trained and competent persons to operate cranes. Operators should know what they are lifting and what it weighs. For example, the rated capacity of mobile cranes varies with the length of the boom and the boom radius. When a crane has a telescoping boom, a load may be safe to lift at a short boom length and/or a short boom radius, but may overload the crane when the boom is extended and the radius increases.

All movable cranes must have boom angle indicators; those cranes with telescoping booms must have some means to determine boom lengths, unless the load rating is independent of the boom length. Load rating charts must be posted in the cab of cab-operated cranes. All mobile cranes do not have uniform capacities for the same boom length and radius in all directions around the chassis of the vehicle.

Always check the crane's load chart to ensure that the crane will not be overloaded for the conditions under which it will operate. Plan lifts before starting them to ensure that they are safe. Take additional precautions and exercise extra care when operating around powerlines.

Some mobile cranes cannot operate with outriggers in the traveling position. When used, the outriggers must rest on firm ground, on timbers, or be sufficiently cribbed to spread the weight of the crane and the load over a large enough area. This will prevent the crane from tipping during use.

Hoisting chains and ropes must always be free of kinks or twists and must never be wrapped around a load. Loads should be attached to the load hook by slings, fixtures, and other devices that have the capacity to support the load on the hook. Sharp edges of loads should be padded to prevent cutting slings. Proper sling angles must be maintained so that slings are not loaded in excess of their capacity.

All cranes must be inspected frequently by persons thoroughly familiar with the crane, the methods of inspecting the crane, and what can make the crane unserviceable. Crane activity, the severity of use, and environmental conditions should determine inspection schedules. Critical parts, such as crane operating mechanisms, hooks, air, or hydraulic system components and other load-carrying components, should be inspected daily for any maladjustment, deterioration, leakage, deformation, or other damage.

22.5.6 Slings

When working with slings, employers must ensure that they are visually inspected before use and during operation, especially if used under heavy stress. Riggers or other knowledgeable employees should conduct or assist in the inspection because they are aware of how the sling is used and what makes it unserviceable. A damaged or defective sling must be removed from service.

Slings must not be shortened with knots or bolts or other makeshift devices; sling legs that have been kinked also are prohibited. Slings must not be loaded beyond their rated capacity. Suspended loads must be kept clear of all obstructions, and crane operators should avoid sudden starts and stops when moving suspended loads. Employees also must remain clear of loads about to be lifted and suspended. All shock loading is prohibited.

22.5.7 Powered Industrial Trucks

Workers who must handle and store materials often use fork trucks, platform lift trucks, motorized hand trucks, and other specialized industrial trucks powered by electrical motors or internal combustion engines. Affected workers, therefore, should be aware of the safety requirements pertaining to fire protection, and the design, maintenance, and use of these trucks.

All new powered industrial trucks, except vehicles intended primarily for earth moving or over-the-road hauling, must meet the design and construction requirements for powered industrial trucks established in the American National Standard for Powered Industrial Trucks. Approved trucks also must bear a label or some other identifying mark indicating acceptance by a nationally recognized testing laboratory.

An owner or user must not make modifications and additions affecting capacity and safe operation of the trucks without the manufacturer's prior written approval. In these cases, capacity, operation, and maintenance instruction plates and tags or decals must be changed to reflect the new information. If the truck is equipped with front-end attachments that are not factory installed, the user should request that the truck be marked to identify these attachments and show the truck's approximate weight, including the installed attachment, when it is at maximum elevation with its load laterally centered.

There are 11 different types of industrial trucks, some having greater safeguards than others. There also are designated conditions and locations under which the vast range of industrial-powered trucks can be used. In some instances, powered industrial trucks cannot be used, and in others, they can only be used if approved by a nationally recognized testing laboratory for fire safety. For example, powered industrial trucks must not be used in atmospheres containing hazardous concentrations of the following substances:

- Acetylene
- Butadiene

- Ethylene oxide
- Hydrogen (or gases or vapors equivalent in hazard to hydrogen, such as manufactured gas)
- Propylene oxide
- Acetaldehyde
- Cyclopropane
- Dimethyl ether
- Ethylene
- Isoprene
- Unsymmetrical dimethyl hydrazine

These trucks are not to be used in atmospheres containing hazardous concentrations of metal dust, including aluminum, magnesium and other metals of similarly hazardous characteristics, or in atmospheres containing carbon black, coal, or coke dust. Where dust of magnesium, aluminum, or aluminum bronze dusts may be present, the fuses, switches, motor controllers, and circuit breakers of trucks must be enclosed with enclosures approved for these substances.

There also are powered industrial trucks that are designed, constructed, and assembled for use in atmospheres containing flammable vapors or dusts. These include industrial-powered trucks equipped with additional safeguards to their exhaust, fuel, and electrical systems; with no electrical equipment, including the ignition; with temperature limitation features; and with electric motors and all other electrical equipment completely enclosed.

These specially designed powered industrial trucks may be used in locations where volatile flammable liquids or flammable gases are handled, processed, or used. The liquids, vapors, or gases should, among other things, be confined within closed containers or closed systems from which they cannot escape.

Some other conditions and/or locations in which specifically designed powered industrial trucks may be used include the following:

- Only powered industrial trucks without any electrical equipment, including the ignition, and that have their electrical motors or other electrical equipment completely enclosed should be used in atmospheres containing flammable vapors or dust.
- Powered industrial trucks that are either powered electrically by liquefied petroleum gas or by a gasoline or diesel engine are used on piers and wharves that handle general cargo.

Safety precautions the user can observe when operating or maintaining powered industrial trucks include:

- That high lift rider trucks be fitted with an overhead guard, unless operating conditions do not permit.
- That fork trucks be equipped with a vertical load backrest extension according to manufacturers' specifications, if the load presents a hazard.
- That battery charging installations be located in areas designated for that purpose.
- That facilities be provided for flushing and neutralizing spilled electrolytes when changing or recharging a battery to prevent fires, to protect the charging apparatus from being damaged by the trucks, and to adequately ventilate fumes in the charging area from gassing batteries.
- That conveyor, overhead hoist, or equivalent materials handling equipment be provided for handling batteries.
- That auxiliary directional lighting be provided on the truck where general lighting is less than 2 lumens per square foot.
- That arms and legs not be placed between the uprights of the mast or outside the running lines of the truck.
- That brakes be set and wheel blocks or other adequate protection be in place to prevent movement of trucks, trailers, or railroad cars when using trucks to load or unload materials onto train boxcars.
- That sufficient headroom be provided under overhead installations, lights, pipes, and sprinkler systems.
- That personnel on the loading platform have the means to shut off power to the truck.
- That dockboards or bridgeplates be properly secured, so they won't move when equipment moves over them.
- That only stable or safely arranged loads be handled, and caution be exercised when handling tools.
- That trucks whose electrical systems are in need of repair have the battery disconnected prior to such repairs.
- That replacement parts of any industrial truck be equivalent in safety to the original ones.

22.5.8 Ergonomic Safety and Health Principles

Ergonomics is defined as the study of work and is based on the principle that the job should be adapted to fit the person, rather than forcing the person to fit the job. Ergonomics focuses on the work environment, such as its design and function, and items such as design and function of work stations, controls, displays, safety devices, tools, and lighting to fit the employees' physical requirements and to ensure their health and well being.

Ergonomics includes restructuring or changing workplace conditions to make the job easier and reducing stressors that cause cumulative trauma disorders and repetitive motion injuries. In the area of materials handling and storing, ergonomic principles may require controls such as reducing the size or weight of the objects lifted, installing a mechanical lifting aid, or changing the height of a pallet or shelf.

Although no approach has been found for totally eliminating back injuries resulting from lifting materials, a substantial number of lifting injuries can be prevented by implementing an effective ergonomics program and by training employees in appropriate lifting techniques.

In addition to using ergonomic controls, there are some basic safety principles that can be employed to reduce injuries resulting from handling and storing materials. These include taking general fire safety precautions and keeping aisles and passageways clear.

In adhering to fire safety precautions, employees should note that flammable and combustible materials must be stored according to their fire characteristics. Flammable liquids, for example, must be separated from other material by a fire wall. Also, other combustibles must be stored in an area where smoking and using an open flame or a spark-producing device is prohibited. Dissimilar materials that are dangerous when they come into contact with each other must be stored apart.

When using aisles and passageways to move materials mechanically, sufficient clearance must be allowed for aisles at loading docks, through doorways, wherever turns must be made, and in other parts of the workplace. Providing sufficient clearance for mechanically-moved materials will prevent workers from being pinned between the equipment and fixtures in the workplace, such as walls, racks, posts, or other machines. Sufficient clearance also will prevent the load from striking an obstruction and falling on an employee.

All passageways used by employees must be kept clear of obstructions and tripping hazards. Materials in excess of supplies needed for immediate operations should not be stored in aisles or passageways, and permanent aisles and passageways must be marked appropriately.

22.5.9 Training and Education

OSHA recommends using a formal training program to allow employees to recognize and avoid materials handling hazards. Instructors should be well-versed in matters that pertain to safety engineering and materials handling and storing. The content of the training should emphasize those factors that will contribute to reducing workplace hazards including the following:

- Alerting the employee to the dangers of lifting without proper training.
- Showing the employee how to avoid unnecessary physical stress and strain.
- Teaching workers to become aware of what they can comfortably handle without undue strain.
- Instructing workers on the proper use of equipment.
- Teaching workers to recognize potential hazards and how to prevent or correct them.

Because of the high incidence of back injuries, safe lifting techniques for manual lifting should be demonstrated and practiced at the worksite by supervisors as well as by employees.

A training program to teach proper lifting techniques should cover the following topics:

- Awareness of health risks to improper lifting- citing organizational case histories.
- Knowledge of the basic anatomy of the spine, the muscles, and the joints of the trunk, and the contributions of intra-abdominal pressure while lifting.
- Awareness of individual body strengths and weaknesses- determining one's own lifting capacity.
- Recognition of the physical factors that might contribute to an accident and how to avoid the unexpected.
- Use of safe lifting postures and timing for smooth, easy lifting and the ability to minimize the load-moment effects.
- Use of handling aids such as stages, platforms, or steps, trestles, shoulder pads, handles, and wheels.
- Knowledge of body responses- warning signals- to be aware of when lifting.

22.6 Crane and Derrick Operations

Using cranes or derricks to hoist personnel poses a significant risk to employees being lifted. To help prevent employee injury or death, the occupational Safety and Health Administration (OSHA) regulation, *Title 29 Code of Federal Regulations 1926.550,* limits the use of personnel hoisting in the construction industry and prescribes the proper safety measures for these operations.

Personnel platforms that are suspended from the load line and used in construction are covered by 29 CFR 1926.550(g). In addition, there is no specific provision for suspended personnel platforms in Part 1910. The governing provision, therefore, is general provision 1910.180(h)(3)(v), which prohibits hoisting, lowering, swinging, or traveling while anyone is on the load or hook. OSHA has determined, however, that when the use of a conventional means of access to any elevated work site would be impossible or more hazardous, a violation of 1910.180(h)(3)(v) will be treated as "de minimis" if the employer has complied with the provisions set forth in 1926.550(g)(3), (4), (5), (6), (7), and (8).

The OSHA rule for hoisting personnel is written in performance-oriented language that allows employers flexibility in deciding how to provide the best protection for their employees against the hazards associated with hoisting operations and how to bring their work sites into compliance with the requirements of the standard.

This section discusses OSHA's requirements for hoisting personnel by crane or derrick in the construction industry, prescribes the measures employers must take to bring their work operations into compliance, and describes safe work practices for employees; but it is not a substitute for the actual OSHA rule.

The OSHA rule prohibits hoisting personnel by crane or derrick except when no safe alternative is possible. Based on the review of the record, OSHA determined that hoisting with crane- or derrick-suspended personnel platforms constitutes a significant hazard to hoisted employees and must not be permitted unless conventional means of transporting employees are not feasible or unless they present greater hazards. OSHA determined that compliance with the provisions of this standard will provide the best available protection for personnel being hoisted by these platforms in those limited situations where such hoisting is necessary.

Where conventional means (e.g., scaffolds, ladders) of access would not be considered safe, personnel hoisting operations, which comply with the terms of this standard, would be authorized. OSHA stresses that employee safety-not practicality or convenience must be the basis for the employer's choice of method.

Cranes and derricks used to hoist personnel must be placed on a firm foundation and the crane or derrick must be uniformly level within 1 percent of level grade.

The crane operator must always be at the controls when the crane engine is running and the personnel platform is occupied. The crane operator also must have full control over the movement of the personnel platform. Any movement of the personnel platform must be performed slowly and cautiously without any sudden jerking of the crane, derrick, or the platform. Wire rope used for personnel lifting must have a minimum safety factor of seven. (This means it must be capable of supporting seven times the maximum intended load.) Rotation resistant rope must have a minimum safety factor of ten.

When the occupied personnel platform is in a stationary position, all brakes and locking devices on the crane or derrick must be set.

The combined weight of the loaded personnel platform and its rigging must not exceed 50 percent of the rated capacity of the crane or derrick for the radius and configuration of the crane or derrick.

22.6.1 Instruments and Components

Cranes and derricks with variable angle booms must have a boom angle indicator that is visible to the operator. Cranes with telescoping booms must be equipped with a device to clearly indicate the boom's extended length, or an accurate determination of the load radius to be used during the lift must be made prior to hoisting personnel. Cranes and derricks also must be equipped with (1) an anti-two-blocking device that prevents contact between the load block and overhaul ball and the boom tip, or (2) a two-block damage-prevention feature that deactivates the hoisting action before damage occurs.

22.6.2 Personnel Platforms

Platforms used for lifting personnel must be designed with a minimum safety factor of five and designed by a qualified engineer or a qualified person competent in structural design. The suspension system must be designed to minimize tipping due to personnel movement on the platform.

Each personnel platform must be provided with a standard guardrail system that is enclosed from the toeboard to the mid-rail to keep tools, materials, and equipment from falling on employees below. The platform also must have an inside grab rail, adequate headroom for employees, and a plate or other permanent marking that clearly indicates the platform's weight and rated load capacity or maximum intended load. When personnel are exposed to falling objects, overhead protection on the platform and the use of hard hats are required.

An access gate, d provided, must not swing outward during hoisting and must have a restraining device to prevent accidental opening. All rough edges on the platform must be ground smooth to prevent injuries to employees.

All welding on the personnel platform and its components must be performed by a qualified welder who is familiar with weld grades, types, and materials specified in the platform design.

22.6.3 Loading

The personnel platform must not be loaded in excess of its rated load capacity or its minimum intended load. Only personnel instructed in the requirements of the standard and the task to be performed-along with their tools, equipment, and materials needed for the job-are allowed on the platform. Materials and tools must be secured and evenly distributed to balance the load while the platform is in motion.

22.6.4 Rigging

When a wire rope bridle is used to connect the platform to the load line, the bridle legs must be connected to a master link or shackle so that the load is evenly positioned among the bridle legs. Bridles and associated rigging for attaching the personnel platform to the hoist line must not be used for any other purpose.

Attachment assemblies such as hooks must be closed and locked to eliminate the hook throat opening; an alloy anchor-type shackle with a bolt, nut, and retaining pin may be used as an alternative. "Mousing" (wrapping wire around a hook to cover the hook opening) is not permitted.

22.6.5 Inspecting and Testing

A trial lift of the unoccupied personnel platform must be made before any employees are allowed to be hoisted. During the trial lift, the personnel platform must be loaded at least to its anticipated lift weight. The lift must start at ground level or at the location where employees will enter the platform and proceed to each location where the personnel platform is to be hoisted and positioned. The trial lift must be performed immediately prior to placing personnel on the platform.

The crane or derrick operator must check all systems, controls, and safety devices to ensure the following:

- They are functioning properly.
- There are no interferences.
- All boom or hoisting configurations necessary to reach work locations will allow the operator to remain within the 50-percent load limit of the hoist's rated capacity.

If a crane or derrick is moved to a new location or returned to a previously used one, the trial lift must be repeated before hoisting personnel.

After the trial lift, the personnel platform must be hoisted a few inches and inspected to ensure that it remains secured and is properly balanced.

Before employees are hoisted, a check must be made to ensure the following:

- Hoist ropes are free of kinks.
- Multiple part lines are not twisted around each other.
- The primary attachment is centered over the platform.
- There is no slack in the wire rope.
- All ropes are properly seated on drums and in sheaves.

Immediately after the trial lift, a thorough visual inspection of the crane or derrick, the personnel platform, and the crane or derrick base support or ground must be conducted by a competent person to determine if the lift test exposed any defects or produced any adverse effects on any component or structure. Any defects found during inspections must be corrected before hoisting personnel.

When initially brought to the job site and after any repair or modification, and prior to hoisting personnel, the platform and rigging must be proof tested to 125 percent of the platform's rated capacity. This is achieved by holding the loaded platform-with the load evenly distributed-in a suspended position for 5 minutes. Then a competent person must inspect the platform and rigging for defects. If any problems are detected, they must be corrected and another proof test must be conducted. Personnel hoisting must not be conducted until the proof testing requirements are satisfied.

22.6.6 Pre-Lift Meeting

The employer must hold a meeting with all employees involved in personnel hoisting operations (crane or derrick operator, signal person(s), employees to be lifted, and the person responsible for the hoisting operation) to review the OSHA requirements and the procedures to be followed before any lift operations are performed. This meeting must be held before the trial lift at each new work site and must be repeated for any employees newly assigned to the operation.

22.6.7 Safe Work Practices

Employees must follow these safe work practices:

- Use tag lines unless their use creates an unsafe condition.
- Keep all body parts inside the platform during raising, lowering, and positioning.
- Make sure a platform is secured to the structure where work is to be performed before entering or exiting it, unless such securing would create an unsafe condition.
- Wear a body belt or body harness system with a lanyard. The lanyard must be attached to the lower load block or overhaul ball or to a structural member within the personnel platform. If the hoisting operation is performed over water, the requirements 29 CFR *1926.106- Working over or near water*-must apply.
- Stay in view of, or in direct communication with, the operator or signal person.

Crane and derrick operators must follow these safe work practices:

- Never leave crane or derrick controls when the engine is running or when the platform is occupied.

- Stop all hoisting operations if there are indications of any dangerous weather conditions or other impending danger.
- Do not make any lifts on another load line of a crane or derrick that is being used to hoist personnel.

22.6.8 Movement of Cranes

Personnel hoisting is prohibited while the crane is traveling except when the employer demonstrates that this is the least hazardous way to accomplish the task or when portal, tower, or locomotive cranes are used.

When cranes are moving while hoisting personnel, the following rules apply:

- Travel must be restricted to a fixed track or runway.
- Travel also must be limited to the radius of the boom during the lift.
- The boom must be parallel to the direction of travel.
- There must be a complete trial run before employees occupy the platform.
- If the crane has rubber tires, the condition and air pressure of the tires must be checked and the chart capacity for lifts must be applied to remain under the 50-percent limit of the hoist's rated capacity. Outriggers may be partially retracted as necessary for travel.

Compliance with the common-sense requirements of the OSHA standard and the determination that no other safe method is available should greatly reduce or eliminate the injuries and accidents that occur too frequently during personnel hoisting operations.

Summary

Construction work can be made safer if good work practices are implemented on the job. The current practices of capping exposed rebar can prevent serious injury of steel and concrete workers. In addition, removal of concrete forms on schedule, proper bracing of masonry walls, and keeping loads off green concrete all contribute to improved safety.

OSHA has continued to enforce the electrical ground-fault protection codes with numerous citations. Daily safety inspections of the construction site are needed to maintain a safe working environment.

Crane and derrick operations continue to cause injuries at construction sites along with other material handling and storage operations. Employers must provide the training and follow-up enforcement of good work practices.

Bibliography

National Safety Council (1991) "Accident Facts", Chicago, IL.

OSHA "Excavations", OSHA 226. Order No. 029-016-00176-1, Department of Labor, Washington, D.C.

OSHA (1991) "Ergonomics: The Study of Work", Publication No. 3125, U.S. Department of Labor, Washington, D.C.

OSHA (1997) "Ground-Fault Protection on Construction Sites", U.S. Department of Labor, Washington, D.C.

OSHA (1998) "Concrete and Masonry Construction" Booklet 3106, U.S. Department of Labor, Washington, D.C.

23

Hazardous Waste Operations

THE DUMPING of hazardous waste poses a significant threat to the environment. The Environmental Protection Agency's (EPA) 1995 data show that EPA managed about 277 million metric tons of hazardous waste at licensed Resource Conservation and Recovery Act (RCRA) sites. Hazardous waste is a serious safety and health problem that continues to endanger human and animal life and environmental quality. Hazardous waste, discarded chemicals that are toxic, flammable or corrosive, can cause fires, explosions, and pollution of air, water, and land. Unless hazardous waste is properly treated, stored, or disposed of, it will continue to do great harm to all living things that come into contact with it now or in the future.

Because of the seriousness of the safety and health hazards related to hazardous waste operations, the Occupational Safety and Health Administration (OSHA) issued its Hazardous Waste Operations and Emergency Response Standard, Title 29 Code of Federal Regulations (CFR) Part 1910.120 (See Federal Register 54 (42): 9294-9336, March 6, 1989) to protect workers in this environment and to help them handle hazardous wastes safely and effectively. The standard covers workers in cleanup operations at uncontrolled hazardous waste sites and at EPA-licensed waste TSD facilities; as well as workers responding to emergencies involving hazardous materials.

State, county, and municipal employees such as police, ambulance workers, and firefighters with local fire departments will be covered by the regulations issued by the states operating their own OSHA-approved safety and health programs. EPA regulations will cover these employees in states without state plans. These regulations will be based on OSHA's standard.

This chapter discusses OSHA's requirements for hazardous waste operations and emergency response at uncontrolled hazardous waste sites and treatment, storage, and disposal (TSD) facilities and summarizes the steps an employer must take to protect the health and safety of workers in these environments.

23.1 Provisions of the Standard

23.1.1 Safety and Health Program

An effective and comprehensive safety and health program is essential in reducing work-related injuries and illnesses and in maintaining a safe and healthful work environment. The standard, therefore, requires each employer to develop and implement a written safety and health program that identifies, evaluates, and controls safety and health hazards and provides emergency response procedures for each hazardous waste site or treatment, storage, and disposal facility. This written program must include specific and detailed information on the following topics:

- an organizational workplan
- site evaluation and control
- a site-specific program
- information and training program
- personal protective equipment program
- monitoring
- medical surveillance program
- decontamination procedures
- emergency response program

The written safety and health program must be periodically updated and made available to all affected employees, contractors, and subcontractors. The employer also must inform contractors and subcontractors, or their representatives, of any identifiable safety and health hazards or potential fire or explosion hazards before they enter the work site. Each of the components of the safety and health program is discussed in the following paragraphs.

23.1.2 Workplan

Planning is the key element in a hazardous waste control program. Proper planning will greatly reduce worker hazards at waste sites. A workplan should support the overall objectives of the control program and provide procedures for implementation and should incorporate the employer's standard operating procedures for safety and health. Establishing a chain of command will specify employer and employee responsibilities in carrying out the safety and health program. For example, the plan should include the following:

- Supervisor and employee responsibilities and means of communication.
- Name of person who supervises all of the hazardous waste operations.
- The site supervisor with responsibility for and authority to develop and implement the site safety and health program and to verify compliance.

In addition to this organizational structure, the plan should define the tasks and objectives of site operation as well as the logistics and resources required to fulfill these tasks. For example, the following topics should be addressed:

- the anticipated clean-up and/or operating procedures
- a definition of work tasks and objectives and methods of accomplishment
- the established personnel requirements for implementing the plan
- procedures for implementing training, informational programs, and medical surveillance requirements

Necessary coordination between the general program and site-specific activities also should be included in the actual operations workplan.

23.1.3 Site Evaluation and Control

Site evaluation, both initial and periodic, is crucial to the safety and health of workers. Site evaluation provides employers with the information needed to identify site hazards so they can select appropriate protection methods for employees.

It is extremely important, and a requirement of the standard, that a trained person conduct a preliminary evaluation of an uncontrolled hazardous waste site before entering the site. The evaluation must include all suspected conditions that are immediately dangerous to life or health or that may cause serious harm to employees (e.g., confined space entry, potentially explosive or flammable situations, visible vapor clouds, etc.). As available, the evaluation must include the location and size of the site, site topography, site accessibility by air and roads, pathways for hazardous substances to disperse, a description of worker duties, and the time needed to perform a given task, as well as the present status and capabilities of the emergency response teams. Periodic reevaluations should also be conducted for treatment, storage, and disposal facilities, as conditions or operations change.

Controlling the activities of workers and the movement of equipment is an important aspect of the overall safety and health program. Effective control of the site will minimize potential contamination of workers, protect the public from hazards, and prevent vandalism. The following information is useful in implementing the site control program: a site map, site work zones, site communication, safe work practices, and the name, location and phone number of the nearest medical assistance.

The use of a "buddy system" also is required as a protective measure to assist in the rescue of an employee who becomes unconscious, trapped, or seriously disabled on site. In the buddy system, two employees must keep an eye on each other and only one should be in a specific dangerous area at one time, so that if one gets in trouble, the second can call for help.

23.1.4 Site-Specific Safety and Health Plan

A site-specific safety and health plan is a complementary program element that aids in eliminating or effectively controlling anticipated safety and health hazards. The site-specific plan must include all of the basic requirements of the overall safety and health program, but with attention to those characteristics unique to the particular site. For example, the site-specific plan may outline procedures for confined space entry, air and personal monitoring and environmental sampling, and a spill containment program to address the particular hazards present at the site.

The site safety and health plan must identify the hazards of each phase of the specific site operation and must be kept at the work site. Pre-entry briefings must be conducted prior to site entry and at other times as necessary to ensure that employees are aware of the site safety and health plan and its implementation. The employer also must ensure that periodic safety and health inspections are made of the site and that all known deficiencies are corrected prior to work at the site.

23.1.5 Information and Training Program

As part of the safety and health program, employers are required to develop and implement a program to inform workers (including contractors and subcontractors) performing hazardous waste operations of the level and degree of exposure they are likely to encounter. Employers also are required to develop and implement procedures for introducing effective new technologies that provide improved worker protection in hazardous waste operations. Examples include foams, absorbents, adsorbents, and neutralizers.

Training makes workers aware of the potential hazards they may encounter and provides the necessary knowledge and skills to perform their work with minimal risk to their safety and health. The employer must develop a training program for all employees exposed to safety and health hazards during hazardous waste operations. Both supervisors and workers must be trained to recognize hazards and to prevent them; to select, care for and use respirators properly as well as other types of personal protective equipment; to understand engineering controls and their use; to use proper decontamination procedures; and to understand the emergency response plan, medical surveillance requirements, confined space entry procedures, spill containment program, and any appropriate work practices. Workers also must know the names of personnel and their alternates responsible for site safety and health. The amount of instruction differs with the nature of the work operations, as indicated in Tables 23. 1 and 23.2.

Employees at all sites must not perform any hazardous waste operations unless they have been trained to the level required by their job function and responsibility and have been certified by their instructor as having completed the necessary training. All emergency responders must receive refresher training, sufficient to maintain or demonstrate competency, annually. Employee training requirements are further defined by the nature of the work, for example, temporary emergency response personnel, firefighters, safety officers, HAZMAT personnel, and incident commanders. These requirements may include recognizing and knowing the hazardous materials and their risks, knowing how to select and use appropriate personal protective equipment, and knowing the appropriate control, containment, or confinement procedures and how to implement them. The specific training and competency requirements for each personnel category are explained fully in the final rule (FR54 42:9294, March 6, 1989). For a brief summary of training requirements, see Tables 23.1 and 23.2.

Employees who receive the training specified (see Table 23.1) must receive a written certificate upon successful completion of that training. That training need not be repeated if the employee goes to work at a new site; however, the employee must receive whatever additional training is needed to work safely at the new site. Employees who worked at hazardous waste sites before 1987 and received equivalent training need not repeat the initial training specified in Table 23.1, if the employer can demonstrate that in writing and certify that the employee has received such training.

TABLE 23.1
Training Requirements at Hazardous Waste Clean-Up Sites

Position		Training Requirement
Staff	Routine Site Employees	40 hours initial 24 field hours 8 hours annual refresher
	Routine site employees (minimal exposure)	24 hours initial 8 hours field 8 hours annual refresher
	Non-routine site employees	24 hours initial 8 hours field 8 hours annual refresher
Supervisors/ Managers of:	Routine site employees	40 hours initial 24 hours field 8 hours hazardous waste management 8 hours annual refresher
	Routine site employees (minimal exposure)	24 hours initial 8 hours field 8 hours hazardous waste management 8 hours annual refresher
	Non-routine site employees	24 hours initial 8 hours field 8 hours hazardous waste management 8 hours annual refresher
Treatment, Storage, and Disposal Sites Staff	General Site employees	24 hours initial or equivalent 8 hours annual refresher
	Emergency response personnel	Trained to level of competency Annual refresher

Note: See 29 CFR 1910.120 (e) and (p)(7).

23.1.6 Personal Protective Equipment Program

The standard further requires the employer to develop a written personal protective equipment program for all employees involved in hazardous waste operations. As mentioned earlier, this program also is part of the site-specific safety and health program. The personal protective equipment program must include an explanation of equipment selection and use, maintenance and storage, decontamination and disposal, training and proper fit, donning and doffing procedures, inspection, in-use monitoring, program evaluation, and equipment limitations.

The employer also must provide and require the use of personal protective equipment where engineering control methods are infeasible to

TABLE 23.2

Training Requirements of Emergency Response Staff

Other Emergency Response Staff	Training Requirement
Level 1 – First responder (awareness) level, i.e., witness or discovers a release of hazardous materials and who are trained to notify the proper authorities	Sufficient training or proven experience in specific competencies. Annual refresher
Level 2 – First responder (operations level) i.e., responds to the releases of hazardous substances in a defensive manner, without trying to stop the releases.	Level 1 competency and 8 hours initial or proven experience in specific competencies. Annual refresher.
Level 3 – HAZMAT technician i.e., responds aggressively to stop the release of hazardous substances.	24 hours of Level 2 and proven experience in specific competencies. Annual refresher.
Level 4 – HAZMAT specialist, i.e., responds in support to HAZMAT technicians, but who have specific knowledge of various hazardous substances.	24 hours of Level 3 and proven experience in specific competencies Annual refresher
Level 5 – On – the – scene incident commander, i.e., assumes control over the incident scene beyond the first-responder awareness level.	24 hours of Level 2 and additional competencies. Annual refresher

Note: See 29 CFR 1910.120 (q) (6).

reduce worker exposures at or below the permissible exposure limit. Personal protective equipment must be selected that is appropriate to the requirements and limitations of the site, the task-specific conditions and duration, and the hazards and potential hazards identified at the site. As necessary, the employer must furnish the employee with positive-pressure self-contained breathing apparatus or positive-pressure airline respirators equipped with an escape air supply, and with totally encapsulating chemical protective suits.

23.1.7 Monitoring

Airborne contaminants can present a significant threat to employee safety and health, thus making air monitoring an important component of an effective safety and health program. The employer must conduct monitoring before site entry at uncontrolled hazardous waste sites to identify conditions immediately dangerous to life and health, such as oxygen-deficient atmospheres and areas where toxic substance exposures are above permissible limits. Accurate information on the identification and quantification of airborne contaminants is useful for the following:

- selecting personal protective equipment
- delineating areas where protection and controls are needed
- assessing the potential health effects of exposure
- determining the need for specific medical monitoring

After a hazardous waste cleanup operation begins, the employer must periodically monitor those employees who are likely to have higher exposures to determine if they have been exposed to hazardous substances in excess of permissible exposure limits. The employer also must monitor for any potential condition that is immediately dangerous to life and health or for higher exposures that may occur as a result of new work operations.

23.1.8 Medical Surveillance

A medical surveillance program will help to assess and monitor the health and fitness of employees working with hazardous substances. The employer must establish a medical surveillance program for the following individuals:

- all employees exposed or potentially exposed to hazardous substances or health hazards above the permissible exposure limits for more than 30 days per year
- workers exposed above the published exposure levels (if there is no permissible exposure limit for these substances) for 30 days or more a year
- workers who wear approved respirators for 30 or more days per year on site
- workers who are exposed to unexpected or emergency releases of hazardous wastes above exposure limits (without wearing appropriate protective equipment) or who show signs, symptoms, or illness that may have resulted from exposure to hazardous substances
- members of hazardous materials (HAZMAT) teams

All examinations must be performed under the supervision of a licensed physician, without cost to the employee, without loss of pay and at a reasonable time and place. Examinations must include a medical and work history with special emphasis on symptoms related to the handling of hazardous substances and health hazards and to fitness for duty including the ability to wear any required personal protective equipment under conditions that may be expected at the work site. These examinations must be given:

- prior to job assignment and annually thereafter (or every 2 years if a physician determines that is sufficient)
- at the termination of employment
- before reassignment to an area where medical examinations are not required
- if the examining physician believes that a periodic follow-up is medically necessary
- as soon as possible for employees injured or becoming ill from exposure to hazardous substances during an emergency, or who develop signs or symptoms of overexposure from hazardous substances

The employer must give the examining physician a copy of the standard and its appendices, a description of the employee's duties relating to his or her exposure, the exposure level or anticipated exposure level, a description of any personal protective and respiratory equipment used or to be used, and any information from previous medical examinations. The employer must obtain a written opinion from the physician that contains the results of the medical examination and any detected medical conditions that would place the employee at an increased risk from exposure, any recommended limitations on the employee or upon the use of personal protective equipment, and a statement that the employee has been informed by the physician of the results of the medical examination. The physician is not to reveal, in the written opinion given to the employer, specific findings or diagnoses unrelated to employment.

23.1.9 Decontamination Procedures

Decontamination procedures are a component of the site-specific safety and health plan and, consequently, must be developed, communicated to employees, and implemented before workers enter a hazardous waste site. As necessary, the site safety and health officer must require and monitor decontamination of the employee or decontamination and disposal of the employee's clothing and equipment, as well as the solvents used for decontamination, before the employee leaves the work area. If an employee's non-impermeable clothing becomes grossly contaminated with hazardous substances, the employee must immediately remove that clothing and take a shower. Impermeable protective clothing must be decontaminated before being removed by the employee.

Protective clothing and equipment must be decontaminated, cleaned, laundered, maintained, or replaced to retain effectiveness. The employer must inform any person who launders or cleans such clothing or equipment of the potentially harmful effects of exposure to hazardous substances.

Employees who are required to shower must be provided showers and change rooms that meet the requirements of 29 CFR 1910.141, Subpart J - General Environmental Controls. In addition, unauthorized employees must not remove their protective clothing or equipment from change rooms unless authorized to do so.

23.1.10 Emergency Response

Proper emergency planning and response are important elements of the safety and health program that help minimize employee exposure and injury. The standard requires that the employer develop and implement a written emergency response plan to handle possible emergencies before performing hazardous waste operations. The plan must include, at uncontrolled hazardous waste sites and at treatment, storage, and disposal facilities, the following elements:

- personnel roles, lines of authority, and communication procedures
- pre-emergency planning
- emergency recognition and prevention
- emergency medical and first-aid treatment
- methods or procedures for alerting onsite employees
- safe distances and places of refuge
- site security and control
- decontamination procedures
- critique of response and follow-up
- personal protective and emergency equipment
- evacuation routes and procedures

In addition to the above requirements, the plan must include site topography, layout, and prevailing weather conditions; and procedures for reporting incidents to local, state, and federal government agencies.

The procedures must be compatible with and integrated into the disaster, fire and/or emergency response plans of the site's nearest local, state, and federal agencies. Emergency response organizations may use the local or state emergency response plans, or both, as part of their emergency response plan to avoid duplication of federal regulations.

The plan requirements also must be rehearsed regularly, reviewed periodically, and amended, as necessary, to keep them current with new or changing site conditions or information. A distinguishable and distinct alarm system must be in operation to notify employees of emergencies. The emergency plan also must be made available for inspection and copying by employees, their representatives, OSHA personnel, and other governmental agencies with relevant responsibilities.

When deemed necessary, employees must wear positive-pressure self-contained breathing apparatus and approved self-contained compressed-air breathing apparatus with approved cylinders. In addition, back-up and first-aid support personnel must be available for assistance or rescue.

As already indicated, as part of an effective safety and health program, the employer must institute control methods and work practices that are appropriate to the specific characteristics of the site. Such controls are essential to successful worker protection. Some control methods are described in the following paragraphs.

23.2 Other Provisions

23.2.1 Engineering Controls and Work Practices

To the extent feasible, the employer must institute engineering controls and work practices to help reduce and maintain employee exposure at or below permissible exposure limits. To the extent not feasible, engineering and work practice controls may be supplemented with personal protective equipment. Examples of suitable and feasible engineering controls include the use or pressurized cabs or control booths on equipment, and/or remotely operated materials handling equipment. Examples of safe work practices include removing all non-essential employees from potential exposure while opening drums, wetting down dusty operations, and placing employees upwind of potential hazards.

23.2.2 Handling and Labeling Drums and Containers

Prior to handling a drum or container, the employer must assure that drums or containers meet the required OSHA, EPA (40 CFR Parts 264-265 and 300), and Department of Transportation (DOT) regulations (49 CFR Parts 171-178), and are properly inspected and labeled. Damaged drums or containers must be emptied of their contents, using a device classified for the material being transferred, and must be properly discarded. In areas where spills, leaks or ruptures occur, the employer must furnish employees with salvage drums or containers, a suitable quantity of absorbent material, and approved fire-extinguishing equipment in the event of small fires.

The employer also must inform employees of the appropriate hazard warnings of labeled drums, the removal of soil or coverings, and the dangers of handling unlabeled drums or containers without prior identification of their contents. To the extent feasible, the moving of drums or containers must be kept to a minimum, and a program must be implemented to contain and isolate hazardous substances being transferred into drums or containers. In addition, an approved EPA ground-penetrating device must be used to determine the location and depth of any improperly discarded drums or containers.

The employer also must ensure that safe work practices are instituted before opening a drum or container. For example, airline respirators and approved electrical equipment must be protected from possible contamination, and all equipment must be kept behind any existing explosion barrier.

Only tools or equipment that prevent ignition shall be used. All employees not performing the operation shall be located at a safe distance and behind a suitable barrier to protect them from accidental explosions. In addition, standing on or working from drums or containers is prohibited. Special care also must be given when an employee handles containers of shock-sensitive waste, explosive materials, or laboratory waste packs. Where an emergency exists, the employer must ensure the following:

- evacuate non-essential employees from the transfer area; protect equipment operators from exploding containers by using a barrier
- make available a continuous means of communication (e.g., suitable radios or telephones), and a distinguishable and distinct alarm system to signal the beginning and end of activities where explosive wastes are handled

If drums or containers bulge or swell or show crystalline material on the outside, they must not be moved onto or from the site unless appropriate containment procedures have been implemented. In addition, lab packs must be opened only when necessary and only by a qualified person. Prior to shipment to a licensed disposal facility, all drums or containers must be properly labeled and packaged for shipment. Staging areas also must be kept to a minimum and provided with adequate access and egress routes.

23.2.3 Sanitation of Temporary Workplaces

Each temporary worksite must have a supply of potable water that is stored in tightly closed and clearly labeled containers and equipped with a tap. Disposable cups and a receptacle for cup disposal also must be provided. The employer also must clearly mark all water outlets that are unsafe for drinking, washing, or cooking. Temporary worksites must be equipped with toilet facilities. If there are no sanitary sewers close to or on the hazardous waste site, the employer must provide the following toilet facilities unless prohibited by local codes:

- privies
- chemical toilets
- re-circulating toilets
- combustion toilets

Heated, well-ventilated, and well-lighted sleeping quarters must be provided for workers who guard the worksite. In addition, washing facilities for all workers must be near the worksite, within controlled work zones, and so equipped to enable employees to remove hazardous substances. The employer also must ensure that food service facilities are licensed.

22.3 Record Keeping

In 1988, OSHA revised the standard requiring employers to provide employees with information to assist in the management of their own safety and health. The standard, Access to Employee Exposure and Medical Records (29 CFR 1910.20), permits direct access to these records by employees exposed to hazardous materials, or by their designated representatives, and by OSHA. The rule applies to, but does not require, medical and exposure records maintained by the employer.

The employer must keep exposure records for 30 years and medical records for at least the duration of employment plus 30 years. Records of employees who have worked for less than 1 year need not be retained after employment, but the employer must provide these records to the employee upon termination of employment. First-aid records of one-time treatment need not be retained for any specified period.

The employer must inform each employee of the existence, location, and availability of these records. Whenever an employer plans to stop doing business and there is no successor employer to receive and maintain these records, the employer must notify employees of their right to access to records at least 3 months before the employer ceases to do business. At the same time, employers also must notify the National Institute for Occupational Safety and Health.

Under the hazardous waste standard, at a minimum, medical records must include the following information:

- employee's name and social security number
- physicians' written opinions
- employee's medical complaints related to exposure to hazardous substances
- information provided to the treating physician

Title III of the Superfund Amendments and Reauthorization Act of 1986 (SARA) requires employers covered by the Hazard Communication Standard (HCS) (29 CFR 1910.1200) to maintain Material Safety Data Sheets (MSDSs) and submit such information to State emergency response commissions, local emergency planning committees, and the local fire department. Under this requirement, employers covered by HCS must provide chemical hazard information to both employees and surrounding communities. Consequently, in the case of an emergency response situation to hazardous substances at a site, the local fire department may already be aware of the chemicals present at the site since data may have been provided through MSDSs.

Summary

Hazardous wastes, when not handled properly, can pose a significant safety and health risk. OSHA recognizes the need to improve the quality of the hazardous waste work environment and has, therefore, issued this standard. This standard provides employers and employees with the information and training necessary to improve workplace safety and health, thereby greatly reducing the number of injuries and illnesses resulting from exposure to hazardous waste.

Bibliography

Hazard Communication Guidelines for Compliance, OSHA 3111.

Department of Transportation (1987) "Emergency Response Guidebook", Publication No. DOT – P – 5800-4.

Martin, William F., J. M. Lippitt, P.J. Webb, (2000) "Hazardous Waste Handbook for Health and Safety", Butterworth-Heinemann Publishers, Boston, MA.

Martin, William F., M. Gochfeld, eds., (2000) "Protecting Personnel at Hazardous Waste Sites", Butterworth-Heinemann Publishers, Boston, MA.

NIOSH, "Occupational Safety and Health Guidance Manual for Hazardous Waste Site Activities", NIOSH/OSHA/USCG/EPA. Publication No. DHHS (NIOSH) No. 85-115.

OSHA (1997) "Hazardous Waster Operations and Emergency Response", U.S. Department of Labor, Washington, D.C.

U.S. Environmental Protection Agency (1995) Office of Solid Waste and Emergency Response, "The Hazardous Waste System", Washington, DC.

24

Noise and Hearing Conservation

EXCESSIVE NOISE in the occupational environment can lead to a variety of undesirable effects. Probably the most obvious and best quantified of these effects is noise induced hearing loss. However, other adverse consequences of occupational noise have been documented including; interference with communication (e.g. masking verbal messages or other audible signals necessary to perform a job), creation of safety problems (e.g. masking audible warning signals), contributions to poor job performance (e.g. producing a sense of apathy or helplessness), and a host of other detrimental effects such as annoyance, fatigue, irritability, and sleeplessness. Additionally, various human studies as well as certain animal studies have implicated high noise levels as a contributor to stress related disorders such as cardiovascular disease.

Obviously, addressing all these difficulties brought on by excessive noise levels in the occupational environment is a matter of utmost importance to you as a small business owner/manager. Of particular concern is the preservation of your employees' hearing. The key to managing these noise-related problems is the development, implementation, and proper management of an effective hearing conservation program.

Noise control and hearing conservation are the topics of discussion in this training module. As with other modules in this series, this subject-matter will be considered in conjunction with the applicable Occupational Health and Safety Administration (OSHA) standard; 29 CFR 1910.95, Occupational Noise Exposure and the corresponding Hearing Conservation amendment.

Although this chapter is presented in the context of the OHSA Noise and Hearing Conservation Standards, it is important to understand that simply complying with this standard will not insure success in preventing noise induced hearing loss in your employees. An effective Hearing Conservation Program (HCP) depends, to a large extent, on the level of commitment displayed by the small business owner/manager. A strong commitment to hearing conservation often makes it necessary to go beyond the minimal requirements set forth in the standard. The complexity of noise measurement usually requires that a small business use an industrial hygienists consultant to set up a hearing conservation program.

24.1 OSHA Standard for Occupational Noise Exposure

The OSHA standard for noise exposure, 29 CFR 1910.95, sets forth maximum sound pressure levels allowable for continuous exposures to noise as a function of the exposure duration within a work day. In accordance with the damage-risk model utilized by OSHA, the maximum continuous sound pressure level to which an unprotected worker may be exposed for eight hours per day is 90 decibels as measured on the A scale of a standard sound level meter (i.e. 90 dBA) at slow response.

The 90 dBA criterion value is continuous noise exposure for eight hours. Most workplaces generate noise exposures that are not constant throughout the day. In these environments the noise level will fluctuate depending upon the processes occurring at a given point in time. For situations such as this, the 90 dBA exposure criterion may be exceeded (up to a maximum level of 115 dBA) for a period of time during the work shift provided that it is off-set by a sufficiently long period of exposure below 90 dBA. In essence, the 90 dBA becomes an average noise exposure over an eight-hour period and represents a specific amount of accumulated noise energy over that period.

Observe that the standard is expressed in terms of both a criterion noise level (90 dBA) and an exposure duration (eight hours). If an unprotected employee's average exposure is greater than 90 dBA for eight hours, you must compensate for the criterion exceedance with a shorter exposure duration. Maximum continuous noise levels for exposure durations shorter than eight hours are provided in the standard. For example, if the average daily noise exposure is 92 dBA, the maximum permissible exposure time for unprotected employees is six hours. At that point this employee has received his/her maximum allowable noise exposure for the day and must spend the remainder of the shift in a quieter environment (i.e. less than 80 dBA).

By the same token, you may also compensate for a longer exposure durations with a reduction in the average level of noise exposure. This is especially important if your workshifts at your facility are in excess of eight hours. If, for example, an unprotected employee works an extended shift of 10. 6 hours, the standard stipulates that the average sound pressure level must not exceed 87 dBA over that period. Again, this 87 dBA is an average noise level over that 10.6 hours with periods with appropriate periods of exposure less than 87 dBA compensating for those periods exceeding 87 dBA. These standards may change due to new research so employers must check the latest 29 CFR 1910.95 annually.

The combination of a 90 dBA noise level and eight hours duration is defined as a 100 % noise dose. All other combinations of noise level and exposure duration are compared to 90 dBA for eight hours to determine the corresponding dose. The dose must be less than or equal to 100 % to achieve compliance.

24.2 Administrative and Engineering Controls

Exposures of employees to average noise levels and duration in excess of 90 dBA for eight hours (or 100% dose) triggers certain actions. When employees are subjected to sound exceeding OSHA guidelines, feasible administrative or engineering controls shall be utilized.

An administrative control is commonly interpreted to mean some change in the affected employees work schedule or job task(s) that would reduce his/her overall exposure. This usually involves a periodic rotation of employees from noisy jobs to less noisy ones such that their overall average daily exposure is reduced below the 90 dBA criterion. Other examples include conducting noisy operations during the second or third shift when fewer people are exposed and providing quiet areas for work breaks. Unfortunately, moving employees from one job to another or conducting noisy operation on off shifts is frequently difficult in a small business setting. Additionally, rotating employees in an out of noisy jobs may reduce the exposure and subsequent hearing loss for any single individual but may result in smaller hearing losses spread across many workers.

A broader interpretation of administrative control includes essentially any managerial decision that would ultimately result in lower noise exposures. This includes maintaining an effective equipment maintenance program, long term planning for noise control purposes (e.g. new equipment purchases, remodeling of existing equipment, or facilities), and enforced noise specifications for new equipment purchases.

For hearing conservation purposes, engineering controls are defined as any modification or replacement of equipment, or related physical change at the noise source or along the transmission path (with the exception of hearing protectors) that reduces the noise level at the employee's ear. Examples of the various categories of engineering controls include the following:

- substitution of noisy machines with quieter machines
- substitution of noisy processes with quieter processes
- vibration dampening
- reducing sound transmission through solids
- reducing sound produced by fluid flow
- isolating noise sources
- isolating the operator

Selection and implementation of engineering controls can be a complex process requiring input from a variety of personnel including equipment operators, engineers, maintenance personnel, and safety and industrial hygiene personnel. Extensive measurement of the noise source(s) including the determination of predominant frequencies is frequently required. For difficult noise control problems, it may be prudent to enlist the services of a noise control consultant to assist in measurement, design, and installation of noise controls.

Another important issue regarding engineering controls is economic feasibility. Engineering controls are technologically feasible for most noise sources but their economic feasibility must be determined on a case-by-case basis. If the noise sources contributing the most significantly to employees' exposures can be identified, the decision must then be made regarding the anticipated effectiveness and cost of the engineering control measures versus the cost of not controlling the source. These decisions are frequently not so clear cut and making them will likely be very challenging to the small business owner/manager.

24.3 Hearing Protective Devices

If the implementation of engineering and/or administrative controls fails to reduce the average shift-long noise exposure of unprotected employees below the 90 dBA/eight hours criterion, the employer is required by the standard to provide and insure the use of personal protective equipment that will reduce the noise exposure to the criterion level. The selection of adequate hearing protective devices, like the selection of any type of personal protective equipment, is not an arbitrary process. The employer must provide plugs, muffs, or a combination of these, for employees based on the specific noise environment in which the protector will be used.

24.4 Hearing Conservation Program

Just as an average exposure of 90 dBA for eight hours triggers mandatory control efforts, the implementation of an effective hearing conservation program (HCP) is required whenever employee noise exposures equal or exceed an 8-hour time-weighted average sound level (TWA) of 85 decibels.

24.4.1 Monitoring

The first component of the Hearing Conservation Program (HCP) is the development and implementation of a noise monitoring program. So much of noise control efforts and the HCP depend on an accurate assessment of the noise levels in the facility and the assessment of the noise exposures of employees. Noise exposure monitoring is conducted to:

- determine whether hazards to hearing exist
- determine whether noise presents a safety hazard by interfering with speech communication or the recognition of audible warning signal
- identify employees for inclusion in the HCP
- classify employees' noise exposures for prioritizing noise control efforts and defining and establishing hearing protection practices
- evaluate specific noise sources for noise control purposes
- evaluate noise control efforts

A variety of measurement techniques and strategies exist utilizing different types of equipment. These can be broken down into two broad categories; surveys conducted for the purpose of assessing an employee's noise exposure (dose) and surveys conducted to obtain information to be utilized for engineering purposes (engineering control determination and preventive maintenance assessment.).

If noise sources are relatively constant and workers are fairly stationary, an estimation of dose can be obtained with a basic sound level meter and a stopwatch. However, the standard explicitly states that "Where circumstances such as high worker mobility, significant variations in sound level, or a significant component of improve noise make area monitoring generally inappropriate, the employer shall use representative personal sampling to comply with the monitoring requirements of this paragraph unless the employer can show that area sampling produces equivalent results." This typically requires the use of an instrument called a dosimeter, an electronic device worn by the worker that automatically averages varying sound levels during a given time period.

An important decision to make is whether to purchase the equipment and develop the expertise necessary to conduct measurements "in house" or whether to hire the services of an outside consultant to assist in this effort. Even if a consultant is hired, the owner/manager must be able to provide information relative to production cycles, processes, exposure duration, numbers exposed, production environment, machinery, and specific operations to the person conducting the monitoring and interpreting the results. It is important to keep the monitoring portion of the HCP in perspective. An inordinate amount of resources should not be spent on the monitoring effort at the expense of other portions of the program. Information should be limited to that which is necessary to make appropriate HCP decisions.

Other requirements regarding the monitoring portion of the HCP stipulated in the standard include:

- Integrate all continuous, intermittent, and impulsive sound levels between 80 and 130 decibels in the noise measurements.
- Calibrate instruments used to measure noise exposure to ensure measurement accuracy.
- Repeat monitoring whenever production, processes, equipment or controls change in such a way as to increase noise exposures to the extent that additional employees may be exposed above the action level.
- Notify each employee exposed above the action level of the monitoring results.
- Allow the affected employee or their representatives the opportunity to observe measurements.

Perhaps the most important information gained from the monitoring effort is a knowledge of the personnel in your facility who are exposed in excess of the action level (85 dBA for 8 hours). These individuals are included in your HCP and will be the focus of several additional activities under this program including periodic hearing tests and special education and training efforts.

24.4.2 Audiometric Testing Program

Noise induced hearing loss occurs gradually over time and is essentially painless. Most likely, employees wouldn't notice any problem at all until a significant hearing loss has accumulated. Unfortunately, at this point the loss that has occurred is irreversible. For this reason, it's important to periodically test worker's hearing to detect early evidence of potential hearing loss. If necessary, prompt action can be taken to correct problems before a significant deterioration of hearing occurs.

The standard requires the small business to establish and maintain an audiometric testing program for all employees in your HCP (i.e. those exposed above the action level) at no cost to them. This program requires periodic hearing tests, called audiograms, to be conducted for these personnel. An audiogram is a test to determine an employee's "hearing threshold' at different frequencies. The hearing threshold is defined as the lowest sound a person with normal hearing can hear in quiet surroundings. A baseline audiogram is obtained for an employee up on initial assignment to a noisy area. Subsequent periodic audiograms can be compared to this baseline for early detection of hearing "threshold shifts". A threshold shift means that sounds must be louder at the various frequencies in order to be heard. This is an indication that hearing sensitivity is being lost at these frequencies and gives warning that it is time to act quickly before significant hearing loss occurs.

It is important to conduct audiograms quickly upon assignment in a noise hazardous area. Evidence has shown that hearing loss, while progressive, is not linear. In other words significant losses can occur early on in the noise exposure experience, even within the first year. Therefore, it is important to conduct audiograms early and then periodically thereafter.

In terms of audiometric testing frequency, OSHA's requirements are outlined as follows:

- Establish a baseline audiogram within six months of an employee's first exposure above the action level. If a mobile test van is used for audiometric services, the time frame for a baseline is extended to one year with the provision that the employee wear HPDs for any period exceeding six months after first exposure until the baseline audiogram is established.
- Conduct periodic audiograms at least annually thereafter for comparison to the baseline.

Audiometric tests must be proceeded by at least 14 hours without exposure to workplace noise. HPDs may be used to meet this requirement.

For maximum protection of the employees, audiograms should be performed on the following five occasions:

- pre-employment
- prior to initial assignment in an noisy work area
- annually as long as the employee is assigned to a noisy job (a time-weighted average exposure level of 85 to 100 dBA), or twice a year for employees with time-weighted average exposures over 100 dBA
- at the time of reassignment out of a noisy job
- at the termination of employment

The standard has additional requirements regarding the qualifications of individuals conducting and interpreting audiograms, evaluation of audiograms, follow-up procedures, standard threshold shifts, and audiometer calibrations.

Remember, the audiometric testing program is, in a sense, the "proof of the pudding" as to the effectiveness of the overall HCP. It indicates whether or not the business is achieving its overall goal of the prevention of significant hearing loss in your employees.

24.4.3 Hearing Protectors

Another vital part of the HCP is the provision of appropriate Hearing Protective Devices (HPDs). The wearing of HPDs is mandatory in certain situations. As noted earlier, HPDs are mandatory for personnel exposed in excess of the criterion of 90 dBA for eight hours. In addition, HPDs are required for any employee exposed to an 8-hour time weighted average of 85 dBA or greater who has been waiting for over six months for the mobile test van visit so that his/her baseline audiogram can be established.

Employees exposed to 85 dBA or above who have experienced a "standard threshold shift must also wear HPDS. A standard threshold shift is defined in the standard as a change in hearing threshold relative to the baseline audiogram of an average of 10 dB or more at 2000, 3000, and 4000 Hz in either ear. The standard also requires that you make HPDs available at no cost to all employees exposed to an 8-hour time weighted average of 85 dBA or greater.

Employees must be given an opportunity to select HPDs for their own use from a variety of suitable devices provided for them. HPDs selected must provide at least enough noise attenuation to reduce exposure to the 90 dBA/eight hours criterion. HPDs worm by employees with a standard threshold shift must attenuate the noise to an 8-hour time-weighted average of 85 dBA or below.

It is the employer's responsibility to ensure proper initial fitting of the devices as well as to supervise their correct use. Employers must train their employees in the use and care of HPDs and replace the devices as necessary. Employers must also periodically re-evaluate the attenuation capabilities of the devices in use to ensure their continued effectiveness and provide more effective devices where necessary.

The importance of the HPD phase of the HCP cannot be overemphasized. HPDs can be important in preventing hearing loss but only if they are properly selected and utilized. Management policies and the enforcement of those policies are critical in the success of the HPD phase of the program.

24.4.4 Training Program

The training program is perhaps the most important of all the HCP elements in ensuring the success of the overall program. Education and motivation is essential in enlisting the cooperation and support of the workers. Employees need to understand why it is in their best interest to cooperate in audiometric testing, in the wearing of HPDs, and in the other aspects of the program that require their participation. Failure here will most certainly lead to failure in the other aspects of the program. Management must emphasize the importance of the educational phase of the HCP by setting a high priority on and requiring attendance at regular hearing conservation training sessions.

The basic requirements set forth in the standard stipulate annual, updated training for each employee in the HCP. This training must include the effects of noise on hearing, the purpose of hearing protectors, the advantages, disadvantages, and attenuation of various types, and instruction on selection, use, and care of HPDS, and the purposes of audiometric testing and an explanation of the test procedures.

A variety of materials are available to assist employers in their education and training efforts. However, training programs are most effective when they contain more than just videos and pamphlets. Live presentations by speakers who are knowledgeable not only of hearing conservation issues in general but also of your company's HCP are usually the most effective and make the best impressions on the attendees.

24.4.5 Recordkeeping

The final aspect of the program addressed in the standard deals with recordkeeping, record retention, and access to records. Basically, employers are required to maintain employee exposure measurement for two years and audiometric test records for the duration of the affected employee's employment. Consult the OSHA Noise and Hearing Conservation Standard for additional recordkeeping details.

Summary

A hearing conservation program (HCP) must be implemented by a business if the work environment has excessive noise. The determination as to the need for a HCP usually requires the services of a person trained in noise measurement.

Hearing loss in employees exposed to levels above 85 dBA can occur early, thus audiograms should be conducted upon assignment. Employers must provide the hearing protection devised plus the training in their use.

Management's full support of a noise reduction program through engineering controls will go a long way toward protecting employee's hearing.

Bibliography

AIHA (1986) "Noise and Hearing Conservation Manual", Fourth Edition, Akron, Ohio.

American Conference of Governmental Industrial Hygienists (1990) Publication No. 9134, "Hearing Conservation Programs", Cincinnati, Ohio, 45240.

H. Pelton (1993) "Noise Control Management", Van Nostrand Reinhold, New York.

NIOSH (1990) "A Practical Guide to Effective Hearing Conservation Programs in the Workplace", USDHHS, CDC, Cincinnati, OH.

25

Confined Spaces

THE HAZARDS encountered and associated with entering and working in confined spaces are capable of causing bodily injury, illness, and death to the worker. Accidents occur among workers because of failure to recognize that a confined space is a potential hazard. It should therefore be considered that the most unfavorable situation exists in every case and that the danger of explosion, poisoning, and asphyxiation will be present at the onset of entry.

Before forced ventilation is initiated, information such as restricted areas within the confined space, voids, the nature of the contaminants present, the size of the space, the type of work to be performed, and the number of people involved should be considered. The ventilation air should not create an additional hazard due to re-circulation of contaminants, improper arrangement of the inlet duct, or by the substitution of anything other than fresh (normal) air (approximately 20.9% oxygen, 79.1% nitrogen by volume). The terms air and oxygen are sometimes considered synonymous. However, this is a dangerous assumption, since the use of oxygen in place of fresh (normal) air for ventilation will expand the limits of flammability and increase the hazards of fire and explosion.

An estimation of the number of workers potentially exposed to confined spaces would be difficult to produce. Based on a rough estimate of the percentage of workers that may work in confined spaces at some time, NIOSH estimates that millions of workers may be exposed to hazards in confined spaces each year.

25.1 Identifying Confined Spaces

If you are required to construct or work in a boiler, cupola, degreaser, furnace, pipeline, pit, pumping station, reaction or process vessel, septic tank, sewage digester, sewer, silo, storage tank, ship's hold, utility vault, vat, or similar type enclosure, you are working in a confined space. Figure

25.1 provides a few examples of confined spaces. A confined space is a space which has any one of the following characteristics:

- limited openings for entry and exit
- unfavorable natural ventilation
- not designed for continuous worker occupancy

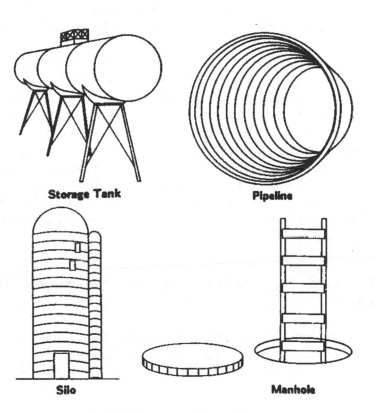

Storage Tank   **Pipeline**

Silo   **Manhole**

Fig. 25.1 Examples of Confined Spaces

Confined space openings are limited primarily by size or location. Openings are usually small in size, perhaps as small as 18 inches in diameter, and are difficult to move through easily. Small openings may make it very difficult to get needed equipment in or out of the spaces, especially protective equipment such as respirators needed for entry into spaces with hazardous atmospheres, or life-saving equipment when rescue is needed. However, in some cases openings may be very large, for example open-topped spaces such as pits, degreasers, excavations, and ships' holds. Access to open-topped spaces may require the use of ladders, hoists, or other devices, and escape from such areas may be very difficult in emergency situations.

Because air may not move freely in and out of confined spaces, the atmosphere inside a confined space can be very different from the atmosphere outside. Deadly gases may be trapped inside, particularly if the space is used to store or process chemicals or organic substances which may decompose. There may not be enough oxygen inside the confined space to support life, or the air could be so oxygen-rich that it is likely to increase the chance of fire or explosion if a source of ignition is present.

Most confined spaces are not designed for workers to enter and work in them on a routine basis. They are designed to store a product, enclose materials and processes, or transport products or substances. Therefore, occasional worker entry for inspection, maintenance, repair, cleanup, or similar tasks is often difficult and dangerous due to chemical or physical hazards within the space.

A confined space may have a combination of these three characteristics, complicating the work in and around them as well as rescue operations during emergencies. If a survey of the work area identifies one or more work spaces with the characteristics listed above, a confined space entry program must be developed.

25.2 Hazards Involved In Entering and Working in Confined Spaces

The atmosphere in a confined space may be extremely hazardous because of the lack of natural air movement. This characteristic of confined spaces can result in oxygen-deficient atmospheres, flammable atmospheres, and/or toxic atmospheres.

25.2.1 Oxygen-Deficient Atmospheres

An oxygen-deficient atmosphere has less than 19.5% available oxygen. Any atmosphere with less than 19.5% oxygen should not be entered without an approved self-contained breathing apparatus (SCBA).

The oxygen level in a confined space can decrease because of work being done, such as welding, cutting, or brazing; or, it can be decreased by certain chemical reactions (rusting) or through bacterial action (fermentation).

The oxygen level is also decreased if oxygen is displaced by another gas, such as carbon dioxide or nitrogen. Total displacement of oxygen by another gas, such as carbon dioxide, will result in unconsciousness, followed by death.

25.2.2 Flammable Atmospheres

Two things make an atmosphere flammable: the oxygen in air; and flammable gas, vapor, or dust in the proper mixture. Different gases have different flammable ranges. If a source of ignition (e.g., a sparking or electrical tool) is introduced into a space containing a flammable atmosphere, an explosion will result.

An oxygen-enriched atmosphere (above 21%) will cause flammable materials, such as clothing and hair, to burn violently when ignited. Therefore, never use pure oxygen to ventilate a confined space. Ventilate with normal air.

25.2.3 Toxic Atmosphere

Most substances (liquids, vapors, gases, mists, solid materials, and dusts) should be considered hazardous in a confined space. Toxic substances can come from the product stored in the space, the work being performed in the space, and the areas adjacent to the space.

The product stored in a confined space can be absorbed into the walls and give off toxic gases when removed or when cleaning out its residue. For example, the removal of sludge from a tank where the decomposed material gives off deadly hydrogen sulfide gas.

Examples of work being performed in a confined space include welding, cutting, brazing, painting, scraping, sanding, and degreasing. Toxic atmospheres are generated from various processes. For example, cleaning solvents are used in many industries for cleaning and degreasing. The vapors from these solvents are very toxic in a confined space.

Toxicants produced by work in the area of confined spaces can enter and accumulate in confined spaces.

25.3 Testing the Atmosphere

It is important to understand that some gases or vapors are heavier than air and will settle to the bottom of a confined space. Also, some gases are lighter than air and will be found around the top of the confined space.

Therefore, it is necessary to test all areas (top, middle, and bottom) of a confined space with properly calibrated testing instruments to determine what gases are present. If testing reveals oxygen-deficiency, or the presence of toxic gases or vapors, the space must be ventilated and re-tested before workers enter. If ventilation is not possible and entry is necessary (for emergency rescue, for example), workers must have appropriate respiratory protection.

Workers should never trust their senses to determine if the air in a confined space is safe. Many toxic gases and vapors are colorless and orderless. It is also impossible to determine the level of oxygen present by using ones senses.

25.4 Ventilation

Ventilation by a blower or fan may be necessary to remove harmful gases and vapors from a confined space. There are several methods for ventilating a confined space. The method and equipment chosen are dependent upon the size of the confined space openings, the gases to be exhausted, and the source of makeup air.

Under certain conditions where flammable gases or vapors have displaced the oxygen level but are too rich to burn, forced air ventilation may dilute them until they are within the explosive range. Also, if inert gases are used in the confined space, the space should be well ventilated and re-tested before workers enter.

A common method of ventilation requires a large hose with one end attached to a fan and the other lowered into a manhole or opening. For example, a manhole would have the ventilating hose run to the bottom to blow out all harmful gases and vapors. The air intake should be placed in an area that will draw in fresh air only. Where possible, ventilation should be continuous because in many confined spaces, the hazardous atmosphere will reform when the flow of air is stopped.

25.5 Isolation

Isolation of a confined space is a process where the space is removed from service by:

- locking out electrical sources, preferably at disconnect switches remote from the equipment

- blanking and bleeding pneumatic and hydraulic lines
- disconnecting belt and chain drives, and mechanical linkages on shaft-driven equipment where possible
- securing mechanical moving parts within confined spaces with latches, chains, chocks, blocks, or other devices

25.6 Respirators

Respirators are devices that can allow workers to safely breathe without inhaling toxic gases or particles. Two basic types are air-purifying, which filter dangerous substances from the air; and air-supplying, which deliver a supply of safe breathing air from a tank or an uncontaminated area nearby. Only air-supplying respirators should be used in confined s paces where there is not enough oxygen (see Figure 25.2).

Selecting the proper respirator for the job, the hazard, and the individual is very important, as is thorough training in the use and limitations of respirators. Questions regarding the proper selection and use of respirators should be addressed to a certified industrial hygienist.

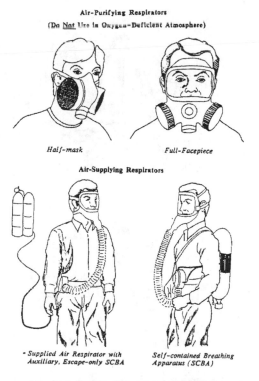

Air-Purifying Respirators
(Do Not Use in Oxygen-Deficient Atmosphere)

Half-mask   *Full-Facepiece*

Air-Supplying Respirators

- Supplied Air Respirator with Auxiliary, Escape-only SCBA   *Self-contained Breathing Apparatus (SCBA)*

FIG. 25.2 Two basic types of respirators.

25.7 Standby/Rescue

A standby person should be assigned to remain on the outside of the confined space and be in constant visual or verbal contact with the workers inside. The standby person should not have any other duties but to serve as standby and know who should be notified in case of emergency. Standby personnel should not enter a confined space until help arrives, and then only with proper protective equipment, life lines, and respirators. Figure 25.3 illustrates confined space entry with hoist and standby personnel.

Over 50% of the workers who die in confined spaces are attempting to rescue other workers. Rescuers must be trained in and follow established emergency procedures and use appropriate equipment and techniques (lifelines, respiratory protection, and standby persons). Steps for safe rescue should be included in all confined space entry procedures. Rescues should be well planned and drills should be frequently conducted on emergency procedures. Unplanned rescues, such as when someone instinctively rushes in to help a downed coworker, can easily result in a double fatality, or even multiple fatalities if there are more than one would-be rescuers.

Fɪɢ. 25.3 Entry with Hoist and Standby Personnel

25.8 General/Physical Hazards

In addition to the areas discussed above, evaluation of a confined space should consider the following potential hazards:

1. Temperature Extremes – Extremely hot or cold temperatures can present problems for workers. For example, if the space has been steamed, it should be allowed to cool before any entry is made.
2. Engulfment Hazards – Loose, granular material stored in bins and hoppers, such as grain, sand, coal, or similar materials, can engulf and suffocate a worker. The loose material can crust or bridge over in a bin and break loose under the weight of a worker.
3. Noise – Noise within a confined space can be amplified because of the design and acoustic properties of the space. Excessive noise can not only damage hearing, but can also affect communication, such as causing a shouted warning to go unheard.
4. Slick/Wet Surfaces – Slips and falls can occur on a wet surface causing injury or death to workers. Also, a wet surface will increase the likelihood for and effect of electric shock in areas where electrical circuits, equipment, and tools are used.
5. Falling Objects – Workers in confined spaces should be mindful of the possibility of falling objects, particularly in spaces which have topside openings for entry, and where work is being done above the water.

25.9 Evaluating the Confined Space

Table 25.1 contains a checklist which should be used to evaluate a confined space. A confined space should not be entered until every question on the checklist has been considered and the space has been determined to be safe.

Summary

Confined spaces present a real risk to workers. Training of employees in confined space identification and proper entry procedures is

TABLE 25.1
Confined Space Evaluation Checklist

Question	YES	NO
Is entry necessary?		
TESTING:		
Are the instruments used in atmosphere testing properly calibrated?		
Was the atmosphere in the confined space tested?		
Was oxygen at 19.5% - not more than 21%?		
Were toxic, flammable, or oxygen-displacing gases/vapors present? Hydrogen Sulfide, Carbon Monoxide, Methane, Carbon Dioxide, or other: _____		
MONITORING:		
Will the atmosphere in the space be monitored while work is going on?		
Continuously?		
Periodically? (If yes, give interval: _____)		
CLEANING:		
Has the space been cleaned before entry is made?		
Was the space steamed?		
If so, was it allowed to cool?		
VENTILATION		
Has the space been ventilated before entry?		
Will ventilation be continued during entry?		
Is the air intake for the ventilation system located in an area that is free of combustible dusts, vapors, and toxic substances?		
If the atmosphere was found unacceptable and then ventilated, was it re-tested before entry?		
ISOLATION:		
Has the space been isolated from other systems?		
Has electrical equipment been locked out?		
Have disconnects been used where possible?		
Has mechanical equipment been blocked, chocked, and disengaged where necessary?		
Have lines under pressure been blanked and bled?		
CLOTHING/EQUIPMENT		
Is special equipment required (e.g., rescue equipment, communications equipment, etc.)? If so, specify: _____		
Is special equipment required (e.g., boots, chemical suits, glasses, etc.)? If so, specify: _____		
Are special tools required (e.g., sparkproof)? If so, specify: _____		
RESPIRATORY PROTECTION		
Are MSHA/NIOSH approved respirators of the type required available at the worksite?		
Is respiratory protection required (e.g., air-purifying, supplied air, self-contained breathing apparatus, etc.)? If so, specify type: _____		
Can the worker get through the opening with a respirator on? (If unsure, find out before attempting entry.)		
TRAINING		
Have you been trained in proper use of a respirator?		
Have you received first aid/CPR training?		
Have you been trained in confined space entry and do you know what to look for?		
STANDBY/RESCUE		
Will there be a standby person on the outside in constant visual or auditory communication with the person on the inside?		
Will the standby person be able to see and/or hear the person inside at all times?		
Has the standby person(s) been trained in rescue procedures?		
Will safety lines and harnesses be required to remove a person?		

TABLE 25.1 (Continued)

Question	YES	NO
Are company rescue procedures available to be followed in the event of an emergency?		
Are you familiar with emergency rescue procedures?		
Do you know who to notify and how in the event of an emergency?		
PERMIT*		
Has a confined space entry permit been issued?		
Does the permit include a list of emergency telephone numbers?		

*The permit is a written authorization stating that the space has been tested by a qualified person; that the space is safe for entry; what precautions, equipment, etc., are required; and the work to be done.

a top priority. Workers must be aware of the three hazards involved in entering and working in confined spaces. These hazards are oxygen-deficient atmospheres, flammable atmospheres, and toxic atmospheres. Testing of the confined space atmosphere and ventilation as required are integral parts of a safe confined space entry program.

Lack of hazard awareness and unplanned rescue attempts led to the following deaths:

- On July 23, 1985, a city worker was removing an inspection plate from a sewer line in a 50-foot deep pump station, when the plate blew off allowing raw sewage to enter the room. Two fellow workers and a policeman attempted to rescue the worker from the sludge filled room and were unsuccessful. All four were dead when removed from the pumping station.
- On February 21, 1986, a self-employed truck driver died after entering the top of a 22-foot high x 15-foot square sawdust bin. He suffocated when the sawdust inside the bin collapsed and buried him.
- On July 16, 1986, a worker entered a septic tank to clean out the residue at the bottom and collapsed shortly afterward. Two workers on the outside went in to rescue the downed worker. All three were dead when removed from the tank.
- On October 10, 1986, a self-employed plumbing contractor entered an underground water line vault to inspect a backflow device. The contractor collapsed shortly after entering the vault. A supervisor noticed the man down, and entered the vault in a rescue attempt. Both men had entered an untested oxygen - deficient atmosphere, and died as a result.
- On February 6, 1987, two workers (father and son) at a wastewater plant were working on a digester that was being

drained. They went on top of the digester and opened a hatch to check the sludge level. To provide light in the digester, they lowered an extension cord with an exposed 200 watt light bulb into the digester. The light broke and caused the methane gas in the digester to explode, killing both men instantly.

If the guidelines in this chapter had been followed, these fatalities would have been prevented.

Bibliography

NIOSH (1979) "Criteria for a Recommended Standard – Working in Confined Spaces", DHHS, Publication No. 80-106, Cincinnati, OH.

NIOSH (1987) "A Guide to Safety in Confined Spaces", DHHS, Publication No. 87-113, Cincinnati, OH.

Rekus, John F. (1994) "Complete Confined Spaces Handbook", National Safety Council, Lewis Publishers, Ann Arbor, MI.

Appendix A - Model Policy Statements

The following statements are examples of policy statements which may be used or modified by employers to help prevent employee injury and illness.

"The Occupational Safety and Health Act of 1970 clearly states our common goal of safe and healthful working conditions. The safety and health of our employees continues to be the first consideration in the operation of this business."

"Safety and health in our business must be a part of every operation. Without question it is every employee's responsibility at all levels."

"It is the intent of this company to comply with all laws. To do this we must constantly be aware of conditions in all work areas that can produce injuries. No employee is required to work at a job he or she knows is not safe or healthful. Your cooperation in detecting hazards and, in turn, controlling them is a condition of your employment. Inform your supervisor immediately of any situation beyond your ability or authority to correct."

"The personal safety and health of each employee of this company is of primary importance. The prevention of occupationally-induced injuries and illnesses is of such consequence that it will be given precedence over operating productivity whenever necessary. To the greatest degree possible, management will provide all mechanical and physical facilities required for personal safety and health in keeping with the highest standards."

"We will maintain a safety and health program conforming to the best practices of organizations of this type. To be successful, such a program must embody the proper attitudes toward injury and illness prevention on the part of supervisors and employees. It also requires cooperation in all safety and health matters, not only between supervisor and employee, but also between each employee and his or her co-workers. Only through such a cooperative effort can a safety program in the best interest of all be established and preserved."

"Our objective is a safety and health program that will reduce the number of injuries and illnesses to an absolute minimum, not merely in keeping with, but surpassing, the best experience of operations similar to ours. Our goal is zero accidents and injuries."

"Our safety and health program will include:
- Providing mechanical and physical safeguards to the maximum extent possible.
- Conducting a program of safety and health inspections to find and eliminate unsafe working conditions or practices, to control health hazards, and to comply fully with the safety and health standards for every job.
- Training all employees in good safety and health practices.
- Providing necessary personal protective equipment and instructions for its use and care.
- Developing and enforcing safety and health rules and requiring that employees cooperate with these rules as a condition of employment.
- Investigating, promptly and thoroughly, every accident to find out what caused it and to correct the problem so that it won't happen again.
- Setting up a system of recognition and awards for outstanding safety service or performance."

"We recognize that the responsibilities for safety and health are shared:
- The employer accepts the responsibility for leadership of the safety and health program, for its effectiveness and improvement, and for providing the safeguards required to ensure safe conditions.
- Supervisors are responsible for developing the proper attitudes toward safety and health in themselves and in those they supervise, and for ensuring that all operations are performed with the utmost regard for the safety and health of all personnel involved, including themselves.
- Employees are responsible for wholehearted, genuine operation with all aspects of the safety and health program including compliance with all rules and regulations-and for continuously practicing safety while performing their duties."

Appendix B – Self-Inspection Check Lists

These check lists are typical for general industry and are not all-inclusive. You should add or delete portions or items that do not apply to your operations. You also will need to refer to OSHA standards for complete and specific standards that may apply to your work situation. Source: OSHA Publication #2209, 1996.

Employer Posting

- Is the required OSHA workplace poster displayed in a prominent location where all employees are likely to see it?

- Are emergency telephone numbers posted where they can be readily found in case of emergency?

- Where employees may be exposed to any toxic substances or harmful physical agents, has appropriate information concerning employee access to medical and exposure records and "Material Safety Data Sheets" been posted or otherwise made readily available to affected employees?

- Are signs concerning "Exiting from buildings," room capacities, floor loading, biohazards, exposures to x-ray, microwave, or other harmful radiation or substances posted where appropriate?

- Is the Summary of Occupational Illnesses and Injuries (OSHA Form 200) posted in the month of February?

Recordkeeping

- Are all occupational injury or illnesses, except minor injuries requiring only first aid, being recorded as required on the OSHA 200 log?

- Are employee medical records and records of employee exposure to hazardous substances or harmful physical agents up-to-date and in compliance with current OSHA standards?

- Are employee training records kept and accessible for review by employees, when required by OSHA standards?

- Have arrangements been made to maintain required records for the legal period of time for each specific type record? (Some records must be maintained for at least 40 years.)

- Are operating permits and records up-to-date for such items as elevators, air pressure tanks, and liquefied petroleum gas tanks?

Safety and Health Program

- Do you have an active safety and health program in operation that deals with general safety and health program elements as well as the management of hazards specific to your worksite?

- Is one person clearly responsible for the overall activities of the safety and health program?

- Do you have a safety committee or group made up of management and labor representatives that meets regularly and report in writing on its activities?

- Do you have a working procedure for handling in-house employee complaints regarding safety and health?

- Are you keeping your employees advised of the successful effort and accomplishments you and/or your safety committee have made in assuring they will have a workplace that is safe and healthful?

- Have you considered incentives for employees or workgroups who have excelled in reducing workplace injury/illnesses?

Medical Services and First Aid

- Is there a hospital, clinic, or infirmary for medical care in proximity of your workplace?

- If medical and first-aid facilities are not in proximity of your workplace, is at least one employee on each shift currently qualified to render first aid?

- Have all employees who are expected to respond to medical emergencies as part of their work:

  (1) received first-aid training; (2) had hepatitis B vaccination made available to them; (3) had appropriate training on procedures to protect them from bloodbome pathogens, including universal precautions; and (4) have available and understand how to use appropriate personal protective equipment to protect against exposure to bloodbome diseases?

- Where employees have had an exposure incident involving bloodbome pathogens, did you provide an immediate post exposure medical evaluation and followup?

- Are medical personnel readily available for advice and consultation on matters of employees' health?

- Are emergency phone numbers posted?

- Are first-aid kits easily accessible to each work area, with necessary supplies available, periodically inspected and replenished as needed?

- Have first-aid kit supplies been approved by a physician, indicating that they are adequate for a particular area or operation?

- Are means provided for quick drenching or flushing of the eyes and body in areas where corrosive liquids or materials are handled?

Fire Protection

- Is your local fire department well acquainted with your facilities, its location, and specific hazards?

- If you have a fire alarm system, is it certified as required?

- If you have a fire alarm system, is it tested at least annually?

- If you have interior stand pipes and valves, are they inspected regularly?

- If you have outside private fire hydrants, are they flushed at least once a year and on a routine preventive maintenance schedule?

- Are fire doors and shutters in good operating condition?

- Are fire doors and shutters unobstructed and protected against obstructions, including their counterweights?

- Are fire door and shutter fusible links in place?

- Are automatic sprinkler system water control valves, air and water pressure checked weekly/ periodically as required?

- Is the maintenance of automatic sprinkler systems assigned to responsible persons or to a sprinkler contractor?

- Are sprinkler heads protected by metal guards, when exposed to physical damage?

- Is proper clearance maintained below sprinkler heads?

- Are portable fire extinguishers provided in adequate number and type?

- Are fire extinguishers mounted in readily accessible locations?

- Are fire extinguishers recharged regularly and noted on the inspection tag?

- Are employees periodically instructed in the use of extinguishers and fire protection procedures?

Personal Protective Equipment and Clothing

- Are employers assessing the workplace to determine if hazards that require the use of personal protective equipment (e.g. head, eye, face, hand, or foot protection) are present or are likely to be present?

- If hazards or the likelihood of hazards are found, are employers selecting and having affected employees use properly fitted personal protective equipment suitable for protection from these hazards?

- Has the employer been trained on Personal Protective Equipment (PPE) procedures, i.e. what PPE is necessary for a job tasks, when they need it, and how to properly adjust it?

- Are protective goggles or face shields provided and worn where there is any danger of flying particles or corrosive materials?

- Are approved safety glasses required to be worn at all times in areas where there is a risk of eye injuries such as punctures, abrasions, contusions or bums?

- Are employees who need corrective lenses (glasses or contacts) in working environments having harmful exposures, required to wear only approved safety glasses, protective goggles, or use other medically approved precautionary procedures?

- Are protective gloves, aprons, shields, or other means provided and required where employees could be cut or where there is reasonably anticipated exposure to corrosive liquids, chemicals, blood, or other potentially infectious materials? See 29 CFR 1910.1030(b) for the definition of other potentially infectious materials."

- Are hard hats provided and worn where danger of failing objects exists?

- Are hard hats inspected periodically for damage to the shell and suspension system?

- Is appropriate foot protection required where there is the risk of foot injuries from hot, corrosive, poisonous substances, falling objects, crushing or penetrating actions?

- Are approved respirators provided for regular or emergency use where needed?

- Is all protective equipment maintained in a sanitary condition and ready for use?

- Do you have eye wash facilities and a quick Drench Shower within the work area where employees are exposed to injurious corrosive materials?

- Where special equipment is needed for electrical workers, is it available?

- Where food or beverages are consumed on the premises, are they consumed in areas where there is no exposure to toxic material, blood, or other potentially infectious materials?

- Is protection against the effects of occupational noise exposure provided when sound levels exceed those of the OSHA noise standard?

- Are adequate work procedures, protective clothing and equipment provided and used when cleaning up spilled toxic or otherwise hazardous materials or liquids?

- Are there appropriate procedures in place for disposing of or decontaminating personal protective equipment contaminated with, or reasonably anticipated to be contaminated with, blood or other potentially infectious materials?

General Work Environment

- Are all worksites clean, sanitary, and orderly?

- Are work surfaces kept dry or appropriate means taken to assure the surfaces are slip-resistant?

- Are all spilled hazardous materials or liquids, including blood and other potentially infectious materials, cleaned up immediately and according to proper procedures?

- Is combustible scrap, debris and waste stored safely and removed from the worksite promptly?

- Is all regulated waste, as defined in the OSHA bloodbome pathogens standard (29 CFR 1910.1030), discarded according to federal, state, and local regulations?

- Are accumulations of combustible dust routinely removed from elevated surfaces including the overhead structure of buildings, etc.?

- Is combustible dust cleaned up with a vacuum system to prevent the dust going into suspension?

- Is metallic or conductive dust prevented from entering or accumulating on or around electrical enclosures or equipment?

- Are covered metal waste cans used for oily and paintsoaked waste?

- Are all oil and gas fired devices equipped with flame controls that will prevent flow of fuel if pilots or main burners are not working?

- Are paint spray booths, dip tanks, etc., cleaned regularly?

- Are the minimum number of toilets and washing facilities provided?

- Are all toilets and washing facilities clean and sanitary?

- Are all work areas adequately illuminated?

- Are pits and floor openings covered or otherwise guarded?

- Have all confined spaces been evaluated for compliance with 29 CFR 1910.146?

Walkways

- Are aisles and passageways kept clear?

- Are aisles and walkways marked as appropriate?

- Are wet surfaces covered with non-slip materials?

- Are holes in the floor, sidewalk or other walking surface repaired properly, covered or otherwise made safe?

- Is there safe clearance for walking in aisles where motorized or mechanical handling equipment is operating?

- Are materials or equipment stored in such a way that sharp projectives will not interfere with the walkway?

- Are spilled materials cleaned up immediately?

- Are changes of direction or elevations readily identifiable?

- Are aisles or walkways that pass near moving or operating machinery, welding operations or similar operations arranged so employees will not be subjected to potential hazards?

- Is adequate headroom provided for the entire length of any aisle or walkway?

- Are standard guardrails provided wherever aisle or walkway surfaces are elevated more than 30 inches (76.20 centimeters) above any adjacent floor or ground?

- Are bridges provided over conveyors and similar hazards?

Floor and Wall Openings

- Are floor openings guarded by a cover, a guardrail, or equivalent on all sides (except at entrance to stairways or ladders)?

- Are toeboards installed around the edges of permanent floor opening (where persons may pass below the opening)?

- Are skylight screens of such construction and mounting that they will withstand a load of at least 200 pounds (90 kilograms)?

- Is the glass in the windows, doors, glass walls, et which are subject to human impact, of sufficient thickness and type for the condition of use?

- Are grates or similar type covers over floor openings such as floor drains of such design that foot traffic or rolling equipment will not be affected b the grate spacing?

- Are unused portions of service pits and pits not actually in use either covered or protected by guardrails or equivalent?

- Are manhole covers, trench covers and similar covers, plus their supports designed to carry a truck rear axle load of at least 20,000 pounds (9000 kilograms) when located in roadways and subject to vehicle traffic?

- Are floor or wall openings in fire resistive construction provided with doors or covers compatible with the fire rating of the structure and provided with a self-closing feature when appropriate?

Stairs and Stairways

- Are standard stair rails or handrails on all stairway having four or more risers?

- Are all stairways at least 22 inches (55.88 centimeters) wide?

- Do stairs have landing platforms not less than 30 inches (76.20 centimeters) in the direction of travel and extend 22 inches (55.88 centimeters) in width at every 12 feet (3.6576 meters) or less of vertical rise?

- Do stairs angle no more than 50 and no less than 30 degrees?

- Are stairs of hollow-pan type treads and landings filled to the top edge of the pan with solid material?

- Are step risers on stairs uniform from top to bottom?

- Are steps on stairs and stairways designed or provided with a surface that renders them slip resistant?

- Are stairway handrails located between 30 (76.20 centimeters) and 34 inches (86.36 centimeters) above the leading edge of stair treads?

- Do stairway handrails have at least 3 inches (7.62 centimeters) of clearance between the handrails and the wall or surface they are mounted on?

- Where doors or gates open directly on a stairway, is there a platform provided so the swing of the door does not reduce the width of the platform to less than 21 inches (53.34 centimeters)?

- Are stairway handrails capable of withstanding a load of 200 pounds (90 kilograms), applied within 2 inches (5.08 centimeters) of the top edge, in any downward or outward direction?

- Where stairs or stairways exit directly into any area where vehicles may be operated, are adequate barriers and warnings provided to prevent employees stepping into the path of traffic?

- Do stairway landings have a dimension measured in the direction of travel, at least equal to the width of the stairway?

- Is the vertical distance between stairway landings limited to 12 feet (3.6576 centimeters) or less?

Elevated Surfaces

- Are signs posted, when appropriate, showing the elevated surface load capacity?

- Are surfaces elevated more than 30 inches (76.20 centimeters) above the floor or ground provided with standard guardrails?

- Are all elevated surfaces (beneath which people or machinery could be exposed to falling objects) provided with standard 4-inch (10.16 centimeters) toeboards?

- Is a permanent means of access and egress provided to elevated storage and work surfaces?

- Is required headroom provided where necessary?

- Is material on elevated surfaces piled, stacked or racked in a manner to prevent it from tipping, falling, collapsing, rolling or spreading?

- Are dock boards or bridge plates used when transferring materials between docks and trucks or rail cars?

Exiting or Egress

- Are all exits marked with an exit sign and illuminated by a reliable light source?

- Are the directions to exits, when not immediately apparent, marked with visible signs?

- Are doors, passageways or stairways, that are neither exits nor access to exits, and which could be mistaken for exits, appropriately marked "NOT AN EXIT," "TO BASEMENT," "STOREROOM", etc.?

- Are exit signs provided with the word "EXIT" in lettering at least 5 inches (12.70 centimeters) high and the stroke of the lettering at least 1/2-inch (1.2700 centimeters) wide?

- Are exit doors side-hinged?

- Are all exits kept free of obstructions?

- Are at least two means of egress provided from elevated platforms, pits or rooms where the absence of a second exit would increase the risk of injury from hot, poisonous, corrosive, suffocating, flammable, or explosive substances?

- Are there sufficient exits to permit prompt escape in case of emergency?

- Are special precautions taken to protect employees during construction and repair operations?

- Is the number of exits from each floor of a building and the number of exits from the building itself, appropriate for the building occupancy load?

- Are exit stairways that are required to be separated from other parts of a building enclosed by at least 2-hour fire-resistive construction in buildings more than four stories in height, and not less than I -hour fire-resistive constructive elsewhere?

- Where ramps are used as part of required exiting from a building, is the ramp slope limited to 1 foot (0.3048 meters) vertical and 12 feet (3.6576 meters) horizontal?

- Where exiting will be through frameless glass doors, glass exit doors, or storm doors are the doors fully tempered and meet the safety requirements for human impact?

Exit Doors

- Are doors that are required to serve as exits designed and constructed so that the way of exit travel is obvious and direct?

- Are windows that could be mistaken for exit doors, made inaccessible by means of barriers or railings?

- Are exit doors openable from the direction of exit travel without the use of a key or any special knowledge or effort when the building is occupied?

- Is a revolving, sliding, or overhead door prohibited from serving as a required exit door?

- Where panic hardware is installed on a required exit door, will it allow the door to open by applying a force of 15 pounds (6.75 kilograms) or less in the direction of the exit traffic?

- Are doors on cold storage rooms provided with an inside release mechanism which will release the latch and open the door even if it's padlocked or otherwise locked on the outside?

- Where exit doors open directly onto any street, alley or other area where vehicles may be operated, are adequate barriers and warnings provided to prevent employees from stepping into the path of traffic?

- Are doors that swing in both directions and are located between rooms where there is frequent traffic, provided with viewing panels in each door?

Portable Ladders

- Are all ladders maintained in good condition, joints between steps and side rails tight, all hardware an fittings securely attached and moveable parts operating freely without binding or undue play?

- Are non-slip safety feet provided on each ladder?

- Arc non-slip safety feet provided on each metal o rung ladder?

- Are ladder rungs and steps free of grease and oil?

- Is it prohibited to place a ladder in front of doors opening toward the ladder except when the door is blocked open, locked, or guarded?

- Is it prohibited to place ladders on boxes, barrels, or other unstable bases to obtain additional height?

- Are employees instructed to face the ladder when ascending or descending?

- Are employees prohibited from using ladders that are broken, missing steps, rungs, or cleats, broken side rails or other faulty equipment?

- Are employees instructed not to use the top step of ordinary stepladders as a step?

- When portable rung ladders are used to gain access to elevated platforms, roofs, etc., does the ladder always extend at least 3 feet (0.9144 meters) above the elevated surface?

- Is it required that when portable rung or cleat type ladders are used, the base is so placed that slipping will not occur, or it is lashed or otherwise held in place?

- Are portable metal ladders legibly marked with signs reading "CAUTION" - Do Not Use Around Electrical Equipment" or equivalent wording?

- Are employees prohibited from using ladders as guys, braces, skids, gin poles, or for other than their intended purposes?

- Are employees instructed to only adjust extension ladders while standing at a base (not while standing on the ladder or from a position above the ladder)?

- Are metal ladders inspected for damage?

- Are the rungs of ladders uniformly spaced at 12 inches, (30.48 centimeters) center to center?

Hand Tools and Equipment

- Are all tools and equipment (both company and employee owned) used by employees at their workplace in good condition?

- Are hand tools such as chisels and punches, which develop mushroomed heads during use, reconditioned or replaced as necessary?

- Are broken or fractured handles on hammers, axes and similar equipment replaced promptly?

- Are worn or bent wrenches replaced regularly?

- Are appropriate handles used on files and similar tools?

- Are employees made aware of the hazards caused by faulty or improperly used hand tools?

- Are appropriate safety glasses, face shields, etc. used while using hand tools or equipment which might produce flying materials or be subject to breakage?

- Are jacks checked periodically to ensure they are in good operating condition?

- Are tool handles wedged tightly in the head of all tools?

- Are tool cutting edges kept sharp so the tool will move smoothly without binding or skipping?

- Are tools stored in dry, secure location where they won't be tampered with?

- Is eye and face protection used when driving hardened or tempered spuds or nails?

Portable (Power Operated) Tools and Equipment

- Are grinders, saws and similar equipment provided with appropriate safety guards?

- Are power tools used with the correct shield, guard, or attachment, recommended by the manufacturer?

- Are portable circular saws equipped with guards above and below the base shoe?

- Are circular saw guards checked to assure they are not wedged up, thus leaving the lower portion of the blade unguarded?

- Are rotating or moving parts of equipment guarded to prevent physical contact?

- Are all cord-connected, electrically operated tools and equipment effectively grounded or of the approved double insulated type?

- Are effective guards in place over belts, pulleys, chains, sprockets, on equipment such as concrete mixers, and air compressors?

- Are portable fans provided with full guards or screens having openings 1/2 inch (1.2700 centimeters) or less?

- Is hoisting equipment available and used for lifting heavy objects, and are hoist ratings and characteristics appropriate for the task?

- Are ground-fault circuit interrupters provided on all temporary electrical 15 and 20 ampere circuits, used during periods of construction?

- Are pneumatic and hydraulic hoses on power operated tools checked regularly for deterioration or damage?

Abrasive Wheel Equipment Grinders

- Is the work rest used and kept adjusted to within 1/8 inch (0.3175 centimeters) of the wheel?

- Is the adjustable tongue on the top side of the grinder used and kept adjusted to within 1/4 inch (0.6350 centimeters) of the wheel?

- Do side guards cover the spindle, nut, and flange and 75 percent of the wheel diameter?

- Are bench and pedestal grinders permanently mounted?

- Are goggles or face shields always worn when grinding9

- Is the maximum RPM rating of each abrasive wheel compatible with the RPM rating of the grinder motor?

- Are fixed or permanently mounted grinders connected to their electrical supply system with metallic conduit or other permanent wiring method?

- Does each grinder have an individual on and off control switch?

- Is each electrically operated grinder effectively grounded?

- Before new abrasive wheels are mounted, are they visually inspected and ring tested?

- Are dust collectors and powered exhausts provided on grinders used in operations that produce large amounts of dust?

- Are splash guards mounted on grinders that use coolant to prevent the coolant reaching employees?

- Is cleanliness maintained around grinders?

Powder-Actuated Tools

- Are employees who operate powder-actuated tools trained in their use and carry a valid operators card?

- Is each powder-actuated tool stored in its own locked container when not being used?

- Is a sign at least 7 inches (17.78 centimeters) by 10 inches (25.40 centimeters) with bold face type reading "POWDER-ACTUATED TOOL IN USE" conspicuously posted when the tool is being used?

- Are powder-actuated tools left unloaded until they are actually ready to be used?

- Are powder-actuated tools inspected for obstructions or defects each day before use?

- Do powder-actuated tool operators have and use appropriate personal protective equipment such as hard hats, safety goggles, safety shoes and ear protectors?

Machine Guarding

- Is there a training program to instruct employees on safe methods of machine operation?

- Is there adequate supervision to ensure that employees are following safe machine operating procedures?

- Is there a regular program of safety inspection of machinery and equipment?

- Is all machinery and equipment kept clean and properly maintained?

- Is sufficient clearance provided around and between machines to allow for safe operations, set up and servicing, material handling and waste removal?

- Is equipment and machinery securely placed and anchored, when necessary to prevent tipping or other movement that could result in personal injury?

- Is there a power shut-off switch within reach of the operator's position at each machine?

- Can electric power to each machine be locked out for maintenance, repair, or security?

- Are the noncurrent-carrying metal parts of electrically operated machines bonded and grounded?

- Are foot-operated switches guarded or arranged to prevent accidental actuation by personnel or failing objects?

- Are manually operated valves and switches controlling the operation of equipment and machines clearly identified and readily accessible?

- Are all emergency stop buttons colored red?

- Are all pulleys and belts that are within 7 feet (2.1336 meters) of the floor or working level properly guarded?

- Are all moving chains and gears properly guarded?

- Are splash guards mounted on machines that use coolant to prevent the coolant from reaching employees?

- Are methods provided to protect the operator and other employees in the machine area from hazards created at the point of operation, ingoing nip points, rotating parts, flying chips, and sparks?

- Are machinery guards secure and so arranged that they do not offer a hazard in their use?

- If special handtools are used for placing and removing material, do they protect the operator's hands?

- Are revolving drums, barrels, and containers required to be guarded by an enclosure that is interlocked with the drive mechanism, so that revolution cannot occur unless the guard enclosures is in place, so guarded?

- Do arbors and mandrels have firm and secure bearings and are they free from play?

- Arc provisions made to prevent machines from automatically starting when power is restored after a power failure or shutdown?

- Are machines constructed so as to be free from excessive vibration when the largest size too] is mounted and run at full speed?

- If machinery is cleaned with compressed air, is air pressure controlled and personal protective equipment or other safeguards utilized to protect operators and other workers from eye and body injury?

- Are fan blades protected with a guard having openings no larger than 1/2 inch (1.2700 centimeters), when operating within 7 feet (2.1336 meters) of the floor?

- Are saws used for ripping, equipped with anti-kick back devices and spreaders?

- Are radial arm saws so arranged that the cutting head will gently return to the back of the table when released?

Lockot/Tagout Procedures

- Is all machinery or equipment capable of movement, required to be de-energized or disengaged and locked-out during cleaning, servicing, adjusting or setting up operations, whenever required?

- Where the power disconnecting means for equipment does not also disconnect the electrical control circuit:
  - Are the appropriate electrical enclosures identified?
  - Is means provided to assure the control circuit can also be disconnected and locked-out?

- Is the locking-out of control circuits in lieu of locking-out main power disconnects prohibited?

- Are all equipment control valve handles provided with a means for locking-out?

- Does the lock-out procedure require that stored energy (mechanical, hydraulic, air, etc.) be released or blocked before equipment is locked-out for repairs?

- Are appropriate employees provided with individually keyed personal safety locks?

- Are employees required to keep personal control of their key(s) while they have safety locks in use?

- Is it required that only the employee exposed to the hazard, place or remove the safety lock?

- Is it required that employees check the safety of the lock-out by attempting a startup after making sure no one is exposed?

- Are employees instructed to always push the control circuit stop button immediately after checking, the safety of the lock-out?

- Is there a means provided to identify any or all employees who are working on locked-out equipment by their locks or accompanying tags?

- Are a sufficient number of accident preventive signs or tags and safety padlocks provided for any reasonably foreseeable repair emergency?

- When machine operations, configuration or size requires the operator to leave his or her control station to install tools or perform other operations, and that part of the machine could move if accidentally activated, is such element required to be separately locked or blocked out?

- In the event that equipment or lines cannot be shut down, locked-out and tagged, is a safe job procedure established and rigidly followed?

Welding, Cutting, and Brazing

- Are only authorized and trained personnel permitted to use welding, cutting or brazing equipment?

- Does each operator have a copy of the appropriate operating instructions and are they directed to follow them?

- Are compressed gas cylinders regularly examined for obvious signs of defects, deep rusting, or leakage?

- Is care used in handling and storing cylinders, safety valves, and relief valves to prevent damage?

- Are precautions taken to prevent the mixture of air or oxygen with flammable gases, except at a burner or in a standard torch?

- Are only approved apparatus (torches, regulators, pressure reducing valves, acetylene generators, manifolds) used?

- Are cylinders kept away from sources of heat?

- Are the cylinders kept away from elevators, stairs, or gangways?

- Is it prohibited to use cylinders as rollers or supports?

- Are empty cylinders appropriately marked and their valves closed?

- Are signs reading: DANGER-NO SMOKING, MATCHES, OR OPENLIGHTS, or the equivalent, posted?

- Are cylinders, cylinder valves, couplings, regulators, hoses, and apparatus kept free of oily or greasy substances?

- Is care taken not to drop or strike cylinders?

- Unless secured on special trucks, are regulators removed and valve-protection caps put in place before moving cylinders?

- Do cylinders without fixed hand wheels have keys, handles, or non-adjustable wrenches on stem valves when in service?

- Are liquefied gases stored and shipped valve-end up with valve covers in place?

- Are provisions made to never crack a fuel gas cylinder valve near sources of ignition?

- Before a regulator is removed, is the valve closed and gas released from the regulator?

- Is red used to identify the acetylene (and other fuel-gas) hose, green for oxygen hose, and black for inert gas and air hose?

- Are pressure-reducing regulators used only for the gas and pressures for which they are intended?

- Is open circuit (No Load) voltage of arc welding and cutting machines as low as possible and not in excess of the recommended limits?

- Under wet conditions, are automatic controls for reducing no load voltage used?

- Is grounding of the machine frame and safety ground connections of portable machines checked periodically?

- Are electrodes removed from the holders when not in use?

- Is it required that electric power to the welder be shut off when no one is in attendance?

- Is suitable fire extinguishing equipment available for immediate use?

- Is the welder forbidden to coil or loop welding electrode cable around his body?

- Are wet machines thoroughly dried and tested before being used?

- Are work and electrode lead cables frequently inspected for wear and damage, and replaced when needed?

- Do means for connecting cable lengths have adequate insulation?

- When the object to be welded cannot be moved an fire hazards cannot be removed, are shields used t confine heat, sparks, and slag?

- Are fire watchers assigned when welding or cutting is performed in locations where a serious fire might develop?

- Are combustible floors kept wet, covered by damp sand, or protected by fire-resistant shields?

- When floors are wet down, are personnel protected from possible electrical shock?

- When welding is done on metal walls, are precautions taken to protect combustibles on the other side?

- Before hot work is begun, are used drums, barrels, tanks, and other containers so thoroughly cleaned that no substances remain that could explode, ignite, or produce toxic vapors?

- Is it required that eye protection helmets, hand shields and goggles meet appropriate standards?

- Are employees exposed to the hazards created by welding, cutting, or brazing operations protected with personal protective equipment and clothing?

- Is a check made for adequate ventilation in and where welding or cutting is performed?

- When working in confined places, are environmental monitoring tests taken and means provided for quick removal of welders in case of an emergency?

Compressors and Compressed Air

- Are compressors equipped with pressure relief valves, and pressure gauges?

- Are compressor air intakes installed and equipped so as to ensure that only clean uncontaminated air enters the compressor?

- Are air filters installed on the compressor intake?

- Are compressors operated and lubricated in accordance with the manufacturer's recommendations?

- Are safety devices on compressed air systems checked frequently?

- Before any repair work is done on the pressure system of a compressor, is the pressure bled off and the system locked-out?

- Are signs posted to warn of the automatic starting feature of the compressors?

- Is the belt drive system totally enclosed to provide protection for the front, back, top, and sides?

- Is it strictly prohibited to direct compressed air towards a person?

- Are employees prohibited from using highly compressed air for cleaning purposes?

- If compressed air is used for cleaning off clothing, is the pressure reduced to less than 10 psi?

- When using compressed air for cleaning, do employees wear protective chip guarding and personal protective equipment?

- Are safety chains or other suitable locking devices used at couplings of high pressure hose lines where a connection failure would create a hazard?

- Before compressed air is used to empty containers of liquid, is the safe working pressure of the container checked?

- When compressed air is used with abrasive blast cleaning equipment, is the operating valve a type that must be held open manually?

- When compressed air is used to inflate auto ties, is a clip-on chuck and an inline regulator preset to 40 psi required?

- Is it prohibited to use compressed air to clean up or move combustible dust if such action could cause the dust to be suspended in the air and cause a fire or explosion hazard?

Compressors Air Receivers

- Is every receiver equipped with a pressure gauge and with one or more automatic, spring-loaded safety valves?

- Is the total relieving capacity of the safety valve capable of preventing pressure in the receiver from exceeding the maximum allowable working pressure of the receiver by more than 10 percent?

- Is every air receiver provided with a drain pipe and valve at the lowest point for the removal of accumulated oil and water?

- Are compressed air receivers periodically drained of moisture and oil?

- Are all safety valves tested frequently and at regular intervals to determine whether they are in good operating condition?

- Is there a current operating permit used by the Division of Occupational Safety and Health?

- Is the inlet of air receivers and piping systems kept free of accumulated oil and carbonaceous materials?

Compressed Gas Cylinders

- Are cylinders with a water weight capacity over 30 pounds (I 3.5 kilograms), equipped with means for connecting a valve protector device, or with a collar or recess to protect the valve?

- Are cylinders legibly marked to clearly identify the gas contained?

- Are compressed gas cylinders stored in areas which are protected from external heat sources such as flame impingement, intense radiant heat, electric arcs, or high temperature lines?

- Are cylinders located or stored in areas where they will not be damaged by passing, or failing objects or subject to tampering by unauthorized persons?

- Are cylinders stored or transported in a manner to prevent them from creating a hazard by tipping, failing or rolling?

- Are cylinders containing liquefied fuel gas, stored or transported in a position so that the safety relief device is always in direct contact with the vapor space in the cylinder?

- Are valve protectors always placed on cylinders when the cylinders are not in use or connected for use?

- Are all valves closed off before a cylinder is moved, when the cylinder is empty, and at the completion of each job?

- Are low pressure fuel-gas cylinders checked periodically for corrosion, general distortion, cracks, or any other defect that might indicate a weakness or render it unfit for service?

- Does the periodic check of low pressure fuel-gas cylinders include a close inspection of the cylinders' bottom?

Hoist and Auxiliary Equipment

- Is each overhead electric hoist equipped with a limit device to stop the hook travel at its highest and lowest point of safe travel?

- Will each hoist automatically stop and hold any load up to 125 percent of its rated load if its actuating force is removed?

- Is the rated load of each hoist legibly marked and visible to the operator?

- Are stops provided at the safe limits of travel for trolley hoist?

- Are the controls of hoist plainly marked to indicate the direction-of travel or motion?

- Is each cage-controlled hoist equipped with an effective warning device?

- Are close-fitting guards or other suitable devices installed on hoist to assure hoist ropes will be maintained in the sheave groves?

- Are all hoist chains or ropes of sufficient length to handle the full ran-e of movement of the application while still maintaining two full wraps on the drum at all times?

- Are nip points or contact points between hoist ropes and sheaves which are permanently located within 7 feet (2.1336 meters) of the floor, ground or working platform, guarded?

- Is it prohibited to use chains or rope slings that are kinked or twisted?

- Is it prohibited to use the hoist rope or chain wrapped around the load as a substitute, for a sling?

- Is the operator instructed to avoid carrying loads over people?

Industrial Trucks – Forklifts

- Are only employees who have been trained in the proper use of hoists allowed to operate them?

- Are only trained personnel allowed to operate industrial trucks?

- Is substantial overhead protective equipment provided on high lift rider equipment?

- Are the required lift truck operating rules posted and enforced?

- Is directional lighting provided on each industrial truck that operates in an area with less than 2 footcandles per square foot of general lighting?

- Does each industrial truck have a warning horn, whistle, gong, or other device which can be clearly heard above the normal noise in the areas where operated?

- Are the brakes on each industrial truck capable of bringing the vehicle to a complete and safe stop when fully loaded?

- Will the industrial truck's parking brake effectively prevent the vehicle from moving when unattended?

- Are industrial trucks operating in areas where flammable gases or vapors, or combustible dust or ignitable fibers may be present in the atmosphere, approved for such locations?

- Are motorized hand and hand/rider trucks so designed that the brakes are applied, and power to the drive motor shuts off when the operator releases his or her grip on the device that controls the travel?

- Are industrial trucks with internal combustion engine, operated in buildings or enclosed areas, carefully checked to ensure such operations do not cause harmful concentration of dangerous gases or fumes?

- Are powered industrial trucks being safely operated?

Spraying Operations

- Is adequate ventilation assured before spray operations are started?

- Is mechanical ventilation provided when spraying operations are done in enclosed areas?

- When mechanical ventilation is provided during spraying operations, is it so arranged that it will not circulate the contaminated air?

- Is the spray area free of hot surfaces?

- Is the spray area at least 20 feet (6.096 meters) from flames, sparks, operating electrical motors and other ignition sources?

- Are portable lamps used to illuminate spray areas suitable for use in a hazardous location?

- Is approved respiratory equipment provided and used when appropriate during spraying operations?

- Do solvents used for cleaning have a flash point to 100 degrees F or more?

- Are fire control sprinkler heads kept clean?

- Are "NO SMOKING" signs posted in spray areas, paint rooms, paint booths, and paint storage areas?

- Is the spray area kept clean of combustible residue?

- Are spray booths constructed of metal, masonry, or other substantial noncombustible material?

- Are spray booth floors and baffles noncombustible and easily cleaned?

- Is infrared drying apparatus kept out of the spray area during spraying operations?

- Is the spray booth completely ventilated before using the drying apparatus?

- Is the electric drying apparatus properly grounded?

- Are lighting fixtures for spray booths located outside of the booth and the interior lighted thro sealed clear panels?

- Are the electric motors for exhaust fans placed outside booths or ducts?

- Are belts and pulleys inside the booth fully enclosed?

- Do ducts have access doors to allow cleaning?

- Do all drying spaces have adequate ventilation?

Entering Confined Spaces

- Are confined spaces thoroughly emptied of any corrosive or hazardous substances, such as acids caustics, before entry?

- Are all lines to a confined space, containing inert, toxic, flammable, or corrosive materials valved off and blanked or disconnected and separated before entry?

- Are all impellers, agitators, or other moving parts and equipment inside confined spaces locked-out they present a hazard?

- Is either natural or mechanical ventilation provide prior to confined space entry?

- Are appropriate atmospheric tests performed to check for oxygen deficiency, toxic substances and explosive concentrations in the confined space before entry?

- Is adequate illumination provided for the work to be performed in the confined space?

- Is the atmosphere inside the confined space frequently tested or continuously monitored during conduct of work?

- Is there an assigned safety standby employee outside of the confined space, when required, whose sole responsibility is to watch the work in progress, sound an alarm if necessary, and render assistance?

- Is the standby employee appropriately trained and equipped to handle an emergency?

- Is the standby employee or other employees prohibited from entering the confined space without lifelines and respiratory equipment if there is any question as to the cause of an emergency?

- Is approved respiratory equipment required if the atmosphere inside the confined space cannot be made acceptable?

- Is all portable electrical equipment used inside confined spaces either grounded and insulated, or equipped with ground fault protection?

- Before gas welding or burning is started in a confined space, are hoses checked for leaks, compressed gas bottles forbidden inside of the confined space, torches lightly only outside of the confined area and the confined area tested for an explosive atmosphere each time before a lighted torch is to be taken into the confined space?

- If employees will be using oxygen-consuming equipment-such as salamanders, torches, and furnaces, in a confined space-is sufficient air provided to assure combustion without reducing the oxygen concentration of the atmosphere below 19.5 percent by volume?

- Whenever combustion-type equipment is used in a confined space, are provisions made to ensure the exhaust gases are vented outside of the enclosure?

- Is each confined space checked for decaying vegetation or animal matter which may produce methane?

- Is the confined space checked for possible industrial waste which could contain toxic properties?

- If the confined space is below the ground and near areas where motor vehicles will be operating, is it possible for vehicle exhaust or carbon monoxide to enter the space?

Environmental Controls

- Are all work areas properly illuminated?

- Are employees instructed in proper first-aid and other emergency procedures?

- Are hazardous substances, blood, and other potentially infectious materials identified, which may cause harm by inhalation, ingestion, or skin absorption or contact?

- Are employees aware of the hazards involved with the various chemicals they may be exposed to in their work environment, such as ammonia, chlorine, epoxies, caustics, etc.?

- Is employee exposure to chemicals in the workplace kept within acceptable levels?

- Can a less harmful method or process be used?

- Is the work area's ventilation system appropriate for the work being performed?

- Are spray painting operations done in spray rooms or booths equipped with an appropriate exhaust system?

- Is employee exposure to welding fumes controlled by ventilation, use of respirators, exposure time, or other means?

- Are welders and other workers nearby provided with flash shields during welding operations?

- If forklifts and other vehicles are used in buildings or other enclosed areas, are the carbon monoxide levels kept below maximum acceptable concentration?

- Has there been a determination that noise levels in the facilities are within acceptable levels?

- Are steps being taken to use engineering controls to reduce excessive noise levels?

- Are proper precautions being taken when handling asbestos and other fibrous materials?

- Are caution labels and signs used to warn of hazardous substances (e.g., asbestos) and biohazards (e.g., bloodborne pathogens)?

- Are wet methods used, when practicable, to prevent the emission of airborne asbestos fibers, silica dust and similar hazardous materials?

- Are engineering controls examined and maintained or replaced on a scheduled basis?

- Is vacuuming with appropriate equipment used whenever possible rather than blowing or sweeping dust?

- Are grinders, saws, and other machines that produce respirable dusts vented to an industrial collector or central exhaust system?

- Are all local exhaust ventilation systems designed and operating properly such as air flow and volume necessary for the application, ducts not plugged or belts slipping?

- Is personal protective equipment provided, used and maintained wherever required?

- Are there written standard operating procedures for the selection and use of respirators where needed?

- Are restrooms and washrooms kept clean and sanitary?

- Is all water provided for drinking, washing, and cooking potable?

- Are all outlets for water not suitable for drinking clearly identified?

- Are employees' physical capacities assessed before being assigned to jobs requiring heavy work?

- Are employees instructed in the proper manner of lifting heavy objects?

- Where heat is a problem, have all fixed work areas been provided with spot cooling or air conditioning?

- Are employees screened before assignment to areas of high heat to determine if their health condition might make them more susceptible to having an adverse reaction?

- Are employees working on streets and roadways where they are exposed to the hazards of traffic, required to wear bright colored (traffic orange) warning vests?

- Are exhaust stacks and air intakes so located that contaminated air will not be recirculated within a building or other enclosed area?

- Is equipment producing ultraviolet radiation properly shielded?

- Are universal precautions observed where occupational exposure to blood or other potentially infectious materials can occur and in all instances where differentiation of types of body fluids or potentially infectious materials is difficult or impossible?

Flammable and Combustible Materials

- Are combustible scrap, debris, and waste materials (oily rags, etc.) stored in covered metal receptacles and removed from the worksite promptly?

- Is proper storage practiced to minimize the risk of fire including spontaneous combustion?

- Are approved containers and tanks used for the storage and handling of flammable and combustible liquids?

- Are all connections on drums and combustible liquid piping, vapor and liquid tight?

- Are all flammable liquids kept in closed containers when not in use (e.g., parts cleaning tanks, pans, etc.)?

- Are bulk drums of flammable liquids grounded and bonded to containers during dispensing?

- Do storage rooms for flammable and combustible liquids have explosion-proof lights?

- Do storage rooms for flammable and combustible liquids have mechanical or gravity ventilation?

- Is liquefied petroleum gas stored, handled, and used in accordance with safe practices and standards?

- Are "NO SMOKING" signs posted on liquefied petroleum gas tanks?

- Are liquefied petroleum storage tanks guarded to prevent damage from vehicles?

- Are all solvent wastes, and flammable liquids kept in fire-resistant, covered containers until they are removed from the worksite?

- Is vacuuming used whenever possible rather than blowing or sweeping combustible dust?

- Are firm separators placed between containers of combustibles or flammables, when stacked one upon another, to assure their support and stability?

- Are fuel gas cylinders and oxygen cylinders separated by distance, and fire-resistant barriers, while in storage?

- Are fire extinguishers selected and provided for the types of materials in areas where they are to be used?
    Class A Ordinary combustible material fires.
    Class B Flammable liquid, gas or grease fires.
    Class C Energized-electrical equipment fires.

- Are appropriate fire extinguishers mounted within 75 feet (2286 meters) of outside areas containing flammable liquids, and within 10 feet (3.048 meters) of any inside storage area for such materials?

- Are extinguishers free from obstructions or blockage?

Are all extinguishers serviced, maintained and tagged at intervals not to exceed 1 year?

- Are all extinguishers fully charged and in their designated places?

- Where sprinkler systems are permanently installed, are the nozzle heads so directed or arranged that water will not be sprayed into operating electrical switch boards and equipment?

- Are "NO SMOKING" signs posted where appropriate in areas where flammable or combustible materials are used or stored?

- Are safety cans used for dispensing flammable or combustible liquids at a point of use?

- Are all spills of flammable or combustible liquids cleaned up promptly?

- Are storage tanks adequately vented to prevent the development of excessive vacuum or pressure as a result of filling, emptying, or atmosphere temperature changes?

- Are storage tanks equipped with emergency venting that will relieve excessive internal pressure caused by fire explosion?

- Are "NO SMOKING" rules enforced in areas involving storage and use of hazardous materials?

Hazardous Chemical Exposure

- Are employees trained in the safe handling practices of hazardous chemicals such as acids, caustics, etc.?

- Are employees aware of the potential hazards involving various chemicals stored or used in the workplace such as acids, bases, caustics, epoxies, and phenols?

- Is employee exposure to chemicals kept within acceptable levels?

- Are eye wash fountains and safety showers provided in areas where corrosive chemicals are handled?

- Are all containers, such as vats, and storage tanks labeled as to their contents, e.g., "CAUSTICS"?

- Are all employees required to use personal protective clothing and equipment when handling chemicals (gloves, eye protection, and respirators)?

- Are flammable or toxic chemicals kept in closed containers when not in use?

- Are chemical piping systems clearly marked as to their content?

- Where corrosive liquids are frequently handled in open containers or drawn from storage vessels or pipe lines, are adequate means readily available for neutralizing or disposing of spills or overflows and performed properly and safely?

- Have standard operating procedures been established, and are they being followed when cleaning up chemical spills?

- Where needed for emergency use, are respirators stored in a convenient, clean, and sanitary location?

- Are respirators intended for emergency use adequate for the various uses for which they may be needed?

- Are employees prohibited from eating in areas where hazardous chemicals are present?

- Is personal protective equipment provided, used, and maintained whenever necessary?

- Are there written standard operating procedures for the selection and use of respirators where needed?

- If you have a respirator protection program, are your employees instructed on the correct usage and limitations of the respirators? Are the respirators NIOSH-approved for this particular application? Are they regularly inspected and cleaned, sanitized and maintained?

- If hazardous substances are used in your processes, do you have a medical or biological monitoring system in operation?

- Are you familiar with the Threshold Limit Values or Permissible Exposure Limits of airborne contaminants and physical agents used in your workplace?

- Have control procedures been instituted for hazardous materials, where appropriate, such as respirators, ventilation systems, and handling practices?

- Whenever possible, are hazardous substances handled in properly designed and exhausted booths or similar locations?

- Do you use general dilution or local exhaust ventilation systems to control dusts, vapors, gases, fumes, smoke, solvents or mists which may be generated in your workplace?

- Is ventilation equipment provided for removal of contaminants from such operations as production grinding, buffing, spray painting, and/or vapor degreasing, and is it operating properly?

- Do employees complain about dizziness, headaches, nausea, irritation, or other factors of discomfort when they use solvents or other chemicals?

- Is there a dermatitis problem? Do employees complain about dryness, irritation, or sensitization of the skin?

- Have you considered the use of an industrial hygienist or environmental health specialist to evaluate your operation?

- If internal combustion engines are used, is carbon monoxide kept within acceptable levels?

- Is vacuuming used, rather than blowing or sweeping dusts whenever possible for clean-up?

- Are materials which give off toxic asphyxiant, suffocating or anesthetic fumes, stored in remote or isolated locations when not in use?

Hazardous Substances Communication

- Is there a list of hazardous substances used in your workplace?

- Is there a current written exposure control plan for occupational exposure to bloodbome pathogens and other potentially infectious materials, where applicable?

- Is there a written hazard communication program dealing with Material Safety Data Sheets (MSDS), labeling, and employee training?

- Is each container for a hazardous substance (i.e., vats, bottles, storage tanks, etc.) labeled with product identity and a hazard warning (communication of the specific health hazards and physical hazards)?

- Is there a Material Safety Data Sheet readily available for each hazardous substance used?

- Is there an employee training program for hazardous substances? Does this program include:

    - An explanation of what an MSDS is and how to use and obtain one?
    - MSDS contents for each hazardous substance or class of substances?
    - Explanation of "Right to Know?"
    - Identification of where an employee can see the employers written hazard communication program and where hazardous substances are present in their work areas?
    - The physical and health hazards of substances in the work area, and specific protective measures to be used?
    - Details of the hazard communication program, including how to use the labeling system and MSDSs?

- Does the employee training program on the bloodbome pathogens standard contain the following elements:

    (1) an accessible copy of the standard and an explanation of its contents; (2) a general explanation of the epidemiology and symptoms of bloodbome diseases; (3) an explanation of the modes of transmission of bloodbome pathogens; (4) an explanation of the employer's

exposure control plan and the means by which employees can obtain a copy of the written plan; (5) an explanation of the appropriate methods for recognizing tasks and the other activities that may involve exposure to blood and other potentially infectious materials; (6) an explanation of the use and limitations of methods that will prevent or reduce exposure including appropriate engineering controls, work practices, and personal protective equipment; (7) information on the types, proper use, location, removal, handling, decontamination, and disposal of personal protective equipment; (8) an explanation of the basis for selection of personal protective equipment; (9) information on the hepatitis B vaccine; (10) information on the appropriate actions to take and persons to contact in an emergency involving blood or other potentially infectious materials; (I 1) an explanation of the procedure to follow if an exposure incident occurs, including the methods of reporting the incident and the medical follow-up that will be made available; (12) information on post-exposure evaluations and follow-up; and (13) an explanation of signs, labels, and color coding?

- Are employees trained in the following:

  - How to recognize tasks that might result in occupational exposure?
  - How to use work practice and engineering controls and personal protective equipment and to know their limitations?
  - How to obtain information on the types, selection, proper use, location, removal, handling, decontamination, and disposal of personal protective equipment?
  - Who to contact and what to do in an emergency?

Electrical

- Do you specify compliance with OSHA for all contract electrical work?

- Are all employees required to report as soon as practicable any obvious hazard to life or property observed in connection with electrical equipment or lines?

- Are employees instructed to make preliminary inspections and/or appropriate tests to determine what conditions exist before starting work on electrical equipment or lines?

• When electrical equipment or lines are to be serviced, maintained or adjusted, are necessary switches opened, locked-out and tagged whenever possible?

• Are portable electrical tools and equipment grounded or of the double insulated type?

• Are electrical appliances such as vacuum cleaners, polishers, and vending machines grounded?

• Do extension cods being used have a grounding conductor?

• Are multiple plug adaptors prohibited?

• Are ground-fault circuit interrupters installed on each temporary 15 or 20 ampere, 120 volt AC circuit at locations where construction, demolition, modifications, alterations or excavations are being performed?

• Are all temporary circuits protected by suitable disconnecting switches or plug connectors at the junction with permanent wiring?

• Do you have electrical installations in hazardous dust or vapor areas? If so, do they meet the National Electrical Code (NEC) for hazardous locations?

• Is exposed wiring and cords with frayed or deteriorated insulation repaired or replaced promptly?

• Are flexible cords and cables free of splices or taps?

• Are clamps or other securing means provided on flexible cords or cables at plugs, receptacles, tools, equipment, etc., and is the cord jacket securely held in place?

• Are all cord, cable and raceway connections intact and secure?

• In wet or damp locations, are electrical tools and equipment appropriate for the use or location or otherwise protected?

- Is the location of electrical power lines and cables (overhead. underground, underfloor, other side of walls) determined before digging, drilling or similar work is begun?

- Are metal measuring tapes, ropes, handlines or similar devices with metallic thread woven into the fabric prohibited where they could come in contact with energized parts of equipment or circuit conductors?

- Is the use of metal ladders prohibited in areas where the ladder or the person using the ladder could come in contact with energized parts of equipment. fixtures or circuit conductors?

- Are all disconnecting switches and circuit breakers labeled to indicate their use or equipment served?

- Are disconnecting means always opened before fuses are replaced?

- Do all interior wiring systems include provisions for grounding metal parts of electrical raceways, equipment and enclosures?

- Are all electrical raceways and enclosures securely fastened in place?

- Are all energized parts of electrical circuits and equipment guarded against accidental contact by approved cabinets or enclosures?

- Is sufficient access and working space provided and maintained about all electrical equipment to permit ready and, safe operations and maintenance?

- Are all unused openings (including conduit knockouts) in electrical enclosures and fittings closed with appropriate covers, plugs or plates?

- Are electrical enclosures such as switches, receptacles, and junction boxes, provided with tight-fitting covers or plates?

- Are disconnecting switches for electrical motors in excess of two horsepower, capable of opening the circuit when the motor is in a stalled condition, without exploding? (Switches must be horsepower rated equal to or in excess or the motor hp rating.)

- Is low voltage protection provided in the control device of motors driving machines or equipment which could cause probable injury from inadvertent starting?

- Is each motor disconnecting switch or circuit breaker located within sight of the motor control device?

- Is each motor located within sight of its controller or the controller disconnecting means capable of being locked in the open position or is a separate disconnecting means installed in the circuit within sight of the motor?

- Is the controller for each motor in excess of two horsepower, rated in horsepower equal to or in excess of the rating of the motor it serves?

- Are employees who regularly work on or around energized electrical equipment or lines instructed in the cardiopulmonary resuscitation (CPR) methods?

- Are employees prohibited from working alone on energized lines or equipment over 600 volts?

Noise

- Are there areas in the workplace where continuous noise levels exceed 85 dBA?

- Is there an ongoing preventive health program to educate employees in: safe levels of noise, exposures; effects of noise on their health; and the use of personal protection?

- Have work areas where noise levels make voice communication between employees difficult been identified and posted?

- Are noise levels being measured using a sound level meter or an octave band analyzer and are records being kept?

- Have engineering controls been used to reduce excessive noise levels? Where engineering controls are determined not feasible, are administrative controls (i.e., worker rotation) being used to minimize individual employee exposure to noise?

- Is approved hearing protective equipment (noise attenuating devices) available to every employee working in noisy areas?

- Have you tried isolating noisy machinery from the rest of your operation?

- If you use ear protectors, are employees properly fitted and instructed in their use?

- Are employees in high noise areas given periodic audiometric testing to ensure that you have an effective hearing protection system?

Fueling

- Is it prohibited to fuel an internal combustion engine with a flammable liquid while the engine is running?

- Are fueling operations done in such a manner that likelihood of spillage will be minimal?

- When spillage occurs during fueling operations, is the spilled fuel washed away completely, evaporated, or other measures taken to control vapors before restarting the engine?

- Are fuel tank caps replaced and secured before
- starting the engine?

- In fueling operations, is there always metal contact between the container and the fuel tank?

- Are fueling hoses of a type designed to handle the specific type of fuel?

- Is it prohibited to handle or transfer gasoline in open containers?

- Are open lights, open flames, sparking, or arcing equipment prohibited near fueling or transfer of fuel operations?

- Is smoking prohibited in the vicinity of fueling operations?

- Are fueling operators prohibited in buildings or other enclosed areas that are not specifically ventilated for this purpose?

- Where fueling or transfer of fuel is done through a gravity flow system, are the nozzles of the self-closing type?

Identification of Piping Systems

- When non-potable water is piped through a facility, are outlets or taps posted to alert employees that it is unsafe and not to be used for drinking, washing or other personal use?

- When hazardous substances are transported through above ground piping, is each pipeline identified at points where confusion could introduce hazards to employees?

- When pipelines are identified by color painting, are all visible parts of the line so identified?

- When pipelines are identified by color painted bands or tapes, are the bands or tapes located at reasonable intervals and at each outlet, valve or connection?

- When pipelines are identified by color, is the color code posted at all locations where confusion could introduce hazards to employees?

- When the contents of pipelines are identified by name or name abbreviation, is the information readily visible on the pipe near each valve or outlet?

- When pipelines carrying hazardous substances are identified by tags, are the tags constructed of durable materials, the message carried clearly and permanently distinguishable and are tags installed at each valve or outlet?

- When pipelines are heated by electricity, steam or other external source, are suitable warning signs o tags placed at unions, valves, or other serviceable parts of the system?

Material Handling

- Is there safe clearance for equipment through aisle and doorways?

- Are aisleways designated, permanently marked, and kept clear to allow unhindered passage?

- Are motorized vehicles and mechanized equipment inspected daily or prior to use?

- Are vehicles shut off and brakes set prior to loading or unloading?

- Are containers of combustibles or flammables, when stacked while being moved, always separate by dunnage sufficient to provide stability?

- Are dock boards (bridge plates) used when loading or unloading operations are taking place between -vehicles and docks?

- Are trucks and trailers secured from movement during loading and unloading operations?

- Are dock plates and loading ramps constructed and maintained with sufficient strength to support imposed loading?

- Are hand trucks maintained in safe operating condition?

- Are chutes equipped with sideboards of sufficient height to prevent the materials being handled from falling off?

- Are chutes and gravity roller sections firmly placed or secured to prevent displacement?

- At the delivery end of the rollers or chutes, are provisions made to brake the movement of the handled materials?

- Are pallets usually inspected before being loaded or moved?

- Are hooks with safety latches or other arrangements used when hoisting materials so that slings or load attachments won't accidentally slip off the hoist hooks?

- Are securing chains, ropes, chockers or slings adequate for the job to be performed?

- When hoisting material or equipment, are provisions made to assure no one will be passing under the suspended loads?

- Are material safety data sheets available to employees handling hazardous substances?

Transporting Employees and Materials

- Do employees who operate vehicles on public thoroughfares have valid operator's licenses?

- When seven or more employees are regularly transported in a van, bus or truck, is the operator's license appropriate for the class of vehicle being driven?

- Is each van, bus or truck used regularly to transport employees equipped with an adequate number of seats?

- When employees are transported by truck, are provisions provided to prevent their falling from the vehicle?

- Are vehicles used to transport employees equipped with lamps, brakes, horns, mirrors, windshields and turn signals and are they in good repair?

- Are transport vehicles provided with handrails, steps, stirrups or similar devices, so placed and arranged that employees can safely mount or dismount?

- Are employee transport vehicles equipped at all times with at least two reflective type flares?

- Is a full charged fire extinguisher, in good condition, with at least 4 B:C rating maintained in each employee transport vehicle?

- When cutting tools or tools with sharp edges are carried in passenger compartments of employee transport vehicles, are they placed in closed boxes or containers which are secured in place?

- Are employees prohibited from riding on top of any load which can shift, topple, or otherwise become unstable?

Control of Harmful Substances by Ventilation

- Is the volume and velocity of air in each exhaust system sufficient to gather the dusts, fumes, mists, vapors or gases to be controlled, and to convey them to a suitable point of disposal?

- Are exhaust inlets, ducts and plenums designed, constructed, and supported to prevent collapse or failure of any part of the system?

- Are clean-out ports or doors provided at intervals not to exceed 12 feet (3.6576 meters) in all horizontal runs of exhaust ducts?

- Where two or more different type of operations are being controlled through the same exhaust system, will the combination of substances being controlled, constitute a fire, explosion or chemical reaction hazard in the duct?

- Is adequate makeup air provided to areas where exhaust systems are operating?

- Is the source point for makeup air located so that only clean, fresh air, which is free of contaminates, will enter the work environment?

- Where two or more ventilation systems are serving a work area, is their operation such that one will not offset the functions of the other?

Sanitizing Equipment and Clothing

- Is personal protective clothing or equipment that employees are required to wear or use, of a type capable of being cleaned easily and disinfected?

- Are employees prohibited from interchanging personal protective clothing or equipment, unless it has been properly cleaned?

- Are machines and equipment, which process, handle or apply materials that could be injurious to employees, cleaned and/or decontaminated before being overhauled or placed in storage?

- Are employees prohibited from smoking or eating in any area where contaminates that could be injurious if ingested are present?

- When employees are required to change from street clothing into protective clothing, is a clean change room with separate storage facility for street and protective clothing provided?

- Are employees required to shower and wash their hair as soon as possible after a known contact has occurred with a carcinogen?

- When equipment, materials, or other items are taken into or removed from a carcinogen regulated area, is it done in a manner that will contaminate non-regulated areas or the external environment?

Tire Inflation

- Where tires are mounted and/or inflated on drop center wheels, is a safe practice procedure posted and enforced?

- Where tires are mounted and/or inflated on wheels with split rims and/or retainer rings, is a safe practice procedure posted and enforced?

- Does each tire inflation hose have a clip-on chuck with at least 24 inches (6.9 centimeters) of hose between the chuck and an in-line hand valve and gauge?

- Does the tire inflation control valve automatically shutoff the air flow when the valve is released?

- Is a tire restraining device such as a cage, rack or other effective means used while inflating tires mounted on split rims, or rims using retainer rings?

- Are employees strictly forbidden from taking a position directly over or in front of a tire while it's being inflated?

Appendix C – Codes of Safe Practices

This is a suggested code. It is general in nature and inclusive of many types of small business activities. It is intended only as a model which can be redrafted to describe particular work environments.

General Policy

1. All employees of this firm shall follow these safe practice rules, render every possible aid to safe operations, and report all unsafe conditions or practices to the supervisor/employer.

2. Supervisors shall insist that employees observe and obey every rule, regulation and order necessary to the safe conduct of the work, and shall take such action necessary to obtain compliance.

3. All employees shall be given frequent accident prevention instructions. Instructions, practice drills and articles concerning workplace safety and health shall be given at least once every working day.

4. Anyone known to be under the influence of alcohol and/or drugs shall not be allowed on the job while in that condition. Persons with symptoms of alcohol and/or drug abuse are encouraged to discuss personal or work-related problems with the supervisor/ employer.

5. No one shall knowingly be permitted or required to work while his or her ability or alertness is impaired by fatigue, illness or other causes that might expose the individual or others to injury.

6. Employees should be alert to see that all guards and other protective devices are in proper places and adjusted, and shall report deficiencies. Approved protective equipment shall be worn in

specified work areas.

7. Horseplay, scuffling and other acts which tend to endanger the safety or well-being of employees are prohibited.

8. Work shall be well-planned and supervised to prevent injuries when working with equipment and handling heavy materials. When lifting heavy objects, employees should bend their knees and use the large muscles of the leg instead of the smaller muscles of the back. Back injuries are the most frequent and often the most persistent and painful type of workplace injury.

9. Workers shall not handle or tamper with any electrical equipment, machinery or air or water lines in a manner not within the scope of their duties, unless they have received instructions from their supervisor/employer.

10. All injuries shall be reported promptly to the supervisor/employer so that arrangements can be made for medical and/or first-aid treatment. First-aid materials are located in _____, emergency, fire, ambulance, rescue squad, and doctor's telephone numbers are located on _____, and fire extinguishers are located at _____.

Suggested Safety Rules

- Do not throw material, tools or other objects from heights (whether structures or buildings) until proper precautions are taken to protect others from the falling object hazard.

- Wash thoroughly after handling injurious or poisonous substances.

- Gasoline shall not be used for cleaning purposes.

- Arrange work so that you are able to face ladder and use both hands while climbing.

Use of Tools and Equipment

- Keep faces of hammers in good condition to avoid flying nails and bruised fingers.

- Files shall be equipped with handles; never use a file as a punch or pry.

- Do not use a screwdriver as a chisel.

- Do not lift or lower portable electric tools by the power cords; use a rope.

- Do not leave the cords of these tools where cars or trucks will run over them.

Machinery and Vehicles

- Do not attempt to operate machinery or equipment without special permission, unless it is one of your regular duties.

- Loose or frayed clothing, dangling ties, finger rings, and similar items must not be worn around moving machinery or other places where they can get caught.

- Machinery shall not be repaired or adjusted while in operation.

Index

W